AF389357

Water Quality Assessments for Urban Water Environment

Water Quality Assessments for Urban Water Environment

Editor

Pankaj Kumar

MDPI • Basel • Beijing • Wuhan • Barcelona • Belgrade • Manchester • Tokyo • Cluj • Tianjin

Editor
Pankaj Kumar
Institute for Global Environmental Strategies
Japan

Editorial Office
MDPI
St. Alban-Anlage 66
4052 Basel, Switzerland

This is a reprint of articles from the Special Issue published online in the open access journal *Water* (ISSN 2073-4441) (available at: https://www.mdpi.com/journal/water/special_issues/urban_water_environment).

For citation purposes, cite each article independently as indicated on the article page online and as indicated below:

LastName, A.A.; LastName, B.B.; LastName, C.C. Article Title. *Journal Name* **Year**, *Volume Number*, Page Range.

ISBN 978-3-0365-1869-5 (Hbk)
ISBN 978-3-0365-1870-1 (PDF)

Contents

About the Editor

Pankaj Kumar is working as senior policy researcher in the field of water resources and climate change adaptation in the Institute for Global Environmental Strategies (IGES), Japan. His research work focused on socio-hydrology, water security, hydrological simulation and scenario modelling, and water-health-food-energy nexus, a transdisciplinary work aimed to give policy relevant solutions to enhance community resilience to global change and sustainable development of water environment and human well-being. In addition, he is actively engaged in capacity development on various numerical tools used for water resource management, intended for local government officials and relevant stakeholders in different Asian countries. Also, he has work experience with both IPCC and IPBES as chapter scientist and scoping expert respectively. He has several peer reviewed articles (>70) in high impact factor journals, one authored book, and one edited book to his credit.

Preface to "Water Quality Assessments for Urban Water Environment"

For a long time, human societies have been witnessing the intrinsic connection between their all-encompassing development and freshwater resources. However, with its increasing demand for the swelling population and growing economies, the availability of these finite resources is further skewed down, both in terms of quantity as well as quantity. Rapid urbanization, land use, land cover change, climate change, and frequent extreme weather conditions further exacerbate the challenges for sustainable water resource management. It is predicted that more than 60% of the world's population will live in urban areas by the year 2050, which will lead to an enormous increase in the need for water. In particular, hasty urbanization, poor governance, data scarcity, and insufficient infrastructure are the key factors for urban water scarcity in the developing nations around the world. This leads to other cascading effects, like changes in the hydrological cycle, food insecurity, health issues, high vulnerability to the frequent extreme weather conditions and stressed human well-being, etc.

Most of the conventional research works are working in silos, like quantitative analysis (e.g., water quality analysis, numerical simulation, use of spatial techniques like remote sensing and GIS for water resources analysis, evaluation of extreme weather conditions like flooding, droughts, and its impact on water resources etc.) or qualitative analysis (e.g., citizen science, perception-based studies, questionnaire-based analysis etc.). There is a big gap in relevant scientific information, needed to build robust water management policies. Since existing management frameworks and plans are inadequate, this means that without a transformational change, most of these burgeoning cities within developing nations will face severe water scarcity. On the other hand, the current COVID-19 pandemic has also affected various socio-environmental processes around the globe.

Therefore, there is an urgent need for transdisciplinary research works, which integrate both the biophysical world with socio-economic dimensions to see how they impact hydrological cycle as well as human well-being. The outcomes of such research works will cater to the policy-planners and decision-makers in terms of building climate resilient strategies for better water/wastewater and land use management policies, which ultimately will have substantial environmental and economic co-benefits for communities.

Finally, the outcomes of these sustainable water management plans will also contribute to national adaptation plans, which can briefly sketch the appropriate local actions for addressing local climate change impacts. In addition, most of the SDG goals are intrinsically linked and a water-related goal, i.e., SDG 6.0, is placed at the center of this, so the sustainable management of water resources not only helps nations to achieve different global goals in a timely manner, but also helps socio-environmental development in a holistic manner.

Based on the aforementioned background information, this Special Issue, titled "Water Quality Assessments for Urban Water Environment", strives to highlight the status quo of the water environment, opportunities and challenges for their sustainable management in the urban space, particularly in developing nations around the world. Moreover, it highlights the extent to which different scientific innovations have contributed significantly to solve these challenging issues. Finally, what is the way forward when it comes to enhancing the science policy interface in a better way to achieve global goals, e.g., SDGs at a local level in a timely manner. This issue has eleven papers, covering a variety of water-related issues around the world.

In summary, this Special Issue gives a holistic picture of opportunities and challenges for managing the urban water environment in terms of quality or quantity in a sustainable manner. It provides valuable information about sustainable water resource management in the urban landscape, which is very much useful for policy-makers, decision-makers, local communities, and other relevant stakeholders. Considering the complex nature of urban water security, emerging approaches like socio-hydrology, landscape ecology, regional-circular-ecological sphere etc., which present a perfect combination of hard (infrastructure) and soft (numerical simulations, spatial technologies, participatory approaches, indigenous knowledge) measures, are the potential solutions to manage this precious water resource in coming future.

Pankaj Kumar
Editor

Editorial

Water Quality Assessments for Urban Water Environment

Pankaj Kumar

Natural Resources and Ecosystem Services, Institute for Global Environmental Strategies, Hayama 240-0115, Kanagawa, Japan; kumar@iges.or.jp

Citation: Kumar, P. Water Quality Assessments for Urban Water Environment. *Water* **2021**, *13*, 1686. https://doi.org/10.3390/w13121686

Received: 15 June 2021
Accepted: 16 June 2021
Published: 18 June 2021

Since ages, human societies have witnessed the intrinsic connection between their all-encompassing development and freshwater resources. However, with its increasing demand for the swelling population and growing economies, availability of these finite resources is further skewed down both in terms of quantity as well quantity. Rapid urbanization, land use land cover change, climate change, and frequent extreme weather conditions further exacerbate the challenges for sustainable water resource management. It is predicted that more than 60% of world population will live in the urban areas by the year 2050, which will lead to enormous increase in the need of water. In particular, hasty urbanization, poor governance, data scarcity, insufficient infrastructure are the key factors for urban water scarcity in the developing nations around the world. This leads to other cascading effects like changes in the hydrological cycle, food insecurity, health issues, high vulnerability to the frequent extreme weather conditions and stressed human well-being etc.

Most of the conventional research works are working in silos like quantitative analysis (e.g., water quality analysis, numerical simulation, use of spatial techniques like remote sensing and GIS for water resources analysis, evaluation of extreme weather conditions like flooding, droughts, and its impact on water resources etc.,) or qualitative analysis (e.g., citizen science, perception-based studies, questionnaire-based analysis etc.,). There is a big gap in relevant scientific information, needed to build robust water management policies. Since, existing management frameworks and plans are inadequate, which means that without a transformational change, most of these burgeoning cities within developing nations will face severe water scarcity. On the other hand, current COVID-19 pandemic has also affected various socio-environmental processes around the globe.

Therefore, there is an urgent need for transdisciplinary research works, which integrates both biophysical world with socio-economic dimensions to see how they impact hydrological cycle as well as human well-being. Outcomes of such research works will cater the policy-planners and decision-makers in terms of building climate resilient strategies for better water/wastewater and land use management policies, which ultimately will have substantial environmental and economic co-benefits for communities.

Finally, outcome of this sustainable water management plans will also contribute to national adaptation plans, which can briefly sketch the appropriate local actions for addressing local climate change impacts. In addition, most of the SDG goals are intrinsically linked and water-related goal i.e., SDG 6.0 is placed in the center of this, so sustainable management of water resources not only help nations to achieve different global goals in a timely manner but also help in socio-environmental development in a holistic manner.

Based on the aforementioned background information, this special issue entitled "Water Quality Assessments for Urban Water Environment," strives to highlight the status quo of water environment, opportunities and challenges for their sustainable management in the urban space particularly in developing nations around the world. Moreover, it highlight the extent by which different scientific innovations have contributed significantly to resolve these challenging issues. Finally, what is the way forward to enhance science-policy interface in a better way to achieve global goals e.g., SDGs at local level in a timely manner. This issue has eleven paper, covering variety of water-related issues around the world.

Molekao et al. [1] studied the groundwater resources in the data-scarce region of Mokopane area of South Africa, considering groundwater as the main source of fresh water supply for this region. More specifically, authors have evaluated factors responsible for groundwater quality evolution with space and time using integrated approach of hydrochemical analysis, spatial analysis, statistical analysis, and calculating the water quality index (WQI). Results suggest that rock-water interaction is the main cause of fluoride enrichment in the groundwater. Although most of the water samples are of good quality, this study stressed on the need for regular monitoring and holistic management plans for the sustainable water resource management.

In the second paper, considering the problem of excessive phosphorus in Yangtze river, Dong et al. [2] studied its distribution characteristics at spatio-temporal scale more particularly in the middle and lower reaches of the Yangtze river using water quality analysis. Result suggests that particulate phosphate (PP) and soluble reactive phosphate (SRP) differs with depth and location because of the difference in transportation and sedimentation of the suspended sediment (SS), as well as differences in the location of urban sewage outlets. Lack of timely management of reactive phosphorus will pose a serious threat to the riparian ecosystem like eutrophication.

Considering the enormous effect of rapid global changes on the surface water qualities, Kumar et al. [3] used a methodology called the "Participatory Watershed Land-use Management" (PWLM) approach, which is a combination of participatory approaches and computer simulation modeling, with the goal to help make land-use and climate change adaptation policies more effective at a local scale. This study consists of four major steps: (a) scenario analysis, (b) impact assessment, (c) developing adaptation and mitigation measures and its integration in local government policies, and (d) improvement of land use plan. They have tested this PWLM approach in the Santa Rosa Sub-watershed of the Philippines, a rapidly urbanizing area outside Metro Manila. The scenario analysis step involved a participatory land-use mapping activity (to understand the likely land-use changes in the future), as well as downscaling the Global Climate Model (GCM) data for precipitation and temperature (to understand the local climate scenarios in the future). For impact assessment, the water evaluation and planning (WEAP) tool was used to simulate future river water quality under a business as usual (BAU) scenario and several alternative future scenarios considering different drivers and pressures to a target year. Result suggests that current status of the river is already moderately to extremely polluted compared to desirable water quality. Moreover, this situation will further deteriorate by the year 2030 under all scenarios. In addition, they have also analyzed contributions from different drivers on water quality deterioration and it was found that population growth has the highest impact on future water quality deterioration, while climate change had the lowest (although not negligible). After the impact assessment, different mitigation measures were suggested in a stakeholder consultation workshop, and of them some were adopted to generate a final scenario including countermeasures. The main benefits of the PWLM approach are its high level of stakeholder involvement (through co-generation of the research) and use of free (for developing countries) software and models, both of which will contribute to an enhanced science-policy interface.

Minh et al. [4] studied the impact of agricultural intensification and urbanization on the water quality in An Giang from Vietnamese Mekong Delta (VMD), one of the most productive agricultural deltas in the world. They have assessed seasonal variation of water quality parameters inside full- and semi-dike systems and outside of the dike system frequently used for rice production using multivariable statistical analysis and weighted arithmetic water quality index (WAWQI). The results show that the value of various water quality parameters exceeded the permissible limit given by WHO for both seasons. More precisely, high concentration of NO_2^- and COD were found in rice-cropping system and urban area respectively. The WAWQI showed that 97.5 and 95.0% of water samples fall into the bad and unsuitable drinking categories, respectively. The main reason behind this is the direct discharge of untreated wastewater from the rice intensification and urban sewerage lines in to the ambient surface water bodies. The finding of this study is critically important for decision-makers to design different mitigation or adaptation measures for water resource management in lieu of rapid global changes in a timely manner in An Giang and the VMD.

In recent decades, rapid urbanization and land use land cover changes are putting natural resources especially water resources under enormous pressures throughout in Africa. With this background, Molekao et al. [5] conducted a spatiotemporal analysis of surface water quality and its relation with the land use and land cover (LULC) pattern in Mokopane, Limpopo province, South Africa. Integrated approach of hydrochemical analysis with spatial techniques was used to identify the factors affecting the water quality in particular heavy metals in Mokopane area dominated by mining activities. Based on the estimated values of heavy metal pollution index (HPI) and heavy metal evaluation index (HEI), water bodies are categorized as low to moderately polluted. On the other hand, all water samples fell under the poor category (>100) and beyond based on the calculated water quality index (WQI). Although water quality is showing deteriorating trend from year 2016 to 2019, this trend shows a sign of improvement in year 2020 because of lowering human activities during the lockdown period imposed by COVID-19. Land use has a significant proportional relationship with surface water quality, and it was evident that built-up land had a more significant negative impact on water quality than the other land use classes. Both natural processes (rock weathering) and anthropogenic activities (wastewater discharge, industrial activities etc.,) were found to be playing a vital role in water quality evolution. At last, this study calls for continuous assessment and monitoring of the spatial and temporal variability of water quality in the area to control pollution and health safety in the future.

Although water quality index (WQI) is one of the most extensively used parameters for determining water quality worldwide, however the traditional approach for applying this technique is very time and labor consuming. With the above background, Agrawal et al. [6], applied and examined performance of artificial intelligence techniques namely particle swarm optimization (PSO), a naive Bayes classifier (NBC), and a support vector machine (SVM) for predicting the water quality index. Furthermore, the results obtained were validated from the observed groundwater water quality collected from Chhattisgarh, India. Study results show high prediction accuracy for the WQI indices and recommends that ensemble machine learning (ML) algorithms can be used to estimate and predict the water quality index with significant accuracy. Hence the proposed framework can be directly used for the prediction of the WQI using the measured field parameters while saving significant time and effort.

Despite inhabiting a large proportion of the world population and the rich biodiversity, coastal ecosystems are extremely vulnerable to changes in hydroclimatic factors, which significantly impacts the local socio-economic dimension. In addition, lack of scientific information presents a big hurdle for sustainable management of coastal ecosystem especially in the developing countries. Halder et al. [7] estimated future groundwater demand (domestic, agricultural, and livestock sector) in the fragile Sundarbans ecosystem considering different human population growth rates (high, low, and current) for the year 2050. Simulation results showed that with an increase of groundwater demand for domestic and agricultural sectors by 17% and 35% for Business as Usual (BAU) and high growth respectively. Sustainability of coastal aquifer-dependent rural livelihood is expected to face great danger in the near future. The impact of increasing groundwater demand was analyzed further to identify any socio-economic shifts in this region.

The novel coronavirus (COVID19) pandemic changed the different socio-economic and environmental factors to various degrees around the world. Considering these additional unprecedented shocks induced by the COVID19 pandemic, the next two papers by Avtar et al. [8] and Usman et al. [9] highlighted the importance of studying various challenges and opportunities to manage water resources under these unknown situations. One of the most important effects of the COVID-19 pandemic is lockdown, which has brought countries around the world to a standstill. People stayed indoor, and industrial activities, transportation etc. operated at a minimum level and hence fewer pollutants also entered our ambient environment.

Hence, Avtar et al. [8] hypothesized that due to these pandemic-induced lockdowns, the hydrological residence time (HRT) has increased in the semi-enclosed or closed water bodies like lakes, which can in turn increase the primary productivity. To validate their hypothesis, they quantitatively estimated the chlorophyll-a (Chl-a) concentrations in different lake bodies in China and India using established Chl-a retrieval algorithm. Result shows that concentration of Chl-a increased for closed lake in Wuhan (China), whereas the concentration did not change for a semi-closed lake in India. This opened a door for both scientific communities as well as decision-makers to design proper countermeasures for sustainable water resource management in coming future.

Usman et al. [9] discussed the existence of SARS-CoV-2 in wastewater, which raises the opportunity of tracking wastewater for epidemiological monitoring of COVID19 related diseases. On the other hand, the existence of this virus in wastewater has raised health concerns regarding the fecal–oral transmission of COVID-19. This study highlights the potential inferences of aerosolized wastewater in transmitting this virus. As aerosolized SARS-CoV-2 could offer a more direct respiratory pathway for human exposure, the transmission of this virus remains a significant possibility in the prominent wastewater-associated bioaerosols formed during toilet flushing, wastewater treatment, and sprinkler irrigation. Hence, implementing wastewater disinfection, exercising precautions, and raising public awareness was advocated for prevention of infection spreading.

After having a broader discussion on different water quality issues in urban landscape, two papers (Poudel et al. [10] and Mishra et al. [11]) talked about urban water-security, its challenges and opportunities.

Considering a strong nexus between human health and water security, Poudel et al. [10] used systematic literature review to investigate how health has been incorporated as a dimension in the existing water-security frameworks in different scientific disclosures. The result shows that 11 distinct dimensions have been used to design the existing water-security framework. Public health aspects were mentioned in half of the reviewed papers, direct and indirect health impacts were considered only by 18% and 33% of the papers respectively. Among direct health impacts, diarrhea is the most prevalent one considered for developing a water-security framework. Among different indirect or mediating factors, poor accessibility and availability of water resources in terms of time and distance are the big determinants for causing mental illnesses, such as stress or anxiety, which are being considered when framing water-security framework, particularly in developing nations. Water scarcity in terms of its quantity is more of a common issue for both developed and developing countries. However, poor water quality and mismanagement of water supply-related infrastructure are the main concerns for developing nations, which proved to be the biggest hurdle for achieving water security. The result of this study sheds light on the existing gaps for different water-security frameworks and provides policy-relevant guidelines for its betterment. Moreover, it stresses that a health-specific water-security framework is imperative for dealing with issues arising from public health.

It is well understood that rapid global changes such as urbanization, population growth, socioeconomic change, evolving energy needs, and climate change put unprecedented pressure on water resources and its related systems. Moreover, achieving water security is vital for achieving all 17 sustainable development goals (SDGs) throughout the world. On the other hand, studies on water security with a holistic view with persistently changing dimensions is in its infancy. Hence, Mishra et al. [11] focuses on narrative review work for giving a comprehensive insight on the concept of water security, its evolution with recent environmental changes, its implications, and different possible sustainable solutions to achieve water security. It emphasizes that water security evolves from ensuring reliable access to enough safe water for every person (at an affordable price where market mechanisms are involved) to leading a healthy and productive life, including that of future generations. The constraints on water availability and water quality threaten secured access to water resources for different uses. Despite recent progress in developing new strategies, practices, and technologies for water resource management, their dissemination and implementation has been limited. A comprehensive sustainable approach to address water-security challenges requires connecting social, economic, and environmental systems at multiple scales. This paper captures the persistently changing dimensions and new paradigms of water security providing a holistic view including a wide range of sustainable solutions to address the water challenges.

Way forward

In summary, this special issue gives a holistic picture of opportunities and challenges for managing the urban water environment in terms of quality or quantity in a sustainable manner. It provides valuable information about sustainable water resource management at the urban landscape, which is very much useful for policy-makers, decision-makers, local communities, and other relevant stakeholders. Considering the complex nature of the urban water security, emerging approaches like socio-hydrology, landscape ecology, regional-circular-ecological sphere etc., which presents a perfect combination of hard (infrastructure) and soft (numerical simulations, spatial technologies, participatory approaches, indigenous knowledge) measures, are the potential solutions to manage this precious water resource in coming future.

Funding: Not available.

Institutional Review Board Statement: Not applicable.

Informed Consent Statement: Not applicable.

Conflicts of Interest: The author declares no conflict of interest in the collection, analyses, or interpretation of data; in the writing of the manuscript, or in the decision to publish the results.

References

1. Molekoa, M.D.; Avtar, R.; Kumar, P.; Minh, H.V.T.; Kurniawan, A.T. Hydrochemical assessment of groundwater quality of Mokopane area, Limpopo, South Africa using statistical approach. *Water* **2019**, *11*, 1891. [CrossRef]
2. Dong, L.; Lin, L.; Tang, X.; Huang, Z.; Zhao, L.; Wu, M.; Li, R. Distribution Characteristics and Spatial Differences of Phosphorus in the Main Stream of the Urban River Stretches of the Middle and Lower Reaches of the Yangtze River. *Water* **2020**, *12*, 910. [CrossRef]
3. Kumar, P.; Johnson, B.A.; Dasgupta, R.; Avtar, R.; Chakraborty, S.; Masayuki, K.; Macandog, D. Participatory approach for enhancing robust water resource management: Case study of Santa Rosa sub-watershed near Laguna Lake, Philippines. *Water* **2020**, *12*, 1172. [CrossRef]
4. Minh, H.V.T.; Avtar, R.; Kumar, P.; Le, K.N.; Kurasaki, M.; Ty, T.V. Impact of rice intensification and urbanization on surface water quality in VMD using a statistical approach. *Water* **2020**, *12*, 1710. [CrossRef]
5. Molekoa, M.D.; Avtar, R.; Kumar, P.; Minh, H.V.T.; Dasgupta, R.; Johnson, B.A.; Sahu, N.; Verma, R.L.; Yunus, A.P. Spatio-temporal analysis of surface water quality in Mokopane area, Limpopo, South Africa. *Water* **2021**, *13*, 220. [CrossRef]
6. Agrawal, P.; Sinha, A.; Kumar, S.; Agarwal, A.; Banerjee, A.; Villuri, V.G.K.; Annavarapu, C.S.R.; Dwivedi, R.; Dera, V.V.R.; Sinha, J.; et al. Exploring Artificial Intelligence Techniques for Groundwater Quality Assessment. *Water* **2021**, *13*, 1172. [CrossRef]
7. Halder, S.; Kumar, P.; Das, K.; Dasgupta, R.; Mukherjee, A. Socio-hydrological approach to explore groundwater-human wellbeing nexus: Case study from Sundarbans, India. *Water* **2021**, *13*, 1635. [CrossRef]
8. Avtar, R.; Kumar, P.; Supe, H.; Jie, D.; Sahu, N.; Mishra, B.K.; Yunus, A.P. Did the COVID-19 lockdown induced hydrological residence time intensified the primary productivity in lakes? Observational results based on satellite remote sensing. *Water* **2020**, *12*, 2573. [CrossRef]
9. Usman, M.; Farooq, M.; Farooq, M.; Anastopoulos, I. Exposure to SARS-CoV-2 in Aerosolized Wastewater: Toilet Flushing, Wastewater Treatment, and Sprinkler Irrigation. *Water* **2021**, *13*, 436. [CrossRef]
10. Paudel, S.; Kumar, P.; Dasgupta, R.; Johnson, B.A.; Avtar, R.; Shaw, R.; Mishra, B.K.; Kanbara, S. Nexus between water security framework and public health: A comprehensive scientific review. *Water* **2021**, *13*, 1365. [CrossRef]
11. Mishra, B.K.; Kumar, P.; Saraswat, C.; Chakraborty, S.; Gautam, A. Water security in a changing environment: Concept, challenges and solutions. *Water* **2021**, *13*, 490. [CrossRef]

Article

Hydrogeochemical Assessment of Groundwater Quality of Mokopane Area, Limpopo, South Africa Using Statistical Approach

Mmasabata Dolly Molekoa [1], Ram Avtar [1,2,*], Pankaj Kumar [3], Huynh Vuong Thu Minh [1] and Tonni Agustiono Kurniawan [4]

[1]   Graduate School of Environmental Science, Hokkaido University, Sapporo 060-0810, Japan
[2]   Faculty of Environmental Earth Science, Hokkaido University, Sapporo 060-0810, Japan
[3]   Natural Resources and Ecosystem Services, Institute for Global Environmental Strategies, Hayama 240-0115, Japan
[4]   Key Laboratory of the Coastal and Wetland Ecosystems (Xiamen University), Ministry of Education College of the Environment and Ecology, Xiamen University, Xiamen 361102, Fujian, China
*   Correspondence: ram@ees.hokudai.ac.jp; Tel.: +81-011-706-2261

Received: 14 July 2019; Accepted: 7 September 2019; Published: 11 September 2019

Abstract: Despite being a finite resource, both the quality and quantity of groundwater are under tremendous pressure due to rapid global changes, viz. population growth, land-use/land-cover changes (LULC), and climate change. The 6th Sustainable Development Goal (SDG) aims to "Ensure availability and sustainable management of water and sanitation for all". One of the most significant dimensions of the SDG agenda is the emphasis on data and governance. However, the lack of good governance coupled with good observed data cannot ensure the achievement of SDG6. Therefore, this study strives to evaluate water quality status and hydrochemical processes governing it in the data-scarce Mokopane area of South Africa. Groundwater is the main source of fresh water supply for domestic usage, intensive agriculture, and mining activities in Mokopane. In this study, hydrogeochemical analysis of groundwater samples was employed to calculate the water quality index (WQI) and evaluate factors governing water quality evolution in the study area. Statistical and spatial analysis techniques were carried out to divide sampling sites into clusters and delineate principal factors responsible for determining water quality of the sampled groundwater. Results suggest that most of the physico-chemical parameters are within permissible limits for drinking water set by the World Health Organization (WHO), except for high fluoride in some samples. Na-HCO$_3$ is the most abundant water type followed by Mg-HCO$_3$, which indicates dominance of Na$^+$, Mg^{2+}, and HCO$_3{}^{\pm}$. Rock-water interaction is the prime factor responsible for fluoride enrichment in water. The alkaline nature of groundwater favors the release of exchangeable F$^-$ from minerals like muscovite. The WQI suggests that 80% of water samples fall into the good and excellent categories. Poor management of untreated domestic sewage and agricultural runoff is a main factor for the bad/very bad categories of water samples. As the area lacks any credible scientific/government work to report water quality and its management aspects, the findings of this study will definitely help both scientific communities and policy makers to do what is needed for sustainable water resource management in a timely manner.

Keywords: groundwater; geospatial analysis; hydrogeochemical assessment; multivariate statistical analysis; water quality index

---

1. Introduction

Groundwater is one of the most important finite natural resources for fresh water on Earth [1–3]. Its importance has increased as people depend on it for agriculture and industry, with domestic use exponentially increasing [4–8]. Groundwater exploration and climate change in recent years have led to accelerated depletion of groundwater levels as well as deterioration of groundwater quality, especially in shallow aquifers [9]. Groundwater quality is as important as its quantity, and therefore it is essential to consider in depth the holistic approach towards its management [10]. According to Subramani et al. [11], ground water quality is relatively more important than surface water as it is less susceptible to contamination/pollution and is suitable for different purposes. The quality or physico-chemical characteristics of groundwater is highly dependent on anthropogenic activities like agriculture, mining, urban settlements, etc.; as well as natural processes like rock–water interaction, geological formations, and aerobic/anaerobic conditions of the aquifers [6,12–14]. It is reported that spatio-temporal variation in groundwater quality is also predominantly governed by the hydrogeochemical process occurring at the recharge area and groundwater–surface water interaction [15]. Also, climate change and frequent extreme weather conditions exaggerate the severity of water scarcity [16–18]. Groundwater quality assessment can indicate its suitability for different purposes like drinking and irrigation. The concentration of different chemical parameters beyond their permissible limit leads to serious health complications such as fluorosis, blue baby syndrome, and arsenicosis from fluoride, nitrate, and arsenic, respectively [19,20]. Weathering and leaching are the biggest drivers for high fluoride concentration in groundwater [21,22], whereas nitrate enrichment is mainly attributed to agricultural runoff with high usage of fertilizers and leaching from untreated sewerage systems [7,23]. Therefore, diligent monitoring is a vital part of groundwater quality management, and thus, for maintenance of healthy aquatic ecosystems and human health [24–26].

Usage of different tools and techniques (e.g., numerical modelling, geographical information systems (GIS)) by both scientific communities and decision makers for groundwater quality assessment and management has rapidly increased over the past few decades [27–30]. A GIS is one of the proven vital tools in studies of groundwater for different purposes (spatial distribution of contaminants, estimating the effect of land-use/land-cover (LULC) changes on water quality, overlay analysis for estimating groundwater potential, etc.) throughout the world [31–35]. In this regard, the use of the water quality index (WQI) can provide simplistic information from complex water quality data to the public and managers. WQI can be used for assessment of surface as well as groundwater quality. Moreover, various WQIs that were developed for irrigation and drinking purposes can help to minimize the adverse effects of water use [6,36]. With the above background, this study focuses on groundwater quality assessment in Mokopane in South Africa using laboratory analysis of water quality parameters well-supported by interpretation through different statistical and GIS tools. This integrated assessment has helped us to understand well the processes governing hydro-geochemical changes in spatial patterns, and hence, the direction for effective management of groundwater resources. The rationale behind choosing this study area is that most of the households in this area rely on groundwater as the sole source of fresh water. However, there are various local reports and complaints relating to groundwater quality deterioration credited to poor sanitation, mining activities, and extensive agriculture in the area [37]. Also, the majority of Mokopane communities rely on pit toilets which have the potential to adversely affect the groundwater quality. Despite the above facts, no detailed scientific study has been conducted to analyze the current status of groundwater quality in the Mokopane. Therefore, this study is an integrated approach (chemical, statistical, spatial analysis) to analyze groundwater quality and processes which are the most likely source of pollutants.

2. Study Area

2.1. Site Description

The study area is located in the northern part of Limpopo province of South Africa. Limpopo borders Namibia, Botswana, Zimbabwe, and Mozambique. Mokopane is a town located in the Waterberg district of Limpopo province (Figure 1) with a population of 328,905 in 2016 [38]. It is located at 24°11′2″ S latitude and 29°0′46″ E longitude. Figure 2 shows the climatic condition of the study area which falls within steppe climate with mean annual maximum and minimum temperatures of 23.4 °C and 13 °C respectively [39]. The average annual rainfall is 490mm which normally occurs in November, December, and January. Mokopane has less rainfall during the winter season. July and August are normally dry due to the winter climate.

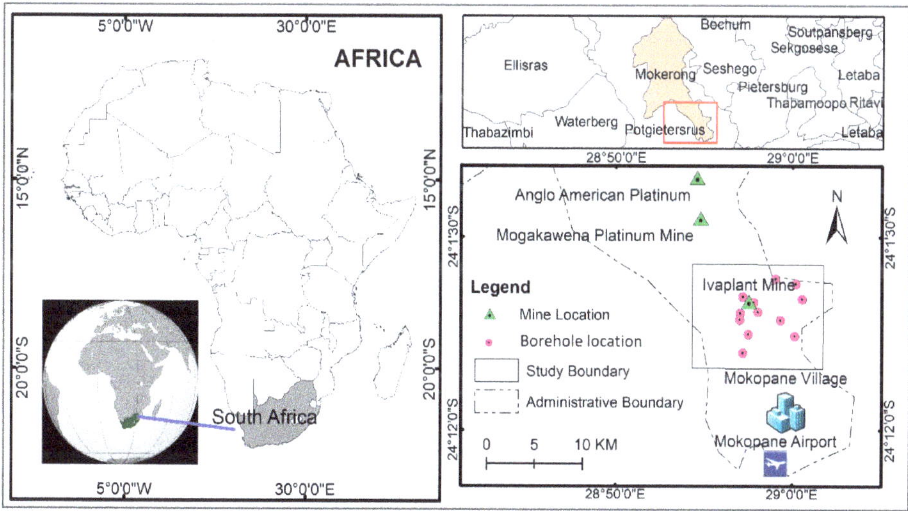

Figure 1. Location of the study area sampling points and mining sites.

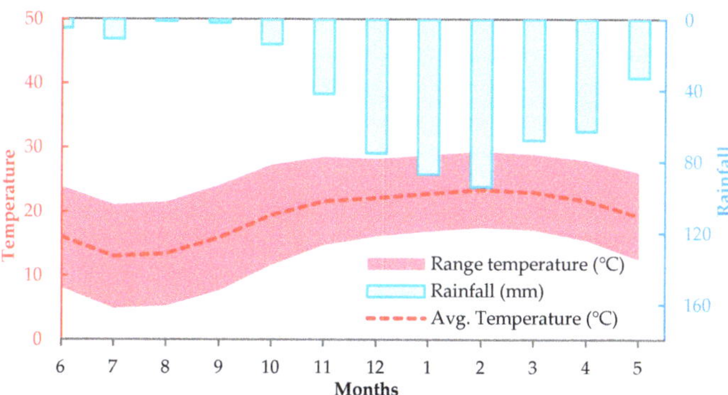

Figure 2. Climate data of average rainfall and temperature in Mokopane.

Economically, Mokopane is known for its intensive agricultural activities until the commencement of Mogalakwena Platinum mine and Ivanplats mine. The mines provide employment to a large portion of the Mokopane population. The Mogalakwena mine is located about 20 km northwest of Mokopane and Ivanplats mine inside Mokopane area. Geologically the study area has deposit of the Bushveld complex. It contains the world's largest reserves of platinum-group minerals (PGMs). The Bushveld Complex comprises the Rustenburg Layered suite, the Lebowa Granites, and the Rooiberg Felsics, that are overlain by the Karoo sediments [40,41]. The development of the mining sector in the study area triggered rapid development in informal settlements and, hence, poor domestic sewerage treatment in the area. Poor settlement and sewerage treatment can cause high risks to human health in the near future if these issues are not taken into consideration.

2.2. Land-Use/Land-Cover Classification

The LULC map of the study area was prepared using the Landsat 8 satellite image acquired on 16th March 2018. The area was classified into five classes using maximum likelihood classification algorithm in the ENVI 5.2 software (Figure 3). Five classes of LULC map were classified, viz. agriculture land, settlement, vegetation, bare land, and water bodies. The selection of classification classes was based on the field work and other existing maps. Figure 3 shows the LULC maps and locations of the water samples and Ivaplats mining area. Results show that most of the study area was covered with agriculture land (53%) followed by settlement (32%), vegetation (13%), bare land (2%), and water bodies (1%). Recently, the settlement area has increased in the study area because of the increase in mining activities resulting in high employment opportunities.

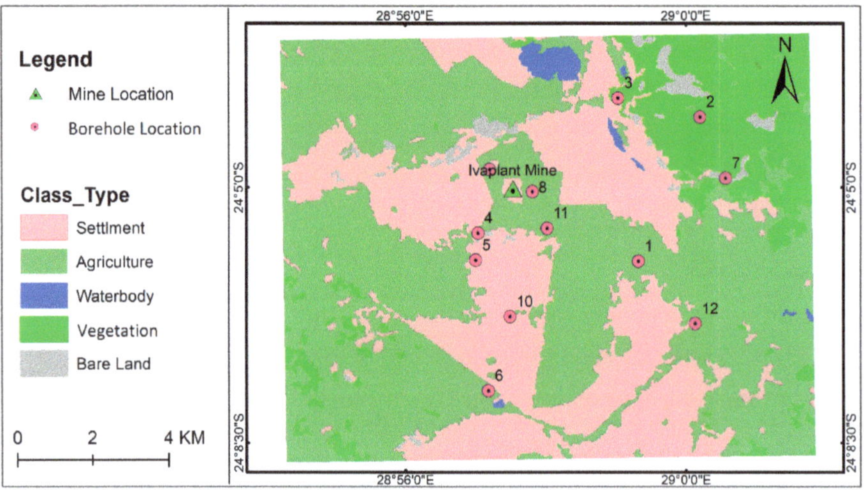

Figure 3. Land-use/land-cover (LULC) map based on Landsat 2018 by using maximum likelihood classification algorithm.

3. Methodology

3.1. Flow Chart of Research Methodology

Figure 4 shows the flowchart of methodology adopted in this study. Analyzed water quality data was further processed using piper diagram, speciation modeling, groundwater quality index (GWQI) development, statistical analysis, and was well supported by spatial image development.

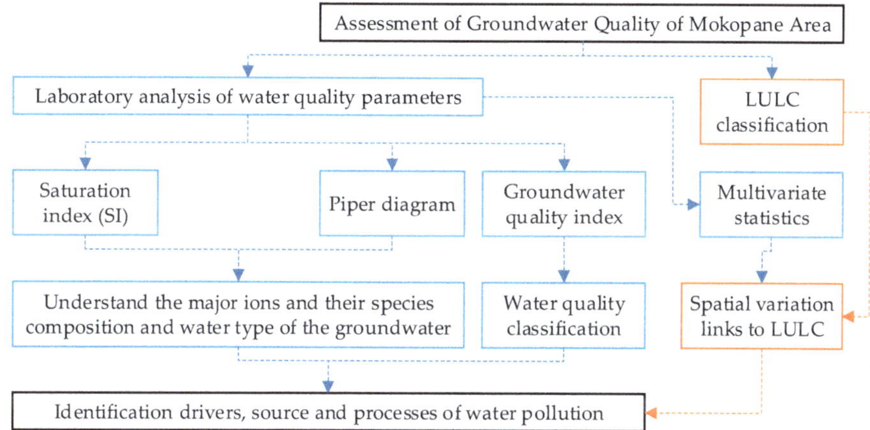

Figure 4. Flowchart of the methodology employed in this study for the assessment of groundwater quality.

3.2. Sample Collections and Hydrochemical Analysis

A total of 12 groundwater samples were collected on 12th June 2018 and analyzed for 21 physico-chemical parameters at Capricorn Veterinary Laboratories. Sampling coordinates were recorded using the Global Positioning System (GPS III, Garmin) as shown in Figure 1. The average well depth of groundwater sampling wells is 103.4 m below ground level (mbgl). Figure 1 shows the locations of the three major mining areas (Ivanplats mine, Mogakawena platinum mine, and Anglo American platinum mine). The inline flow cell that ensures the exclusion of minimized fluctuations and atmospheric contamination was used on-site in order to measure the electrical conductivity (EC) and pH. Precision of > 5% achieved with the Model Beverly, MA, 01.915, known as the transportable Orion Thermo Water analyzing kit, was used for all kinds of on-site measurements. Collected groundwater samples were stored in thoroughly rinsed polyethylene bottles. In order to stabilize trace metals (pH~2), the collected samples for major cation analysis were acidified by 1% HNO_3; while samples collected for nitrate were acidified with H_3BO_3. The concentration of HCO_3^- was analyzed by acid titration (using Metrohm Multi-Dosimat); while other anions Cl^-, NO_3^-, SO_4^{2-}, and PO_4^{3-} were analyzed by DIONEX ICS-90 ion chromatograph with an error percentage of less than 2% using duplicates. Inductively coupled optical emission spectrometry (ICP-OES) was used to evaluate major cations using duplicates, with a precision of less than 2%. One replicate was used to check the accuracy of the instruments after the analysis of every five samples for each instrument. The analytical precision was checked by the normalized inorganic charge balance (NICB) for major ions. The fractional difference between the total cations and anions is defined as $[(Tz^+ - Tz^-)/(Tz^+ + Tz^-)]$. Here, Tz^+ and Tz^- represent the total milli-equivalent of cations and anions, respectively. The observed charge balance supports the quality of the data points, which is better than ±8%, and this charge imbalance was generally in favour of positive charge.

3.3. Statistical Analysis

Different multivariate statistical tools, viz. cluster analysis (CA) using agglomerative hierarchical clustering (AHC) and principal component analysis (PCA) were applied on the analyzed dataset to find out the interrelationship between the groundwater quality variables and to determine the major factors influencing the groundwater quality [7,12,14,42–46]. In addition, the Piper diagram was also used to classify water samples into water types [46,47].

3.3.1. Saturation Index for Estimating Groundwater Mineralization

The saturation index (SI) was calculated in order to understand the origin of groundwater mineralization in the study area. According to Garrels and Christ [48], mineral saturation states with respect to selected minerals were computed as shown by Equation (1).

$$SI = log_{10}(K_{IAP}/K_{SP}) \tag{1}$$

where SI is the saturation index, K_{IAP} is the ion activity product of a particular solid phase, and K_{SP} is the solubility product of the phase. Based on the SI values, the saturation states were recognized as saturated (equilibrium; SI = 0), unsaturated (dissolution; SI < 0), and oversaturated (precipitation; SI > 0).

3.3.2. Water Quality Index

This study has used the water quality index (WQI), which has been considered as one of the most reliable tools to classify water contamination levels for both ground and surface water [49]. WQI is based on a scale of 0–300 whereby lower values indicate good water quality, whereas higher values are an indication of contaminated water. Based on the WQI values of < 50, 50–100, 100–200, 200–300, and > 300, water samples were categorized as excellent, good, poor, very poor, and unsuitable for drinking, respectively. The following steps were taken in order to calculate WQI:

1. Calculating relative weight: this was calculated using Equation (2).

$$W_i = \frac{w_i}{\sum_i^n w_i} \tag{2}$$

   where W_i represents the relative weight of each parameter sampled, w_i represents the weight of each parameter, and n represents the total number of parameters. All calculated values of W_i are shown in Table 1.

2. Calculating Q value: this was calculated using Equation (3).

$$Q_i = \frac{C_i \times 100}{S_i} \tag{3}$$

   where Q_i is quality rating, C_i is concentration of each parameter (mg/L), and S_i is derived from the WHO water quality standard.

3. Finally, WQI was calculated using Equation (4).

$$WQI = \sum W_i \times Q_i \tag{4}$$

Water quality classification and ratings were designed by Singh et al. [50]. In their paper, they have shown that excellent ≤ 50, good water = 50–100, poor water = 100–200, very poor = 200–300, and water unsuitable for drinking purpose ≥ 300. Table 1 shows the relative weights of chemical parameters.

3.3.3. Spatial Analyses

In general, asymmetrical distribution of water quality occurred due to water samples being collected from many sites and at different time [51]. To reduce this asymmetry, statistical analysis is one of the methods often used to evaluate water quality variation in many studies in Vietnam, India, and Japan [7,12,52]. The various multiple statistical tools, namely, cluster analysis (CA), PCA, factor analysis, and the Piper diagram were useful to differentiate changes in water quality in a spatial pattern [7,12,52].

Table 1. The relative weight of chemical parameters.

Parameter	Parameter Standard	Q_i	Weighted Factor (w_i)	W_i
pH	9.5	89.47	4	0.07
EC	150	49.67	4	0.07
TDS	1000	48.4	4	0.07
SS	25	44	4	0.07
Ca	150	31.6	2	0.04
K	15	64.8	2	0.04
Mg	200	34.34	2	0.04
Na	200	27.38	3	0.06
Cl	200	5.14	4	0.07
F	1.5	16	5	0.09
NH_4	0.2	130	5	0.09
NO_3	1	34	5	0.09
SO_4	400	0.58	2	0.04
Si	9.2	22.39	3	0.06
PO_4	10	0.50	2	0.04
COD	75	24	3	0.06
Sum			54	1

3.4. Mapping Spatial Distribution of Groundwater Quality

Inverse distance weighting (IDW) interpolation is a geospatial method to see the spatial distribution of groundwater quality parameters. This interpolation method was developed by Shepard, Bartier, and Keller [53,54]. Although the IDW method contains a "bull's eye" which has an effect on concentric regions of the same value near the known points [55], IDW is based on the distance-weighting exponent which can precisely control the influence of the distance of the points [54]. The IDW interpolation method was used to see the spatial distribution of groundwater quality in Mokopane using ArcGIS 10.3 software.

4. Results and Discussion

4.1. General Groundwater Chemistry

A statistical summary of the groundwater quality is presented in Table 2. The pH values of the groundwater samples varied from 7.31 to 9.24 with an average value of 8.30, suggesting the alkaline nature of groundwater. The alkaline nature of groundwater also suggests enhanced dissolution due to the interaction between soil and rainwater [56]. In addition, the alkaline aquifer condition depicts the influence of an arid/semiarid climate with relatively low rainfall and high evaporation. The electrical conductivity of groundwater varied from 728.0–1895.0 µS/cm with an average value of 1022.17 µS/cm indicating high ionic activity in the area. This also suggests soil mineralization or salt deposition due to climatic conditions as supported by high pH. Looking into the analyzed water quality data for major ions, anionic abundance was in the order of $HCO_3^- > Cl^- > SO_4^{2-} > CO_3^{2-} > NO_3^- > F^- > PO_4^{3-}$, while cationic abundance was found in the order of $Na^+ > Mg^{2+} > Ca^{2+} > K^+$. The highest average concentration of HCO_3^- among anions is due to the presence of carbonaceous sandstones in the aquifers and weathering of carbonate minerals in rain water followed by subsequent precipitation of HCO_3^- along with other cations. Higher Cl^- and SO_4^{2-} concentration also categorized as secondary salt mainly came from the leaching of sewage effluent, especially due to poor sewage management in the informal settlements. In addition, higher concentrations of SO_4^{2-} can also be attributed to the leaching of organic matter and agricultural runoff carrying unutilized SO_4^{2-}. Concentration of NO_3^- and PO_4^{3-} is not of concern as it is well below the permissible WHO limits for all groundwater samples.

Concentration of F^- varied from 0.24 to 3.39 mg/L. Sporadic fluoride contamination, i.e., >1.5 mg/L permissible limit set by WHO, for some of the groundwater samples is a matter of concern and puts the local people at moderate risk of fluorosis (tooth decay particularly in children and bone deformation) if consumed for longer periods of time and not given timely attention for its remediation. For cations, $Na^+ > Mg^{2+} > Ca^{2+} > K^+$ and the average milli-equivalent ratio of $Na^+ + K^+/Mg^{2+} + Ca^{2+}$ was found to be 1.08, indicating the prominence of both alumina-silicate and carbonaceous weathering in the study area.

Table 2. Statistical summary of groundwater quality in the study area.

Parameter	Range	Average	St Dev	WHO Permissible Limit (WHO, 2011)
pH	7.31–9.24	8.30	0.79	-
Depth (mbgl)	60.0–169.0	103.42	33.3	-
Temp. (°C)	25.40–26.3	25.8	0.25	-
TH (mg/L)	81.62–440.46	269.52	105.0	50
EC (µs/cm)	728.0–1895	1022.17	333.54	-
TDS (mg/L)	473.00–1232	664.25	216.84	-
SS (mg/L)	10.00–42	21.50	12.11	-
HCO_3^- (mg/L)	162.50–544.6	346.67	106.81	-
CO_3^{2-} (mg/L)	0.00–119.2	41.32	43.96	-
Ca^{2+} (mg/L)	2.50–53.02	19.65	19.00	200
K^+ (mg/L)	1.46–9.72	3.45	2.35	200
Mg^{2+} (mg/L)	7.64–100.54	53.75	24.86	150
Na^+ (mg/L)	47.32–329.76	133.58	75.15	200
Mn (mg/L)	0.01–0.48	0.11	0.15	1
Cl^- (mg/L)	7.21–188.19	83.24	50.76	600
F^- (mg/L)	0.24–3.39	1.35	0.98	1.5
NH_4^+ (mg/L)	0.19–0.89	0.48	0.32	25
NO_3^- (mg/L)	0.06–19.96	4.99	8.09	50
PO_4^{3-} (mg/L)	0.01–0.01	0.01	1.81E-18	50
SO_4^{2-} (mg/L)	2.01–203.18	44.79	58.44	400
Si (mg/L)	0.43–36.46	10.12	12.75	-

TH—Total hardness, EC—Electric Conductivity, TDS—Total dissolved solid, SS—suspended solid, mbgl—meter below ground level.

The concentrations of most of the parameters were found to be well within the permissible limits set by the WHO (2011) except for F^- and Na^+ for few samples. High concentration of Na^+ in groundwater samples causes only scale formation in the pipelines due to high hardness of the water but does not cause any health effect to the consumers. Potential sources of high Na^+ are alumino-silicate weathering like schist, granitic rocks containing Na^+, and cation-exchange Na^+/K^+ with Ca^{2+}/Mg^{2+}. On the other hand, as high concentration of F^- in the groundwater (i.e., >1.5 mg/L) causes fluorosis (dental, skeletal), further investigation was made about it. It was found that 75% of the water samples contain F^- concentration less than 1.5 mg/L, hence the water is safe to drink or use for cooking. The rest of the water samples were categorized under low risk (17% with F^- concentration between 1.5 to 2.9 mg/L) and high risk (8% with F^- concentration between 3.0–5.0 mg/L).

The fluoride ion (F^-) normally remains non-reactive for groundwater with pH ≤ 7.0 because of its low solubility and remains absorbed in clay or exchangeable materials. However, once the pH of groundwater goes beyond 7.0 (i.e., alkaline medium), the hydroxyl group (OH^-) group replaces the exchangeable fluoride from silicate minerals like biotite, muscovite etc. resulting in fluoride enrichment. The whole chemical process can be explained for muscovite using Equation (5).

$$KAl_2[AlSi_3O_{10}]F_2 + 2OH^- \rightarrow KAl_2[AlSi_3O_{10}][OH_2] + 2F^- \tag{5}$$

Another possible chemical reaction, where hydrolysis of alumina-silicate minerals produces bicarbonate ions, which can enhance fluoride dissolution, is calculated using Equations (6) and (7).

$$NaAlSi_3O_8 + H_2CO_3 + H_2O \rightarrow Na^+ + HCO_3^- + 2H_4SiO_4 + 2Al_2Si_2O_5(OH)_4 \tag{6}$$

$$CaF_2 + 2HCO_3^- \rightarrow CaCO_3 + 2F^- + H_2O + CO_2 \tag{7}$$

IDW interpolation was used to show the spatial distribution of SO_4^{2-} and F^- as shown in Figure 5. The high concentration of F^- was observed in boreholes 2, 3, and 7, where it exceeded the permissible limit of 1.5 mg/L. On the other hand, high SO_4^{2-} concentration was observed in boreholes 8, 9, and 11. To further support its origin, speciation modelling for minerals related to fluoride was conducted. This is discussed later in the manuscript.

To further strengthen our findings, the depth profile of F^- and SO_4^{2-} is plotted and shown in Figure 6. It is found that F^- concentration has a significant association with groundwater well depth. Its concentration is generally higher for groundwater samples from higher well depths. This supports the ideas of geogenic origin as presented above. On the other hand, the higher concentration of SO_4^{2-} is found in groundwater samples with shallow well depths. As anthropogenic activities are also potential sources for SO_4^{2-}, it can be expected that leaching of untreated and unmanaged domestic sewage will contaminate shallow aquifers first, thus firmly supporting the findings.

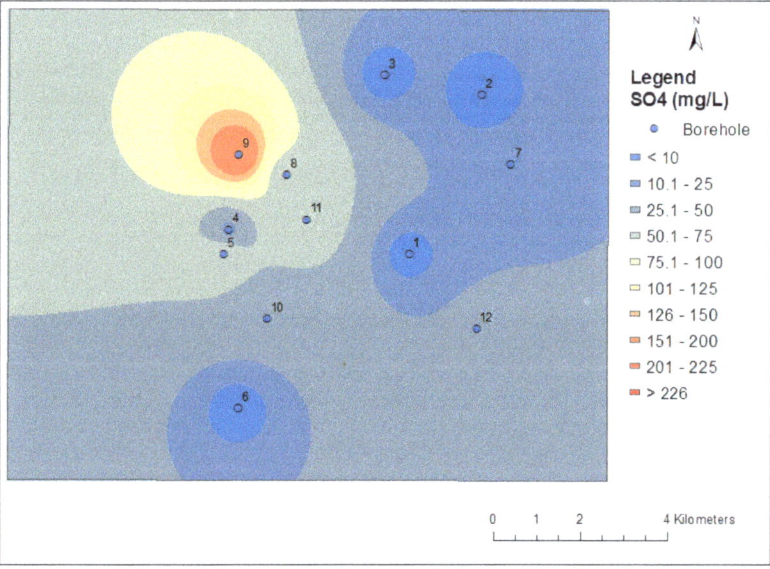

Figure 5. *Cont.*

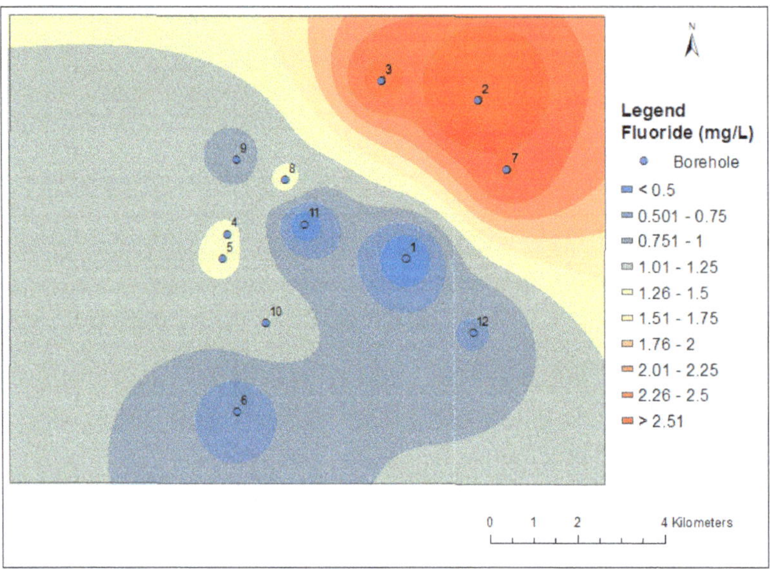

Figure 5. Spatial distribution of sulphate and fluoride using inverse distance weighting (IDW) interpolation in the study area.

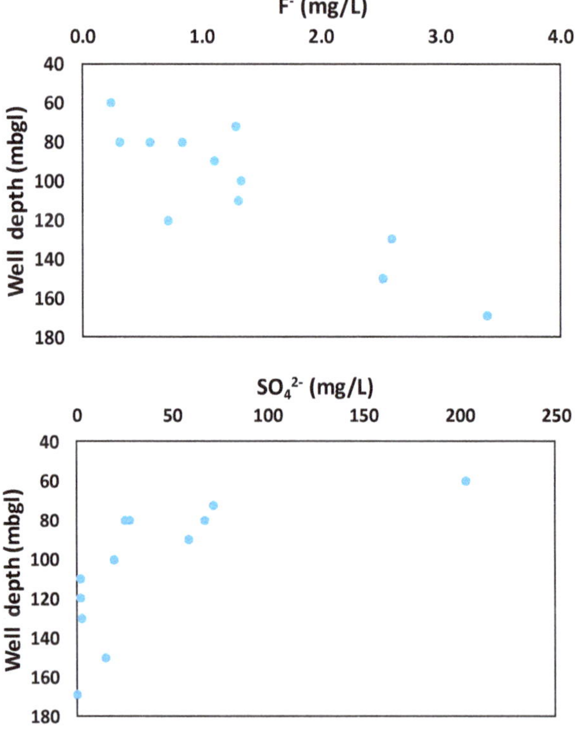

Figure 6. Scatter plot showing depth profile for F^- and SO_4^{2-}.

4.2. Graphical Presentations of Water Quality Data

In this study, physico-chemical data were further analyzed using the Piper diagram to get a deep insight into hydro-chemical processes controlling water quality of groundwater samples. The water type was calculated by plotting major cations and anions in the Piper diagram as shown in Figure 7. All water samples fell under three water facies, i.e., Na-HCO^3 (50%), Mg-HCO^3 (42%), and Na-Cl (8%). These water facies show chemical reactions during rock-water interaction within lithological context. Results indicate the dominance of weak acids (i.e., HCO^3) over hard acids (i.e., SO^4 and Cl). Also, ion exchange, dissolution of magnesite, semi-arid/arid climate, alkaline conditions as well as alumina-silicate/carbonate weathering are responsible for high concentrations of Na$^+$, Mg^{2+} and HCO$_3^-$. However, with poor management of pollution sources like untreated sewerage and agricultural runoff, there is a migration tendency from bicarbonate pole to the chloride pole. In addition, the effect of pit toilets still in practice for several local communities in the study area cannot be neglected. Henceforth, a further detailed investigation of microbial colonies (in particular, *E. coli* or fecal coliform) is a matter of research priority.

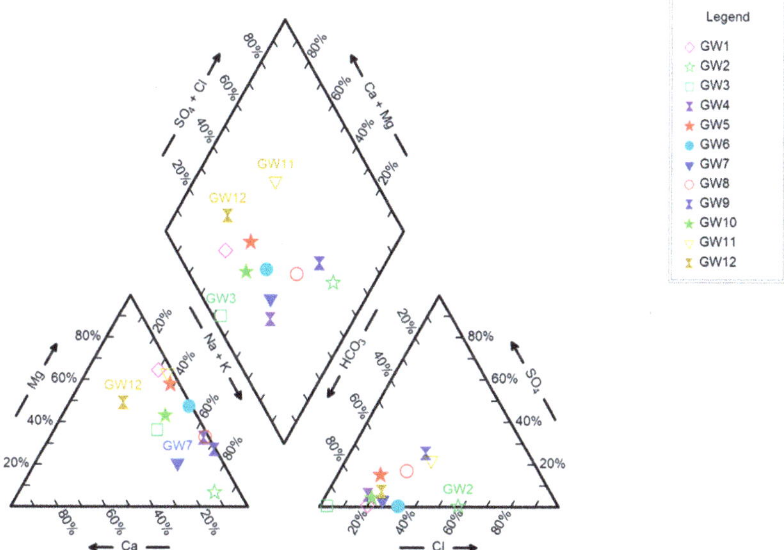

Figure 7. Piper diagram of 12 groundwater samples in the study area.

4.3. Speciation Modelling for Selected Minerals

Speciation modelling using the PHREEQC code [57] was performed to calculate the saturation index (SI) of selected minerals in the groundwater samples. The result is shown in Figure 8. A positive value of SI means water samples are saturated with respect to that mineral, while negative value indicates under-saturation of water samples with respect to those minerals. Here, six different minerals namely hydroxyapetite, fluorite, calcite, anhydrite, dolomite, and aragonite were selected which could possibly control the water chemistry in the study area. The results show that most of the groundwater samples are over-saturated with respect to the calcite, dolomite and aragonite minerals clearly depicting the source of HCO$_3^-$ through carbonate weathering. Looking in fluoride and hydroxyapatite, it was found that water samples were normally under-saturated except for a few samples. Looking for the saturation index for anhydrite, it was found that all water samples were unsaturated with respect to this mineral, indicating that both Ca^{2+} and SO$_4^{2-}$ are not coming from anhydrite. Also, SO$_4^{2-}$ is mainly coming from anthropogenic sources rather than geogenic origin.

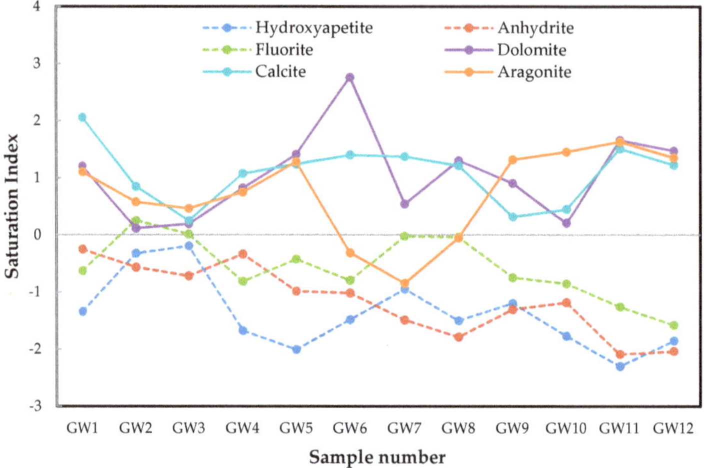

Figure 8. Saturation index for selected minerals related to fluoride in groundwater.

4.4. Groundwater Quality Index

WQI was calculated for groundwater samples. The results are shown in Figure 9 and Table A1. Computed WQI values ranged from 41.97 to 239.66 with an average value of 80.37, which can be placed in four categories, namely very poor water (1), poor water (1), good water (5), and excellent water (5). Still, more than 80% of the water samples fell into the good and excellent water categories. The very poor water type of only one groundwater sample (i.e., GW5) is mainly due to the combined effect of relatively higher concentrations of TDS, EC, and NO_3^- as well as high salt values. Also, looking at GW7 and GW10, WQI values were 77.78 and 176.77, respectively. These are marginal values and are susceptible to worsening into the poor and very poor categories, respectively. Therefore, timely actions for their management is very much advisable.

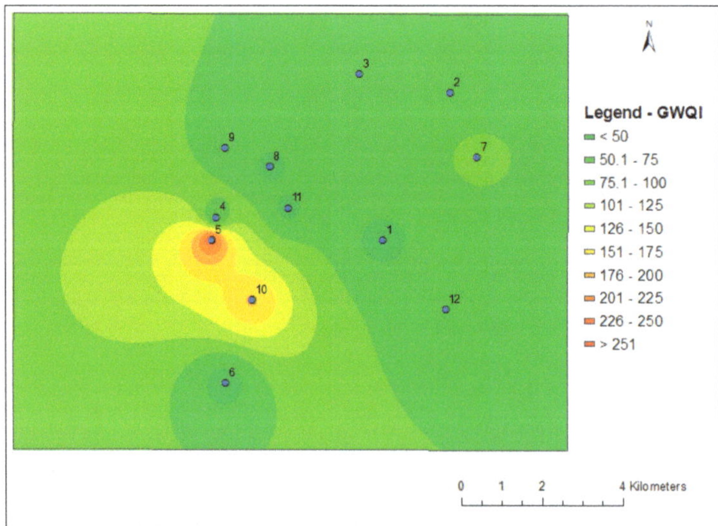

Figure 9. Spatial distribution of groundwater quality index (GWQI) using IDW interpolation in the study area.

4.5. Spatial Analysis of Groundwater Quality

Cluster analysis was used in order to classify water quality and the pollution sources for different samples. Here, with regard to the dendrogram cross section, all groundwater samples were divided into three groups (Figure 10a,b). Cluster 1 includes four groundwater samples i.e., GW2, GW3, GW7, and GW12. Cluster 2 includes five samples i.e., GW1, GW4, GW6, GW8, and GW11. Finally, cluster 3 includes the rest of the three samples including GW5, GW9, and GW10. The first group, which has four samples, is characterized in terms of where water quality is mainly governed by vegetation, agricultural activities, and close vicinity to settlements. We can compare the locations of water samples and the LULC map shown in Figure 3. It can be seen that group 1 samples are near vegetation, agricultural activities, and in close vicinity to settlements. Similarly, looking into group 2, water quality is mainly governed by hydro-chemical reactions going on both in agricultural land and mining areas, whereas water quality for group 3 is mainly being governed by hydrochemical processes going on in settlement areas.

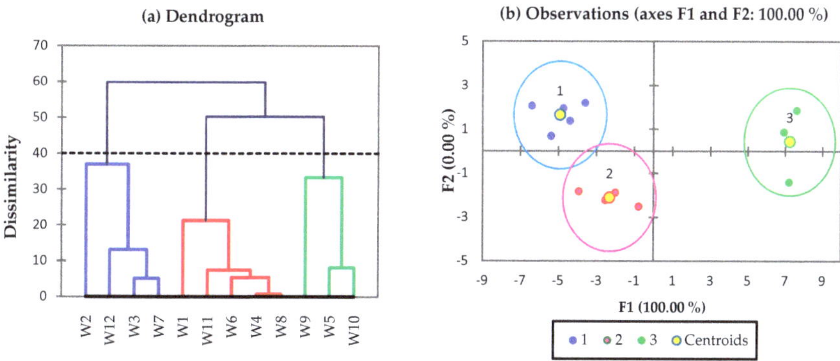

Figure 10. (**a**) Dendrogram showing spatial similarities of groundwater sampling locations using cluster analysis, (**b**) results from discriminant analysis crusted in 3 groups based on the distance metric. Ward agglomeration method and dissimilarity of euclidean distance was applied.

In addition to the dendrogram, similarities and dissimilarities among three groups were evaluated with the help of factor analysis. The result is shown in Figure 5. Here, the similarity distance among clusters indicates that the groundwater samples in different clusters with lesser gaps between them had similar characteristics with respect to groundwater quality due to the same background source and type of pollution sources. Results suggest that there is a significant difference between cluster 3 and the other clusters.

The result of PCA is shown in Figure 11. Here, both x and y axes represented by F1 and F2, explained 31.06% and 22.15% of the total variance, respectively. High variation for parameters like pH, EC, TH, CO_3^{2-}, HCO_3^-, K^+, Si, and Ca^{2+} were reported. Cluster 1 groundwater samples normally have high pH, CO_3^{2-}, SS, and K; whereas cluster 2 has higher values of SS, Si, and CaH. Finally, cluster 3 has high EC, TDS, CO_3^{2-}, HCO_3^-, and Si.

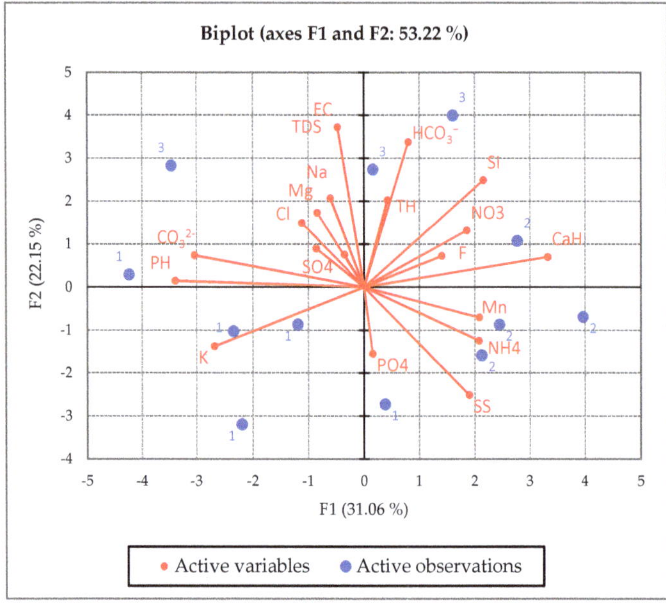

Figure 11. Result for biplot diagram for three clusters in the study area.

5. Conclusions

The integrated approach using statistical analysis, speciation modeling, and graphical analysis was applied to examine the hydrochemical characteristics of groundwater quality in the Mokopane area, South Africa. The study suggests that groundwater quality is in good condition, except for a few boreholes due to sporadic pollution from F^- and SO_4^{2-} in the study area. Mainly F^- pollution is of major concern due to very high concentration at sporadic locations. Results of speciation modelling for six selected minerals also suggest that most of the groundwater samples are over-saturated with respect to the calcite, dolomite, and aragonite minerals clearly depicting the source of HCO_3^- through carbonate weathering. Looking into fluoride and hydroxyapatite, it was found that water samples are normally under-saturated except for few samples. The mineral composition as suggested by speciation modelling and the alkaline nature of groundwater supports the chemical reaction for fluoride enrichment through weathering. The depth profile of SO_4^{2-} indicates that leaching of untreated sewage is the prime source of high SO_4^{2-} concentration especially in shallow aquifers. Overall, both natural processes (i.e., rock–water interaction, silicate weathering) and poor management of anthropogenic activities (i.e., mining, leaching from untreated sewage and agricultural runoff) have a combined effect on the evolution of water quality. Computed WQI values suggest that more than 80% of the water samples fell into the good and excellent water categories. Although few samples stood in the poor and very poor categories and the water quality may further worsen if timely actions for their management are not taken. In summary, this study gives a snapshot of the status quo of water quality which is of high scientific merit despite being based on a small number of samples in the data-scarce region of Mokopane area. This study will be useful as a baseline for the planning of sustainable water resources management in the mining area in the near future to help in implementation of SDGs. The limitation of this study is the limited number of borehole samples for investigation. Hence, a detailed examination of water quality along with soil samples on a larger spatio-temporal scale can be a subject of future study.

Author Contributions: Conceptualization, M.D.M., R.A., P.K., H.V.T.M.; methodology, M.D.M., R.A., P.K., H.V.T.M.; investigation, D.M., R.A.; writing, M.D.M., R.A., P.K., H.V.T.M.; writing—review and editing, M.D.M., R.A., P.K., H.V.T.M., T.A.K.

Funding: This research received no external funding.

Acknowledgments: The author is grateful to the Environmental Department of Ivanplats mine for providing data for this research paper, the Faculty of Environmental Earth Science (Hokkaido University) for facilities, and anonymous reviewers for their comments. The author further extends gratitude to the JICA ABE Initiative scholarship for the opportunity to study in Hokkaido University.

Conflicts of Interest: The authors declare no conflict of interest.

Appendix A

Table A1. Calculated water quality index (WQI) for each borehole water sample.

Boreholes	WQI	Water Quality
1	43.18	Excellent water
2	64.47	Good water
3	66.85	Good water
4	44.78	Excellent water
5	239.66	Very poor water
6	41.97	Excellent water
7	77.78	Good water
8	44.04	Excellent water
9	63.76	Good water
10	176.77	Poor water
11	44.81	Excellent water
12	56.34	Good water

References

1. USGS. How Important Is Groundwater? Available online: https://www.usgs.gov/faqs/how-important-groundwater?qt-News_science_products=0#qt-news_science_products (accessed on 11 July 2019).
2. Avtar, R.; Tripathi, S.; Kumar Aggarwal, A. Assessment of Energy–Population–Urbanization Nexus with Changing Energy Industry Scenario in India. *Land* **2019**, *8*, 124. [CrossRef]
3. Avtar, R.; Tripathi, S.; Aggarwal, A.K.; Kumar, P. Population-Urbanization-Energy Nexus: A Review. *Resources* **2019**, *8*, 136. [CrossRef]
4. Ahmad, M.; Bastiaanssen, W.G.; Feddes, R. Sustainable use of groundwater for irrigation: A numerical analysis of the subsoil water fluxes. *Irrig. Drain. J. Int. Comm. Irrig. Drain.* **2002**, *51*, 227–241. [CrossRef]
5. Avtar, R.; Kharrazi, A. Exploring Renewable Energy Resources Using Remote Sensing and GIS—A Review. *Resources* **2019**, *8*, 149. [CrossRef]
6. Minh, H.V.T.; Avtar, R.; Kumar, P.; Tran, D.Q.; Ty, T.V.; Behera, H.C.; Kurasaki, M. Groundwater Quality Assessment Using Fuzzy-AHP in an Giang Province of Vietnam. *Geosciences* **2019**, *9*, 330. [CrossRef]
7. Minh, H.V.T.; Kurasaki, M.; Van Ty, T.; Tran, D.Q.; Le, K.N.; Avtar, R.; Rahman, M.; Osaki, M. Effects of Multi-Dike Protection Systems on Surface Water Quality in the Vietnamese Mekong Delta. *Water* **2019**, *11*, 1010. [CrossRef]
8. Le, N.K.; Jha, K.M.; Jeong, J.; Gassman, W.P.; Reyes, R.M.; Doro, L.; Tran, Q.D.; Hok, L. Evaluation of Long-Term SOC and Crop Productivity within Conservation Systems Using GFDL CM2.1 and EPIC. *Sustainability* **2018**, *10*, 2665. [CrossRef]
9. Zahedi, S. Modification of expected conflicts between Drinking Water Quality Index and Irrigation Water Quality Index in water quality ranking of shared extraction wells using Multi Criteria Decision Making techniques. *Ecol. Indic.* **2017**, *83*, 368–379. [CrossRef]
10. Neisi, A.; Mirzabeygi Radford, M.; Zeyduni, G.; Hamzezadeh, A.; Jalili, D.; Abbasnia, A.; Yousefi, M.; Khodadadi, R. Data on fluoride concentration levels in cold and warm season in City area of Sistan and Baluchistan Province, Iran. *Data Brief.* **2018**, *18*, 713–718. [CrossRef]
11. Subramani, T.; Elango, L.; Damodarasamy, S.R. Groundwater quality and its suitability for drinking and agricultural use in Chithar River Basin, Tamil Nadu, India. *Environ. Geol.* **2005**, *47*, 1099–1110. [CrossRef]

12. Avtar, R.; Kumar, P.; Singh, C.K.; Sahu, N.; Verma, R.L.; Thakur, J.K.; Mukherjee, S. Hydrogeochemical Assessment of Groundwater Quality of Bundelkhand, India Using Statistical Approach. *Water Qual. Expo. Health* **2013**, *5*, 105–115. [CrossRef]

13. Kumar, P.; Ram, A. Chapter 4: Integrating major ion chemistry with statistical analysis for geochemical assessment of groundwater quality in coastal aquifer of Saijo plain, Ehime prefecture, Japan. In *Water Quality: Indicators, Human Impact and Environmental Health*; Nova Publication: Haryana, India, 2013; pp. 99–108. ISBN 978-1-62417-111-6.

14. Avtar, R.; Kumar, P.; Surjan, A.; Gupta, L.; Roychowdhury, K. Geochemical processes regulating groundwater chemistry with special reference to nitrate and fluoride enrichment in Chhatarpur area, Madhya Pradesh, India. *Environ. Earth Sci.* **2013**, *70*, 1699–1708. [CrossRef]

15. Matthess, G. *The Properties of Groundwater*; Department of Earth Science, J. Wiley and Sons Inc.: New York, NY, USA, 1982; p. 406.

16. Kumar, P.; Kumar, M.; Ramanathan, A.L.; Tsujimura, M. Tracing the factors responsible for arsenic enrichment in groundwater of the middle Gangetic Plain, India: A source identification perspective. *Environ. Geochem. Health* **2010**, *32*, 129–146. [CrossRef] [PubMed]

17. Dat, T.Q.; Kanchit, L.; Thares, S.; Trung, N.H. Modeling the Influence of River Discharge and Sea Level Rise on Salinity Intrusion in Mekong Delta. In Proceedings of the 1st Environment Asia International Conference, Bangkok, Thailand, 23–26 March 2011; Volume 35, pp. 685–701.

18. Ty, T.V. Scenario-based Impact Assessment of Land Use/Cover and Climate Changes on Water Resources and Demand: A Case Study in the Srepok River Basin, Vietnam—Cambodia. *Water Res. Manag.* **2012**, *26*, 1387–1407. [CrossRef]

19. World Health Organization (WHO). *Guidelines for Drinking-Water Quality*, 3rd ed.; WHO: Geneva, Switzerland, 2006; Volume 1, Available online: http://www.who.int/water_sanitation_health/dwq/gdwq0506.pdf (accessed on 8 July 2019).

20. UNICEF. *UNICEF Handbook on Water Quality*; United Nations Childrens Fund: New York, NY, USA, 2008.

21. Alagumuthu, G.; Rajan, M. Chemometric studies of water quality parameters of Sankarankovil block of Tirunelveli, Tamilnadu. *J. Environ. Biol.* **2010**, *31*, 581–586.

22. Singh, C.K.; Rina, K.; Singh, R.; Shashtri, S.; Kamal, V.; Mukherjee, S. Geochemical modeling of high fluoride concentration in groundwater of Pokhran area of Rajasthan, India. *Bull. Environ. Contam. Toxicol.* **2011**, *86*, 152–158. [CrossRef] [PubMed]

23. Kumar, P.; Kumar, A.; Singh, C.K.; Saraswat, C.; Avtar, R.; Ramanathan, A.; Herath, S. Hydrogeochemical evolution and appraisal of groundwater quality in Panna District, Central India. *Expo. Health* **2016**, *8*, 19–30. [CrossRef]

24. Chapman, D.V. *Water Quality Assessments: A Guide to the Use of Biota, Sediments and Water in Environmental Monitoring*; CRC Press: Boca Raton, FL, USA, 1996; ISBN 0-419-21590-5.

25. Olajire, A.A.; Imeokparia, F.E. Water Quality Assessment of Osun River: Studies on Inorganic Nutrients. *Environ. Monit. Assess.* **2001**, *69*, 17–28. [CrossRef] [PubMed]

26. Vasanthavigar, M.; Srinivasamoorthy, K.; Prasanna, M.V. Evaluation of groundwater suitability for domestic, irrigational, and industrial purposes: A case study from Thirumanimuttar river basin, Tamilnadu, India. *Environ. Monit. Assess.* **2012**, *184*, 405–420. [CrossRef] [PubMed]

27. Hua, A.K. Land use land cover changes in detection of water quality: A study based on remote sensing and multivariate statistics. *J. Environ. Public Health* **2017**, *2017*. [CrossRef]

28. Khan, A.; Khan, H.H.; Umar, R. Impact of land-use on groundwater quality: GIS-based study from an alluvial aquifer in the western Ganges basin. *Appl. Water Sci.* **2017**, *7*, 4593–4603. [CrossRef]

29. Narany, T.S.; Aris, A.Z.; Sefie, A.; Keesstra, S. Detecting and predicting the impact of land use changes on groundwater quality, a case study in Northern Kelantan, Malaysia. *Sci. Total Environ.* **2017**, *599*, 844–853. [CrossRef] [PubMed]

30. Machiwal, D.; Cloutier, V.; Güler, C.; Kazakis, N. A review of GIS-integrated statistical techniques for groundwater quality evaluation and protection. *Environ. Earth Sci.* **2018**, *77*, 681. [CrossRef]

31. Gustafsson, P. High resolution satellite data and GIS as a tool for assessment of groundwater potential of semi-arid area. In Proceedings of the IXth Thematic Conference on Geologic Remote Sensing, Pasadena, CA, USA, 8–11 February 1993.

32. Dabral, S.; Sharma, N. An Integrated Geochemical and Geospatial Approach for Assessing the Potential Ground Water Recharge Zones in Mahi-Narmada Inter Stream Doab Area, Gujarat, India. *J. Environ. Earth Sci.* **2013**, *3*, 134–144.

33. Demir, Y.; Erşahin, S.; Güler, M.; Cemek, B.; Günal, H.; Arslan, H. Spatial variability of depth and salinity of groundwater under irrigated ustifluvents in the Middle Black Sea Region of Turkey. *Environ. Monit. Assess.* **2009**, *158*, 279–294. [CrossRef] [PubMed]

34. Baalousha, H. Assessment of a groundwater quality monitoring network using vulnerability mapping and geostatistics: A case study from Heretaunga Plains, New Zealand. *Agric. Water Manag.* **2010**, *97*, 240–246. [CrossRef]

35. Dash, J.; Sarangi, A.; Singh, D. Spatial variability of groundwater depth and quality parameters in the national capital territory of Delhi. *Environ. Manag.* **2010**, *45*, 640–650. [CrossRef] [PubMed]

36. Abbasnia, A.; Yousefi, N.; Mahvi, A.H.; Nabizadeh, R.; Radfard, M.; Yousefi, M.; Alimohammadi, M. Evaluation of groundwater quality using water quality index and its suitability for assessing water for drinking and irrigation purposes: Case study of Sistan and Baluchistan province (Iran). *Hum. Ecol. Risk Assess. Int. J.* **2019**, *25*, 988–1005. [CrossRef]

37. Busari, O. Groundwater use in parts of the Limpopo Basin, South Africa. In Proceedings of the 3rd International Conference of Energy and Development-Environment-Biomedicine, Athens, Greece, 29–31 December 2009; pp. 13–18.

38. Statistics. South Africa Statistics. South Africa (Web) 2016. Available online: http://www.statssa.gov.za/ (accessed on 8 July 2019).

39. Climate Data. Climate Mokopane. Available online: https://en.climate-data.org/africa/south-africa/limpopo/mokopane-953/ (accessed on 1 July 2019).

40. Mineral Council in South Africa. Chamber of Mines. Available online: https://www.mineralscouncil.org.za/ (accessed on 18 August 2019).

41. USGS. *Platinum-Group Elements in Southern Africa—Mineral. Inventory and an Assessment of Undiscovered Mineral Resources*; US Department of the Interior: Reston, VA, USA, 2010; p. 126.

42. Xin, X.; Li, K.; Finlayson, B.; Yin, W. Evaluation, prediction, and protection of water quality in Danjiangkou Reservoir, China. *Water Sci. Eng.* **2015**, *8*, 30–39. [CrossRef]

43. Minh, H.V.T.; Ngoc, D.T.H.; Ngan, H.Y.; Men, H.V.; Van, T.N.; Ty, T.V. Assessment of Groundwater Level and Quality: A Case Study in O Mon and Binh Thuy Districts, Can Tho City, Vietnam. *Naresuan Univ. Eng. J.* **2016**, *11*, 25–33.

44. Lever, J.; Krzywinski, M.; Altman, N. Principal component analysis. *Nat. Methods* **2017**, *14*, 641. [CrossRef]

45. Shrestha, S.; Kazama, F. Assessment of surface water quality using multivariate statistical techniques: A case study of the Fuji river basin, Japan. *Environ. Model. Softw.* **2007**, *22*, 464–475. [CrossRef]

46. Busico, G.; Cuoco, E.; Kazakis, N.; Colombani, N.; Mastrocicco, M.; Tedesco, D.; Voudouris, K. Multivariate statistical analysis to characterize/discriminate between anthropogenic and geogenic trace elements occurrence in the Campania Plain, Southern Italy. *Environ. Pollut.* **2018**, *234*, 260–269. [CrossRef] [PubMed]

47. Fetter, C.W. *Applied Hydrogeology*, 3rd ed.; Macmillan College Publishing Company: New York, NY, USA, 1994.

48. Garrels, R.M.; Christ, C.L. *Solutions, Minerals, and Equilibria*; Harper & Row: New York, NY, USA, 1965.

49. Tiwari, T.; Mishra, M. A preliminary assignment of water quality index of major Indian rivers. *Indian J. Environ. Prot.* **1985**, *5*, 276–279.

50. Singh, S.; Hussian, A. Water quality index development for groundwater quality assessment of Greater Noida sub-basin, Uttar Pradesh, India. *Cogent Eng.* **2016**, *3*, 1177155. [CrossRef]

51. Fu, L.; Wang, Y.-G. Statistical Tools for Analyzing Water Quality Data. In *Water Quality Monitoring and Assessment*; InTech: London, UK, 2012; ISBN 978-953-51-0486-5.

52. Singh, C.K.; Shashtri, S.; Mukherjee, S.; Kumari, R.; Avatar, R.; Singh, A.; Singh, R.P. Application of GWQI to Assess Effect of Land Use Change on Groundwater Quality in Lower Shiwaliks of Punjab: Remote Sensing and GIS Based Approach. *Water Res. Manag.* **2011**, *25*, 1881–1898. [CrossRef]

53. Shepard, D. A two-dimensional interpolation function for irregularly-spaced data. In Proceedings of the 23rd ACM National Conference, Las Vegas, NV, USA, 27–29 August 1968; ACM: New York, NY, USA, 1968; pp. 517–524.

54. Bartier, P.M.; Keller, C.P. Multivariate interpolation to incorporate thematic surface data using inverse distance weighting (IDW). *Comput. Geosci.* **1996**, *22*, 795–799. [CrossRef]
55. Geographic Information Technology Tranning Alliance Distance-Based Interpolation. Available online: http://www.gitta.info/ContiSpatVar/en/html/Interpolatio_learningObject2.xhtml (accessed on 4 September 2019).
56. Subramanian, V.; Saxena, K. Hydrogeochemistry of groundwater in the Delhi region of India. In Proceedings of the Hamburg Symposium, Hamburg, Germany, 18–19 August 1983.
57. Parkhurst, D.L.; Appelo, C. User's guide to PHREEQC (Version 2): A computer program for speciation, batch-reaction, one-dimensional transport, and inverse geochemical calculations. *Water Resour. Investig. Rep.* **1999**, *99*, 312.

Article

Distribution Characteristics and Spatial Differences of Phosphorus in the Main Stream of the Urban River Stretches of the Middle and Lower Reaches of the Yangtze River

Lei Dong [1,2], Li Lin [1,2,*], Xianqiang Tang [1,2,*], Zhuo Huang [1,2], Liangyuan Zhao [1,2], Min Wu [1,2] and Rui Li [1,2]

[1] Basin Water Environmental Research Department, Changjiang River Scientific Research Institute, Wuhan 430010, China; dongleigushi@163.com (L.D.); ckyshj@163.com (Z.H.); zhaoliangyuannew@163.com (L.Z.); xiaomai2100@126.com (M.W.); Leeruiwh@outlook.com (R.L.)
[2] Key Lab of Basin Water Resource and Eco-Environmental Science in Hubei Province, Wuhan 430010, China
*   Correspondence: linli1229@hotmail.com (L.L.); ckyshj@126.com (X.T.)

Received: 26 December 2019; Accepted: 20 March 2020; Published: 23 March 2020

Abstract: Excessive phosphorus is the main problem of water pollution in the main stream of the Yangtze River, while it is not clear about the distribution characteristics and spatial differences of phosphorus in the urban river stretches of the middle and lower reaches of the Yangtze River. In this study, a field survey in June 2014 revealed that the average particulate phosphorus (PP) concentration ranged from 0.195 mg/L to 0.105 mg/L from Wuhan (WH) in the middle reaches of the Yangtze River to Shanghai (SH, 1081 km from WH) in the lower reaches of the Yangtze River, and the average PP-to-the total phosphorus (TP) ratio decreased from 85.71% in WH to 45.65% in SH, while the average soluble reactive phosphate (SRP) concentration ranged from 0.033 to 0.125 mg/L, and the average SRP-to-total dissolved phosphorus (TDP) ratio increased from 60.73% in WH to 88.28% in SH. In general, PP was still an important form of TP in the middle and lower reaches of the Yangtze River. The concentrations of PP and SRP at different sampling locations and water depths in the same monitoring section showed differences, which might be related to the transportation and sedimentation of suspended sediment (SS) and differences in the location of urban sewage outlets. Historical data showed that the concentration and particle size of the SS decreased over time, while the discharge of wastewater also increased over time in the Yangtze River Basin. The measured results showed that there was a significant positive correlation between SS and PP. As a result, the concentration of SRP might increase in the middle and lower reaches of the Yangtze River. If the SRP concentration is not properly controlled, the degree of eutrophication of water body could significantly increase in the Yangtze River estuary, the riparian zone of the urban river stretches, the tributary slow-flow section, and the corresponding lakes connected with the Yangtze River.

Keywords: middle and lower reaches of the Yangtze River; urban river stretches; phosphorus; spatial distribution; bioavailability; suspended sediment (SS)

---

1. Introduction

With changing natural conditions and economic and social development in the Yangtze River Basin, the hydrological characteristics and the pollutants caused by human activities have changed; thus, altering the concentrations of major pollutants. In 2006, concentrations of chemical oxygen demand (COD), ammonia nitrogen, total phosphorus (TP), and total nitrogen (TN) exceed the standard of Grade III based on the National Surface Water Environmental Quality Standards (GB 3838-2002) in China (China 2002) [1]. The contribution rates of COD and ammonia nitrogen from point sources accounted for 54.2% and 55.9% of contribution, respectively, indicating that COD and ammonia nitrogen mainly came from point source pollution, which is closely related to the large-scale discharge of industrial and domestic wastewater. However, the contribution rates of TN and TP from point sources were only 21.4% and 24.5%, respectively, indicating that nitrogen and phosphorus were mainly from nonpoint source pollution [1]. In the past ten years, under the guidance of the water pollution prevention and control policy aimed at reducing COD and ammonia nitrogen in the aquatic environment, the treatment effect of ammonia nitrogen and COD was remarkable in the Yangtze River Basin, but the prevention and control of TP were relatively weak. Thus, TP had become the primary excess pollutant in the main stream of the Yangtze River. Meanwhile, the eutrophication of rivers due to hydrodynamic reasons are generally less severe than those of relatively static water bodies, such as lakes and reservoirs, but the accumulation of nutrients increases the risk of eutrophication of rivers. TN/TP ratio can change the limiting characteristics of nutrients in water body and become one of the key factors regulating the growth of phytoplankton growth.

At present, various phosphorus studies and reports have mainly concentrated on phosphorus forms and its distribution characteristics in the Three Gorges Reservoir [2–5] and phosphorus in the sediment and water bodies of shallow lakes in the middle and lower reaches of the Yangtze River [6–8]. However, a previous study of the spatial distribution of phosphorus in the Yangtze River often used the phosphorus concentration of the sampling point to represent the phosphorus concentration of a monitoring section, which did not truly reflect the phosphorus concentration of the monitoring section. Since phosphorus and suspended sediment (SS) are closely related in the water body, phosphorus could be adsorbed onto the surface of SS and settle to the bottom, resulting in the uneven vertical distribution of phosphorus; additionally, the spatial layout and location of the sewage outlets of each city were also different, which revealed that phosphorus was inconsistent in different positions of the same monitoring section. However, few reports have analyzed and evaluate phosphorus pollution in the main urban river stretches along the middle and lower reaches of the Yangtze River [9,10]. It is well known that the middle and lower reaches of the Yangtze River are economically developed with large population, but fragile ecological environment. Yangtze River Basin has a large amount of pollutants discharged, with many hidden risks, and the pressure of drinking water safety is high [9]. Therefore, in order to protect the water environment quality of the Yangtze River, it is necessary to understand distribution characteristics and spatial differences of phosphorus in the middle and lower reaches of the Yangtze River.

The middle and lower reaches of the Yangtze River enter the flood season in June each year. Nutrients, such as phosphorus, enter the Yangtze River with urban sewage outlets and surface runoff, which leads to an increase in phosphorus content in the water body. The objectives of this study were to (1) determine the concentration levels of phosphorus in water body along main stream of the urban river stretches of the middle and lower reaches of the Yangtze River in the flood season (June 2014); (2) investigate the spatial distributions of particulate phosphorus (PP) and soluble reactive phosphate (SRP) in the main urban river stretches; (3)analyze the influence of cities on phosphorus distribution in water body in the main stream; and (4) study the possible changes in PP and SRP under the current situation. This work is expected to contribute to the understanding of the distribution behaviors of phosphorus in the main stream of the urban river stretches of the middle and lower reaches of the Yangtze River and provide a reference for water environment management and protection. In addition, our results can also be used as reference levels for future phosphorus monitoring programs of water body from the Yangtze River.

2. Sampling, Testing and Data Analysis

2.1. Sampling Sites and Sample Collection

In July 2014, water samples was collected at 14 representative sites (Figure 1), namely the upper reach of Wuhan (UWH), the lower reach of Wuhan (LWH), the upper reach of Jiujiang (UJJ), the lower reach of Jiujiang (LJJ), the upper reach of Anqing (UAQ), the lower reach of Anqing (LAQ), the upper reach of Nanjing (UNJ), the lower reach of Nanjing (LNJ), the upper reach of Zhenjiang (UZJ), the lower reach of Zhenjiang (LZJ), the upper reach of Nantong (UNT), the lower reach of Nantong (LNT), the upper reach of Shanghai (USH), and the lower reach of Shanghai (LSH). In the main stream of the urban river stretches of the middle and lower reaches of the Yangtze River, UWH, LWH, UJJ, LJJ, UAQ, LAQ, UNJ, LNJ, UZJ, LZJ, UNT, LNT, USH, and LSH were 616, 684, 879, 916, 1030, 1062, 1318, 1368,1399, 1436, 1567, 1600, 1651, and 1697 km from the Three Gorges Dam.

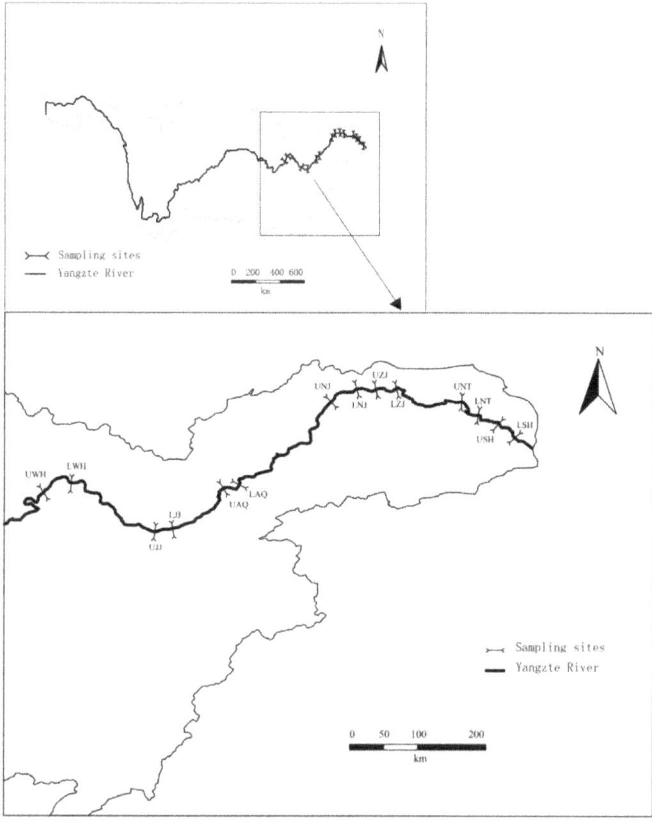

Figure 1. Location of sampling sites along the middle and lower reaches of the Yangtze River.

The upper reach of the urban river stretch included the central line, whereas the lower reach of the urban river stretch included the left and right shore and the central line, in which the central line was midstream in the river, and the left and right vertical lines were 50–100 m from the shores.

At each sampling site, water samples from the surface (0.2 times the water depth), middle (0.5 times the water depth), and bottom (0.8 times the water depth) of the column were obtained to analyze the concentrations of nitrogen and phosphorus nutrients. Some of the water samples were directly processed by digestion to measure TP and TN; the other water samples were immediately filtered through 0.45 μm cellulose acetate membranes to measure total dissolved phosphorus (TDP) and soluble reactive phosphate (SRP). Another 1 L water sample of each sampling site was immediately filtered through a 0.45 μm cellulose acetate membranes and then entrapped and dried to a constant weight of solid material at 103–105 °C to measure suspended sediment (SS). Moreover, the temperature (T), pH, electrical conductivity (EC), dissolved oxygen (DO), turbidity (TURB), chlorophyll-*a* (chl-*a*), and oxidation-reduction potential (ORP) were field measured with a multiparameter water quality sonde (USA, YSI EXO2). All water samples were collected on a hydrological ship.

2.2. Sample Measurement

Concentrations of TP and TDP were determined by potassium persulfate digestion and the ammonium molybdate spectrophotometric method (GB 11893-89). The PP concentration was measured by subtracting the TDP concentration from the TP concentration. The SRP in each sample was determined by the molybdenum blue/ascorbic acid method [4]. Concentrations of TN were determined by using a national standard entitled the alkaline potassium persulfate digestion-UV spectrophotometric method (HJ 636-2012). SS was determined by using the gravimetric national standard method (GB 11901-89). All water samples were measured on hydrological ship. After the water samples were collected, for every 10 water samples tested, one of the water samples should be randomly selected for parallel sample analysis. The relative error of the measured value of the parallel sample must be less than 5%. In addition, quality control samples were also added when testing water samples, and the determination results of quality control samples must be within the range of their standard values, so as to ensure the accuracy of the test results of each collected water sample.

2.3. Data Collection and Analysis

The concentrations of TN, TP, TDP, PP, and SRP in upper and lower reaches of the urban river stretches along the middle and lower reaches of Yangtze River were obtained from a field survey in July 2014. Hydrologic and sediment data were collected from the Changjiang Sediment Bulletin [11], while wastewater discharge in the Yangtze River Basin was obtained from the Water resources bulletin of the Yangtze River Basin and the southwestern rivers [12]. Correlation analysis of PP and SS were carried out using SPSS (IBM SPSS Statistics 22.0).

3. Results and Discussion

3.1. Water Physicochemical Properties

Major physical and chemical parameters at different sampling points are summarized in Table 1. The physical and chemical parameters of main urban river stretches were the average values of the upstream and downstream of the city. During the flood season, the water was weakly alkaline in general, and the average pH values ranged from 7.98 to 8.13. The average DO concentrations ranged from 6.85 to 7.77 mg/L, which satisfied the standard of class I-II water as outlined in the "Environmental Quality Standards for Surface Water" (GB 3838-2002). Moreover, along the main stream of the urban river stretches from the middle to lower reaches of the Yangtze River, the average turbidity level observably increased from 30.88 nephelometric turbidity units (NTU) in Wuhan (WH) to 64.40 NTU in Nanjing (NJ) and then decreased to 51.28 NTU in Shanghai (SH), which increased by 66.1% overall. The average temperature range was 21.8–23.9 °C, which tended to slightly decrease from the middle reach to the lower reach of the Yangtze River. The average EC level ranged from 151.5 to 243.0 μs/cm, and the average chlorophyll-a level ranged from 3.15 to 4.15 μg/L. The average ORP level moderately decreased from 533.4 mv in WH to 418.2 mv in SH, which decreased by 21.60% overall (Table 1).

Table 1. Water physicochemical parameters of the main urban river stretches along the middle and lower reaches of Yangtze River.

Sampling Sites	Parameters						
	pH (-)	DO (mg/L)	TURB (NTU)	T (°C)	EC (µs/cm)	chl-a (µg/L)	ORP (mv)
Wuhan (WH)	7.98 ± 0.05	7.77 ± 0.17	30.88 ± 5.81	23.9 ± 0.4	228.0 ± 35.1	4.15 ± 0.44	533.4 ± 37.1
Jiujiang (JJ)	8.06 ± 0.05	6.85 ± 0.14	48.38 ± 7.41	23.0 ± 0.3	243.0 ± 43.2	3.63 ± 0.45	500.8 ± 37.6
Anqing (AQ)	8.08 ± 0.03	6.92 ± 0.19	59.55 ± 5.72	23.4 ± 0.6	210.3 ± 54.1	4.00 ± 0.18	496.4 ± 33.5
NanJing (NJ)	8.13 ± 0.01	7.21 ± 0.13	64.40 ± 7.72	23.7 ± 0.4	151.5 ± 45.7	3.88 ± 0.41	483.6 ± 47.6
Zhengjiang (ZJ)	8.00 ± 0.02	7.67 ± 0.24	60.98 ± 4.67	23.5 ± 0.1	170.0 ± 49.6	4.25 ± 0.19	451.5 ± 42.5
Nantong (NT)	8.04 ± 0.07	6.78 ± 0.21	59.13 ± 3.37	22.3 ± 0.4	228.5 ± 40.9	3.53 ± 0.28	414.1 ± 27.8
Shanghai (SH)	8.01 ± 0.13	6.22 ± 0.04	51.28 ± 2.58	21.8 ± 0.5	174.5 ± 47.8	3.15 ± 0.34	418.2 ± 35.3

3.2. Distribution Characteristics of Phosphorus in the Main Urban River Stretches

Phosphorus is the first controlling element that restricts algae growth in most fresh water, which is also an important factor that causes water eutrophication [13–15]. The form of phosphorus in the water body is TDP and PP. The main existing form of TDP is SRP, while PP are mainly adsorbed on the surface of sediment particles as well as combined with living cells and organic clastic molecules. Therefore, it is crucially important to accurately monitor the quantity of various forms of phosphorus concentration and its change in water bodies, which is conducive for the comprehensive assessment of the state of the water environment.

3.2.1. Distribution of TP, PP, and Percent of PP to TP

TP is one of the important indicators for water quality evaluation. The TP and PP concentrations of main urban river stretches were the average values of the upstream and downstream of the city. As shown in Figure 2, the average TP concentration range from WH to SH was 0.161–0.245 mg/L, which exhibited an upward trend from JJ to ZJ along the Yangtze River. The average TP concentrations in the river sections of JJ and AQ were up to the standard class III, and the average TP concentrations in other urban river sections were class IV, with TP concentration analysis referring to the environmental quality standards for surface water (GB 3838-2002). Overall, as a result of intensive human activities and developed economies, water pollution affected the ecological environment of the middle and lower reaches of the Yangtze River [16,17]. Phosphorus mainly exists in the particle state in surface water [2], and PP is the main existing form of phosphorus in river systems and estuarine areas [18]. PP is also known as potential bioavailable phosphorus, which is the principal repository of bioavailable phosphorus (BAP) [19]. PP contributed a greater proportion of TP in the Yangtze River basin [4,5,20]. A survey from 1997 to 1998 showed that the ratio of PP-to-TP was approximately 93% in the wet season in the middle and lower reaches of the Yangtze River [21], and a 1998 and 1999 survey showed that the ratio of PP-to-TP reached up to 70%–90% [22].

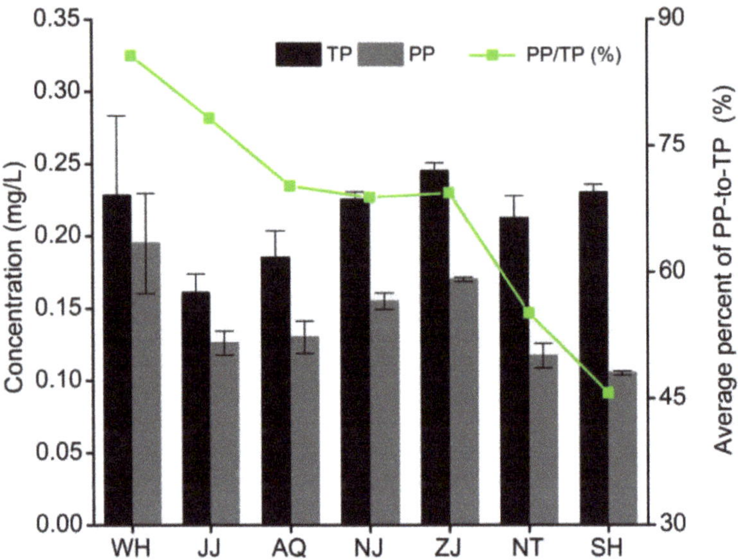

Figure 2. Distribution of total phosphorus (TP), particulate phosphorus (PP), and PP-to-TP (%) of water samples in the main urban river stretches along the middle and lower reaches of Yangtze River.

Taking this into account in this survey, the average PP concentration range from WH to SH was 0.105–0.195 mg/L, with the minimum and maximum values in SH and WH, respectively (Figure 2). When comparing the PP concentrations in WH to those in SH, more than 46.18% of PP was trapped and deposited in the main urban river stretches of the middle and lower reaches of the Yangtze River. The average PP-to-TP ratio decreased from 85.71% to 45.65% from WH to SH. Although PP was still the main form of TP in the middle and lower reaches of the Yangtze River, the PP-to-TP ratio exhibited a decreasing trend when compared with the results of historical literature [21,22].

The SS and PP concentrations of main urban river stretches were the average values of the upstream and downstream of the city. As shown in Figure 3, the average SS and PP concentrations decreased from 151.42 mg/L and 0.195 mg/L in WH to 90.62 mg/L and 0.105 mg/L in SH, respectively, exhibited similar decreases of 40.15% and 46.15%, respectively. SS and PP showed a significant positive correlation (r = 0.793, P < 0.01). Therefore, combining with the related literature [4,5], it was concluded that SS is the most important carrier of PP.

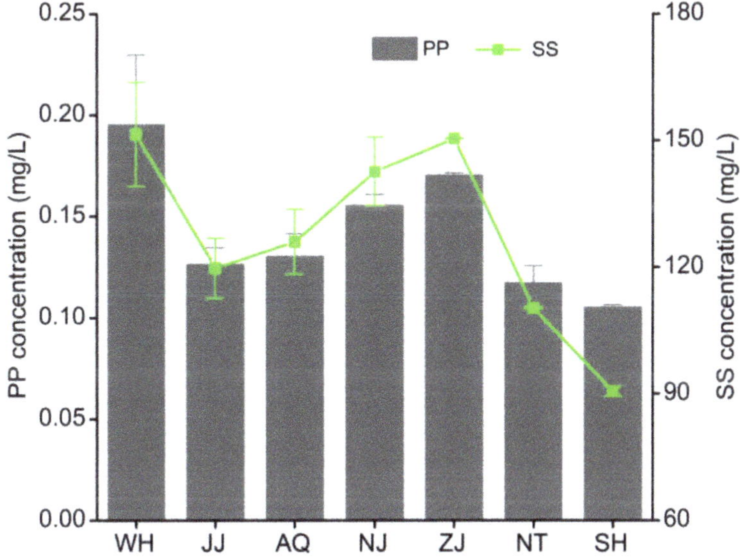

Figure 3. Distribution of PP and suspended sediment (SS) of water samples in the main urban river stretches along the middle and lower reaches of Yangtze River.

Overall, the average PP-to-TP ratio showed a decreasing trend from WH to SH, possibly because the average velocity of the river had a tendency to slow along the middle and lower reaches of the Yangtze River [23] and the SS settled gradually with the PP, the reducing the PP-to-TP ratio overall.

3.2.2. Distribution of TDP, SRP, and Percent of SRP to TDP

TDP is an important component of phosphorus in the water body [24,25]. The TDP and SRP concentrations of the urban section are the average values of the upstream and downstream of the city. As shown in Figure 4, the average TDP concentration range from WH to SH was 0.033–0.125 mg/L. The average TDP concentration along the middle and lower reaches of the Yangtze River demonstrated a gradually increasing trend, which reached its maximum in SH, approximately 3.8 times that observed in WH.

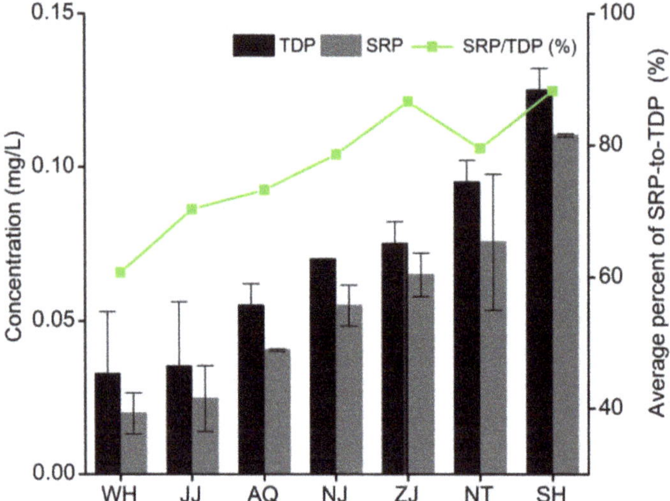

Figure 4. Distribution of total dissolved phosphorus (TDP), soluble reactive phosphate (SRP), and SRP to TDP (%) of water samples in the main urban river stretches along the middle and lower reaches of the Yangtze River.

SRP is a directly accessible phosphorus source for aquatic organisms, and it can be directly absorbed and used by aquatic plants and animals, such as phytoplankton, which causes eutrophication problems [25]. SRP gradually becomes one of the main factors for researching the phosphorus cycle and the biological effect of phosphorous in the aquatic environment. The average SRP concentration was 0.020-0.110 mg/L from WH to SH, which showed an overall increasing trend (Figure 4). The average SRP-to-TDP ratio increased from 60.73% in WH to 88.28% in SH, which indicated that inorganic phosphorus existed in the main form of SRP, which was consistent with results of previous literature [20].

From the natural point of view, the average velocity of the river had a tendency to slow [23], resulting in sedimentation of PP and an increase in the SRP concentration. From the perspective of human activities, according to the "Emission Standards for Pollutants in Urban Sewage Treatment Plants" (GB 18918-2002), even if the TP emission concentration reached the Class A standard (0.5 mg/L), it would not satisfy the class V water standard (GB 3838-2002), and phosphorus emission concentrations in urban river sections were still very high. Therefore, the SPR concentration exhibited an increasing trend along the middle and lower reaches of the Yangtze River.

3.2.3. Effect of Phosphorus Distribution on Water Eutrophication

Different TN/TP ratios of water body had certain effects on the growth and reproduction of phytoplankton [26–29]. The nutrient restriction criteria for eutrophication assessment are as follows: when the TN/TP ratio in water is <7, nitrogen is a limiting nutrient element; when the TN/TP ratio in water is >30, phosphorus will become a limiting factor for algae growth; when the TN/TP ratio is between 7 and 30, it represents the suitable proportion of algae growth; when phytoplankton are physiologically and biochemically active, they absorb TN and TP at a ratio of 16:1 [30,31].

The spatial distribution of TP and TN in the middle and lower reaches of the Yangtze River is given in Figure 5. The TN and TP concentrations of the upper reach of the urban river stretch are the average values of the surface, middle and bottom water samples in the central line of the upstream, whereas the TN and TP concentrations of the lower reaches of the urban river stretch are the average values of the water samples in different water layers (surface, middle and bottom) in the left shore, the central line and right shore of the downstream. The average nitrogen and phosphorus concentrations of water samples ranged from 1.42–2.30 mg/L to 0.152–0.267 mg/L, respectively, whereas the average range of the average ratio of TN/TP was 6.4–11.9 in the reaches of major cities along the middle and lower Yangtze River. The maximum and minimum values appeared in the UZJ and UJJ reaches of the Yangtze River, respectively.

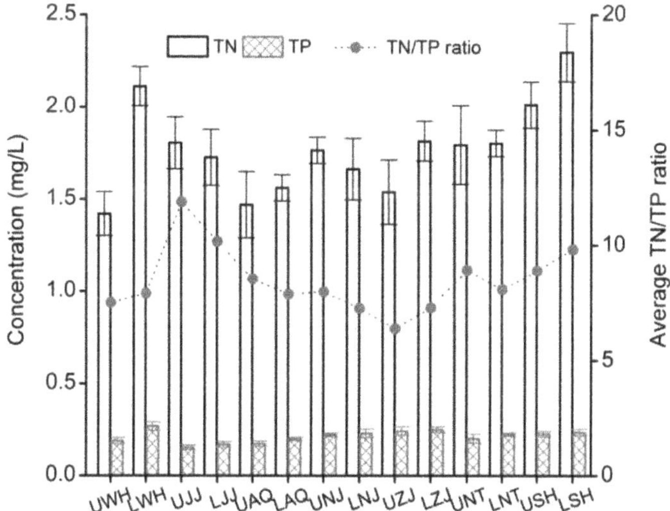

Figure 5. Comparison of total nitrogen (TN), TP, and TN/TP ratio in the upper reaches and the lower reaches of the urban river stretches along the middle and lower reaches of the Yangtze River.

Overall, in the flood season, the TN/TP ratios of water bodies were adapted for algae growth in the middle and lower reaches of the Yangtze River. However, due to its short residence time, strong exchange capacity and high flow velocity, the main stream of the middle and lower reaches of the Yangtze River had a low risk of large-scale cyanobacterial blooms.

3.3. The Impact of Cities on Phosphorus Distribution

The PP and SRP concentrations of the upper reach of the urban river stretch are the average values of the surface, middle and bottom water samples in the central line of the upstream, whereas the PP and SRP concentrations of the lower reaches of the urban river stretch are the average values of the water samples in different water layers (surface, middle and bottom) in the left shore, the central line and right shore of the downstream. As shown in Table 2, the average PP concentrations in the upper reaches of the urban sections were 0.106–0.171 mg/L, while the average PP concentrations in the lower reaches of the urban sections were 0.104–0.220 mg/L. Overall, the average PP concentrations of the lower reaches were relatively higher than those of the upper reaches. In particular, the average PP concentration in the lower reach of the WH section was 28.6% higher than that in the corresponding upper reach.

Moreover, the average SRP concentrations in the upper reaches of the urban sections were 0.015–0.111 mg/L, while the average SRP concentrations in the lower reach of urban sections were 0.024–0.110 mg/L. The average PP concentrations of the lower reach were also relatively higher than those of the upper reach.

Table 2. The PP and SRP concentrations from the upper reaches and the lower reaches of the urban river stretches along the middle and lower reaches of the Yangtze River (mg/L).

The Urban River Stretch	PP Concentration		SRP Concentration	
	The Upper Reach	**The Lower Reach**	**The Upper Reach**	**The Lower Reach**
WH	0.171 ± 0.012	0.220 ± 0.006	0.015 ± 0.001	0.024 ± 0.001
JJ	0.132 ± 0.002	0.120 ± 0.005	0.017 ± 0.001	0.032 ± 0.001
AQ	0.122 ± 0.009	0.138 ± 0.002	0.041 ± 0.006	0.040 ± 0.001
NJ	0.151 ± 0.007	0.159 ± 0.009	0.050 ± 0.003	0.060 ± 0.003
ZJ	0.171 ± 0.010	0.169 ± 0.011	0.060 ± 0.007	0.070 ± 0.004
NT	0.111 ± 0.007	0.123 ± 0.009	0.060 ± 0.006	0.091 ± 0.006
SH	0.106 ± 0.011	0.104 ± 0.013	0.111 ± 0.010	0.110 ± 0.013

In summary, the average PP and SRP concentrations of the lower reaches were relatively higher than those of the upper reaches. It can be concluded that urban sewage discharge contributes a large amount of phosphorus in the Yangtze River Basin. Therefore, the phosphorus emissions from the sewage treatment plants in major cities along the Yangtze River should also be monitored and controlled.

3.4. Spatial Differences of Phosphorus

3.4.1. PP and SRP in Surface and Bottom Layer Water

Figure 6 shows the PP distribution of the surface water and the bottom water from the central line in the upper and lower reaches of the urban river stretches. Generally, the PP concentrations of the surface and the bottom layers of the lower reaches at each urban section were higher than that of the corresponding layers of the upper reaches. Comparing vertically, the PP concentration distribution of each sampling section was uneven at each layer; on the whole, the PP concentration of the bottom layer was relatively higher than the surface concentration. Spatial differences in the SRP concentration were similar to those of the PP concentration.

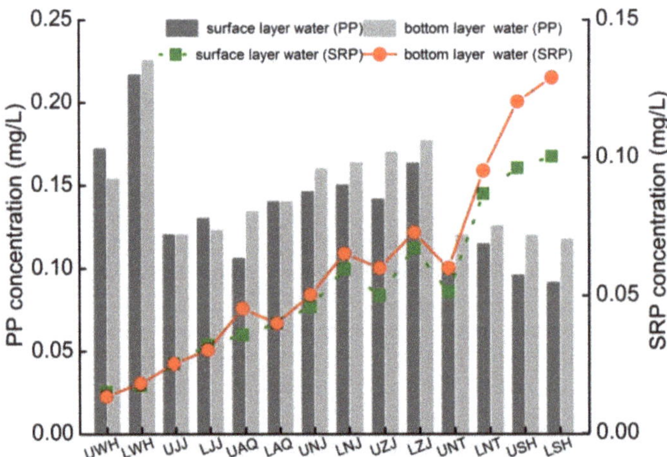

Figure 6. Comparison of PP and SRP concentrations of the surface water and bottom water from the central line in the upper and lower reaches of the urban river stretches.

Based on the monitoring data, there were certain differences between the TP and SRP concentrations in the surface water and the bottom water of each sampling section. Because the distribution of phosphorus was closely related to that of the SS in the water body, phosphorus could have been adsorbed onto the surface of SS and settled to the bottom, resulting in the uneven vertical distribution of phosphorus.

3.4.2. PP and SRP in the Left and Right Shores and the Central Line of Urban Lower Reaches

When comparing the PP concentrations in the left and right shore samples and in the central line from LWH to LSH, the PP concentrations of the left and right shores were higher than that of the central line in surface water samples (Figure 7). Spatial differences in the SRP concentration were similar to those of the PP concentration. Therefore, there were certain differences between the TP and SRP concentrations in the left and right shores and the central line of each sampling section. The main possible reason was that phosphorus could not be quickly diluted with slow water flow to the shore, which easily led to phosphorous accumulating near the shores; on the other hand, the concentration differences may be related to the location of the sewage outlets in the urban river stretches.

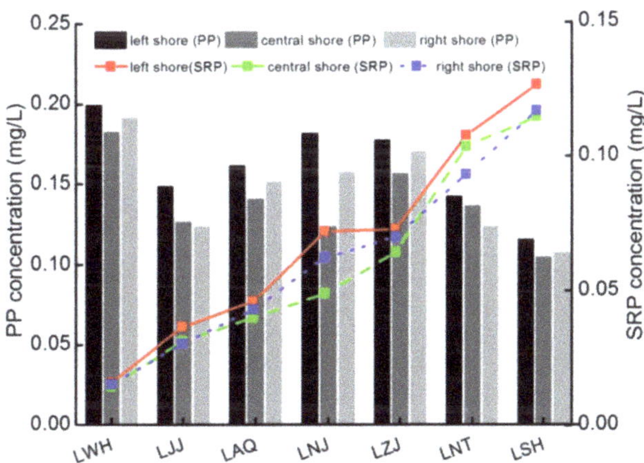

Figure 7. Comparison of PP and SRP concentrations in surface water samples from the lower reaches of urban river stretches. Legend: the lower reach of Wuhan (LWH), the lower reach of Jiujiang (LJJ), the lower reach of Anqing (LAQ), the lower reach of Nanjing (LNJ), the lower reach of Zhenjiang (LZJ), the lower reach of Nantong (LNT), and the lower reach of Shanghai (LSH).

3.5. The Possible Changes in PP and SRP under the Current Situation

3.5.1. The Possible Changes of PP

Based on historical monitoring data [11], the annual total discharged SS loads and the corresponding mean concentrations of Hankou Station decreased from 111 million tons and 149 mg/L in 2010 to 69.8 million tons and 94 mg/L in 2017, which were reduced by 37.18% and 36.91%, respectively, whereas the annual total discharged SS loads and the corresponding mean concentrations of Datong Station decreased from 185 million tons and 181 mg/L in 2010 to 104 million tons and 111 mg/L in 2017, which were reduced by 43.78% and 38.67%, respectively (Figure 8). The discharged SS volume and the corresponding mean concentrations tended to decrease in the main stream of the middle and lower reaches of the Yangtze River. The field survey results (June 2014) indicated that the SS in the middle and lower reaches of the Yangtze River was positively correlated with PP (Figure 3). The sedimentation of SS led to a decrease in PP along the middle and lower reaches of the Yangtze River. Therefore, it could be inferred that the PP concentration would show a decreasing trend with the decrease of the discharged SS volume and the corresponding mean concentrations in the middle and lower reaches of the Yangtze River.

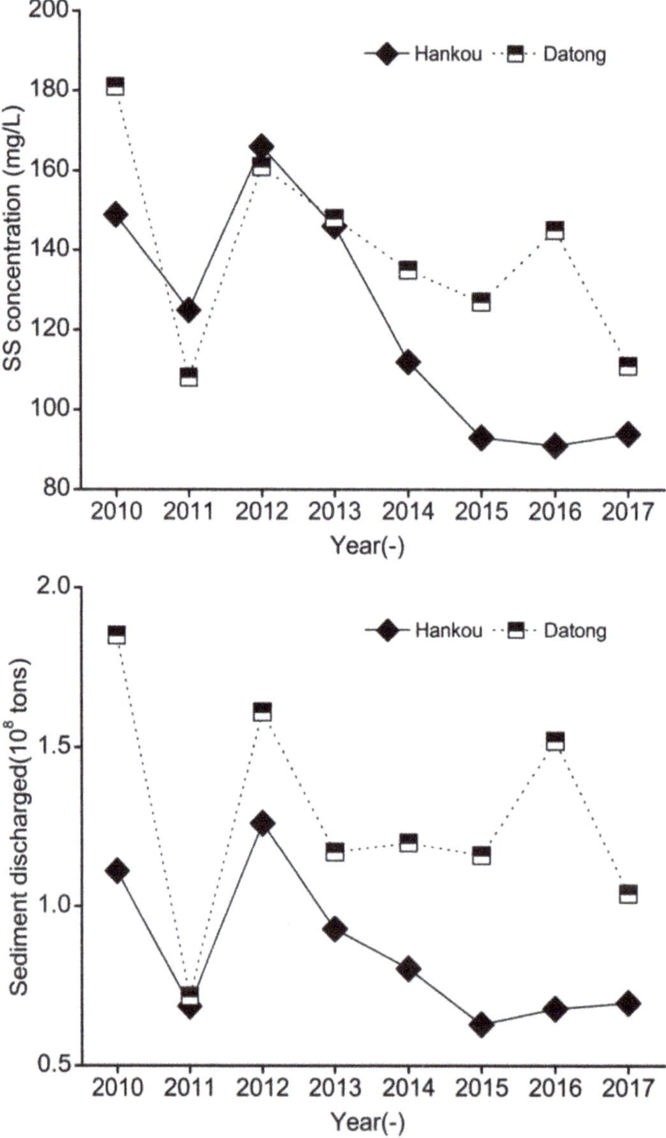

Figure 8. Variations in the annual mean SS concentrations and total sediment discharged volumes between 2010 and 2017.

3.5.2. The Possible Changes in SRP

The historical statistical results indicated that the annual median diameter and total runoff volume of Hankou changed from 13 μm and 747.2 billion tons in 2010 to 19 μm and 737.2 billion tons in 2017, respectively, whereas the annual median diameter and total runoff volume of Datong changed from 13 μm and 1022.0 billion tons in 2010 to 16 μm and 937.8 billion tons in 2017, respectively (Figure 9). Therefore, after the Three Gorges Reservoir (TGR) was fully operational in 2010, based on historical monitoring data from Hankou Station and Datong Station [11], the runoff volume of the middle and lower reaches of the Yangtze River did not change much, but the median diameter of the SS increased overall. Predictably, under the condition of stable runoff volume and TP concentration in the middle and lower reaches of the Yangtze River, the insufficient supply of discharged SS and the corresponding increase in median particle size would cause the high ratio of SRP discharge along the middle and lower reaches of Yangtze River. If the current trend continues to develop, SRP will not be well controlled, and the SRP concentration will continue to increase.

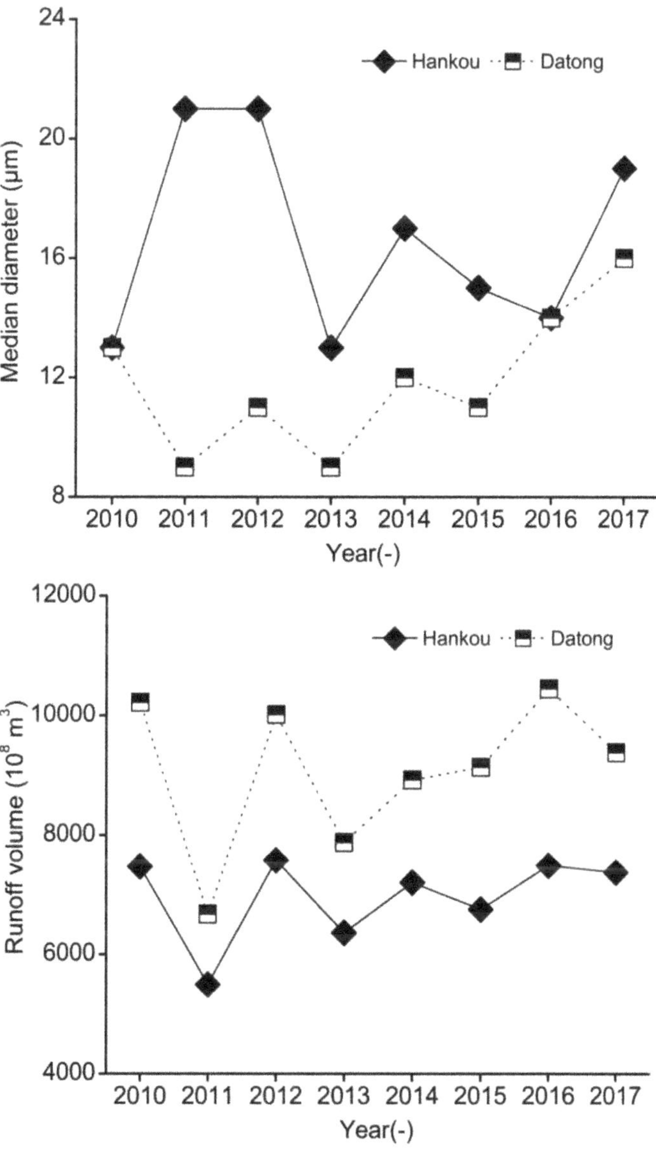

Figure 9. Variation in the annual mean median particle size of SS and total runoff volume between 2010 and 2017.

In summary, although there was no eutrophication problem along the middle and lower reaches of the Yangtze River, the phosphorus pollution problem was still very prominent. As shown in Figure 10, in general, the SS concentration had a decreasing trend in the middle and lower reaches of the Yangtze River; moreover, the discharge of wastewater in the Yangtze River Basin had also increased overall with the development of the social economy [12], with the result that the SRP had an increasing trend overall. It could be deduced that the pollution of phosphorus will still be adverse in the middle and lower reaches of the Yangtze River in the future. Pertinent research has also showed that phosphorus was in the cumulative state in the Yangtze River Basin, and the accumulation might continue in comparison with the experience of large river basins in Europe and the United States [32]. The continuous increase of phosphorus in the Yangtze River might first cause eutrophication of the water body in sensitive areas, even the risk of blooms; secondly, phosphorus can directly cause harm to human skin, causing various skin inflammations, and vomiting, diarrhea, headaches, and even poisoning.

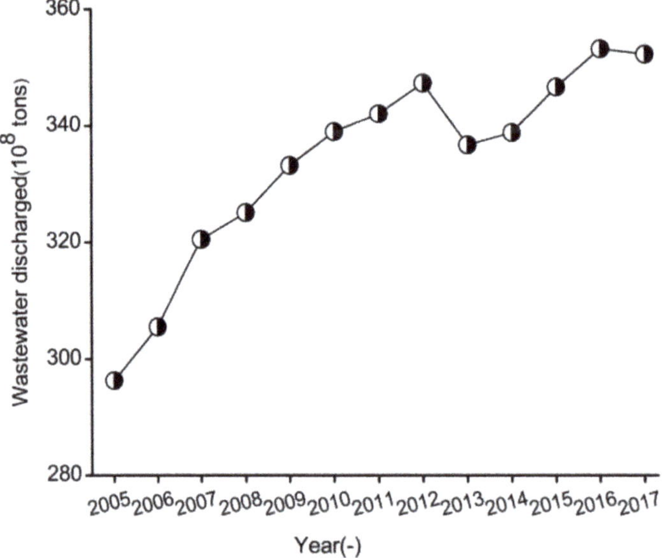

Figure 10. Wastewater discharge in the Yangtze River Basin between 2005 and 2017.

4. Conclusions and Suggestions

Based on monitoring data from June 2014, this paper studied the distribution characteristics and spatial differences of phosphorus in the main urban stretches of the Yangtze River. Combined with historical statistical results, the possible changes in PP and SRP were analyzed under the current situation. Moreover, the following suggestions were concluded with regard to the monitoring and control of phosphorus along the middle and lower reaches of the Yangtze River. The main conclusions and suggestions are as follows:

(1) Based on onsite data analysis in June 2014, the average PP changed from 0.195 mg/L to 0.105 mg/L from WH to SH, and the average PP-to-TP ratio decreased from 85.71% in WH to 45.65% in SH, which showed a gradual decrease, while the average SRP changed from 0.033 to 0.125 mg/L, and the average SRP-to-TDP ratio increased from 60.73% in WH to 88.28% in SH, showing a gradual upward trend. In general, PP was still the principal form of TP in the middle and lower reaches of the Yangtze River; however, the PP-to-TP ratio had a decreasing trend compared with that of historical data.

(2) Overall, the average PP and SRP concentrations of the lower reach of each city were relatively higher than those of the upper reach, which was highly likely to be associated with the discharge of wastewater in urban river stretches along the Yangtze River Basin. Therefore, phosphorus emissions of municipal wastewater treatment plants should also be monitored and analyzed in the main urban river stretches, and the water environmental capacity of phosphorus of the main urban river stretches would also need to be calculated and analyzed to strictly control phosphorus emissions of urban river stretches along the middle and lower reaches of the Yangtze River.

(3) Actual monitoring data showed that the concentrations of the PP and SRP of different sampling locations and water depths in same monitoring section showed differences. As a result, the phosphorus concentration of the sampling point did not represent the phosphorus concentration of the monitoring section. Therefore, the phosphorus flux of the monitoring section needs to be monitored.

(4) Combined with an evaluation of long-term monitoring data and field surveys, the SRP had an increasing tendency on the whole. The risk of water eutrophication was not optimistic, especially in the Yangtze River estuary, the riparian zone of the urban river stretches, the tributary slow-flowing river section and in the corresponding naturally connected lakes. Therefore, the research in the middle and lower reaches of the Yangtze River should focus on strengthening the emergency monitoring and rapid warning of phosphorus pollution in sensitive areas.

This work could help people understand the distribution behavior of phosphorus in the urban rivers in the middle and lower reaches of the Yangtze River and provided a reference for water environment management and protection. The current research results could also provide a reference level for phosphorus monitoring of the Yangtze River. Based on historical monitoring data, combined with model prediction, the migration and transformation of phosphorus in the middle and lower reaches of the Yangtze River should be focused on in the future, so as to provide decision-making for management departments.

Author Contributions: Conceptualization, L.D.; methodology, L.L. and X.T.; investigation, Z.H. and L.L., L.D., and L.Z.; resources, M.W. and R.L.; L.D. wrote the manuscript and all authors contributed to improving the paper. All authors have read and agreed to the published version of the manuscript.

Funding: This research was funded by the Strategic Priority Research Program of the Chinese Academy of Sciences (Grant No. XDA23040303), the State-level Public Welfare Scientific Research Institutes Basic Scientific Research Business Project of China (Grant No. CKSF2019380SH, CKSF2017062/SH), the Young Elite Scientist Sponsorship Program by CAST (Grant No. 2015QNRC001).

Acknowledgments: We thank the reviewers for their useful comments and suggestions.

Conflicts of Interest: The authors declare no conflict of interest.

References

1.   CWRC (Changjiang Water Resource Commission). *The Comprehensive Planning of the Yangtze River Water Resources*; Changjiang Press: Wuhan, China, 2008. (In Chinese)

2.   Duan, S.; Liang, T.; Zhang, S.; Wang, L.; Zhang, X.; Chen, X. Seasonal changes in nitrogen and phosphorus transport in the lower Changjiang River before the construction of the Three Gorges Dam. *Estuar. Coast. Shelf Sci.* **2008**, *79*, 239–250. [CrossRef]

3.   Sun, C.; Shen, Z.; Liu, R.; Xiong, M.; Ma, F.; Zhang, O. Historical trend of nitrogen and phosphorus loads from the upper Yangtze River basin and their responses to the Three Gorges Dam. *Environ. Sci. Pollut. Res.* **2013**, *20*, 8871–8880. [CrossRef] [PubMed]

4.   Tang, X.; Wu, M.; Li, R. Distribution, sedimentation, and bioavailability of particulate phosphorus in the mainstream of the Three Gorges Reservoir. *Water Res.* **2018**, *140*, 44–55. [CrossRef] [PubMed]

5.   Tang, X.; Wu, M.; Li, R. Phosphorus distribution and bioavailability dynamics in the mainstream water and surface sediment of the Three Gorges Reservoir between 2003 and 2010. *Water Res.* **2018**, *145*, 321–331. [CrossRef] [PubMed]

6. Jin, X.; Wang, S.; Bu, Q.; Wu, F. Laboratory Experiments on Phosphorous Release from the Sediments of 9 Lakes in the Middle and Lower Reaches of Yangtze River Region, China. *Water Air Soil Pollut.* **2006**, *176*, 233–251. [CrossRef]

7. Jin, X.C.; Wang, S.R.; Chu, J.Z.; Wu, F.C. Organic Phosphorus in Shallow Lake Sediments in Middle and Lower Reaches of the Yangtze River Area in China. *Pedosphere* **2008**, *18*, 394–400. [CrossRef]

8. Wu, P.; Qin, B.; Yu, G. Estimates of long-term water total phosphorus (TP) concentrations in three large shallow lakes in the Yangtze River basin, China. *Environ. Sci Pollut Res. Int.* **2016**, *23*, 4938–4948. [CrossRef]

9. Liu, R.; Chen, Y.; Sun, C.; Zhang, P.; Wang, J.; Yu, W. Uncertainty analysis of total phosphorus spatial-temporal variations in the Yangtze River Estuary using different interpolation methods. *Mar. Pollut Bull.* **2014**, *86*, 68–75. [CrossRef]

10. Zhou, J.; Zhang, M.; Lu, P. The effect of dams on phosphorus in the middle and lower Yangtze river. *Water Resour Res.* **2013**, *49*, 3659–3669. [CrossRef]

11. CWRC (Changjiang Water Resource Commission). *Changjiang Sediment: Bulletin*; Changjiang Press: Wuhan, China; pp. 2010–2017. (In Chinese)

12. CWRC (Changjiang Water Resource Commission). *Water Resources Bulletin of the Yangtze River Basin and the Southwestern Rivers*; Changjiang Press: Wuhan, China, 2005–2017. (In Chinese)

13. Han, C.N.; Zheng, B.H.; Qin, Y.W.; Ma, Y.Q.; Yang, C.C.; Liu, Z.C.; Cao, W.; Chi, M.H. Impact of upstream river inputs and reservoir operation on phosphorus fractions in water-particulate phases in the Three Gorges Reservoir. *Sci. Total Environ.* **2018**, *610*, 1546–1556. [CrossRef]

14. Heffernan, J.; Barry, J.; Devlin, M.; Fryer, R. A simulation tool for designing nutrient monitoring programmes for eutrophication assessments. *Environmetrics* **2010**, *21*, 3–20. [CrossRef]

15. King, K.W.; Balogh, J.C.; Agrawal, S.G.; Tritabaugh, C.J.; Ryan, J.A. Phosphorus concentration and loading reductions following changes in fertilizer application and formulation on managed turf. *J. Environ. Monit.* **2012**, *14*, 2929–2938. [CrossRef] [PubMed]

16. Huang, J.; Wang, X.; Xi, B.; Xu, Q.; Tang, Y.; Jia, K. Long-term variations of TN and TP in four lakes fed by Yangtze River at various timescales. *Environ. Earth Sci.* **2015**, *74*, 3993–4009. [CrossRef]

17. Huang, Q.; Shen, H.; Wang, Z.; Liu, X.; Fu, R. Influences of natural and anthropogenic processes on the nitrogen and phosphorus fluxes of the Yangtze Estuary, China. *Reg. Environ. Chang.* **2006**, *6*, 125–131. [CrossRef]

18. Sonzogni, W.C.; Chapra, S.C.; Armstrong, D.E.; Logan, T.J. Bioavailability of PhosphorusI nputs to Lakes. *J. Environ. Qual.* **1982**, *11*, 555–563. [CrossRef]

19. Ellison, M.E.; Brett, M.T. Particulate phosphorus bioavailability as a function of stream flow and land cover. *Water Res.* **2006**, *40*, 1258–1268. [CrossRef]

20. Shen, Z. Phosphorus and Silicate Fluxes in the Yangtze River. *Acta Oceanol. Sin.* **2006**, *61*, 741–751.

21. Duan, S.; Zhang, S.; Chen, X.; Zhang, X.; Wang, L.; Yan, W. Concentrations of nitrogen and phosphorus and nutrient transport to estuary of the Yangtze River. *Environ. Sci.* **2000**, *21*, 53–57. (In Chinese)

22. Xu, K.; Hayashi, S.; Murakami, S.; Maki, H.; Xu, B.; Watanabe, M. Characteristics of Water Quality in the ChangjiangRiver: Observations Conducted in 1998 and 1999. *Acta Geogr. Sin.* **2004**, *59*, 118–124.

23. Guo, L.; He, Q. Freshwater flocculation of suspended sediments in the Yangtze River, China. *Ocean. Dynam.* **2011**, *61*, 371–386. [CrossRef]

24. Paerl, H.W.; Xu, H.; McCarthy, M.J.; Zhu, G.; Qin, B.; Li, Y. Controlling harmful cyanobacterial blooms in a hyper-eutrophic lake (Lake Taihu, China): The need for a dual nutrient (N & P) management strategy. *Water Res.* **2011**, *45*, 1973–1983. [PubMed]

25. Xu, H.; Paerl, H.W.; Qin, B.; Zhu, G.; Gao, G. Nitrogen and phosphorus inputs control phytoplankton growth in eutrophic Lake Taihu, China. *Limnol. Oceanogr.* **2010**, *55*, 420–432. [CrossRef]

26. Bergström, A.-K. The use of TN: TP and DIN: TP ratios as indicators for phytoplankton nutrient limitation in oligotrophic lakes affected by N deposition. *Aquat. Sci.* **2010**, *72*, 277–281. [CrossRef]

27. Cao, X.; Wang, J.; Liao, J.; Sun, J.; Huang, Y. The threshold responses of phytoplankton community to nutrient gradient in a shallow eutrophic Chinese lake. *Ecol. Indic.* **2016**, *61*, 258–267. [CrossRef]

28. Carstensen, J.; Klais, R.; Cloern, J.E. Phytoplankton blooms in estuarine and coastal waters: Seasonal patterns and key species. *Estuar. Coast. Shelf Sci.* **2015**, *162*, 98–109. [CrossRef]

29. Paudel, B.; Velinsky, D.; Belton, T.; Pang, H. Spatial variability of estuarine environmental drivers and response by phytoplankton: A multivariate modeling approach. *Ecol. Inform.* **2016**, *34*, 1–12. [CrossRef]

Water **2020**, *12*, 910

30. Kim, B.; Choi, K.; Kim, C.; Lee, U.H.; Kim, Y.H. Effects of the summer monsoon on the distribution and loading of organic carbon in a deep reservoir, Lake Soyang, Korea. *Water Res.* **2000**, *34*, 3495–3504. [CrossRef]
31. Nie, Z.Y.; Liang, X.Q.; Xing, B.; Ye, Y.S.; Qian, Y.C.; Yu, Y.W. The current water trophic status in Tiaoxi River of Taihu Lake watershed and corresponding coping strategy based on N/P ratio analysis. *Acta Ecol. Sin.* **2012**, *32*, 44–55.
32. Powers, S.M.; Bruulsema, T.W.; Burt, T.P. Long-term accumulation and transport of anthropogenic phosphorus in three river basins. *Nat. Geosci.* **2016**, *9*, 353–356. [CrossRef]

Article

Participatory Approach for More Robust Water Resource Management: Case Study of the Santa Rosa Sub-Watershed of the Philippines

Pankaj Kumar [1,*], Brian Alan Johnson [1], Rajarshi Dasgupta [1], Ram Avtar [2], Shamik Chakraborty [3], Masayuki Kawai [1] and Damasa B. Magcale-Macandog [4]

[1]   Natural Resources and Ecosystem Services, Institute for Global Environmental Strategies, Hayama, Kanagawa 240-0115, Japan; johnson@iges.or.jp (B.A.J.); dasgupta@iges.or.jp (R.D.); kawai@iges.or.jp (M.K.)
[2]   Faculty of Environmental Earth Science, Hokkaido University, Sapporo 060-0810, Japan; ram@ees.hokudai.ac.jp
[3]   Faculty of Sustainability Studies, Hosei University, Tokyo 102-8160, Japan; shamik.chakraborty.76@hosei.ac.jp
[4]   Institute of Biological Sciences, University of the Philippines, Los Banos, Laguna 4031, Philippines; dmmacandog@up.edu.ph
*   Correspondence: kumar@iges.or.jp; Tel.: +81-046-855-3858

Received: 27 March 2020; Accepted: 16 April 2020; Published: 19 April 2020

Abstract: Due to the cumulative effects of rapid urbanization, population growth and climate change, many inland and coastal water bodies around the world are experiencing severe water pollution. To help make land-use and climate change adaptation policies more effective at a local scale, this study used a combination of participatory approaches and computer simulation modeling. This methodology (called the "Participatory Watershed Land-use Management" (PWLM) approach) consist of four major steps: (a) Scenario analysis, (b) impact assessment, (c) developing adaptation and mitigation measures and its integration in local government policies, and (d) improvement of land use plan. As a test case, we conducted PWLM in the Santa Rosa Sub-watershed of the Philippines, a rapidly urbanizing area outside Metro Manila. The scenario analysis step involved a participatory land-use mapping activity (to understand future likely land-use changes), as well as GCM precipitation and temperature data downscaling (to understand the local climate scenarios). For impact assessment, the Water Evaluation and Planning (WEAP) tool was used to simulate future river water quality (BOD and E. coli) under a Business as Usual (BAU) scenario and several alternative future scenarios considering different drivers and pressures (to 2030). Water samples from the Santa Rosa River in 2015 showed that BOD values ranged from 13 to 52 mg/L; indicating that the river is already moderately to extremely polluted compared to desirable water quality (class B). In the future scenarios, we found that water quality will deteriorate further by 2030 under all scenarios. Population growth was found to have the highest impact on future water quality deterioration, while climate change had the lowest (although not negligible). After the impact assessment, different mitigation measures were suggested in a stakeholder consultation workshop, and of them (enhanced capacity of wastewater treatment plants (WWTPs), and increased sewerage connection rate) were adopted to generate a final scenario including countermeasures. The main benefit of the PWLM approach are its high level of stakeholder involvement (through co-generation of the research) and use of free (for developing countries) software and models, both of which contribute to an enhanced science-policy interface.

Keywords: water quality; WEAP; climate change adaptation; urbanization; domestic wastewater management; sustainable development goals

---

1. Introduction

Population growth, rapid urbanization and climate change are continuously depleting finite fresh water resources, affecting both its quality and quantity [1]. Since institutional capacity are often limited, the threat of poor water quality, as well as unavailability of water, looms large on the horizon, particularly for developing countries [2,3]. To deal with future uncertainties in water quality and predict its future situation, scenario analysis through numerical quantification (using Integrated Water Resource Models) are of great significance, which may address key decision choices and facilitate targeted intervention. Nonetheless, water quality scenarios, as in other cases, requires trans-disciplinary approaches, including hydrological science, climate science, social science and local policies. It is imperative to ensure that local policy planners can understand, relate and utilize the hydrological modelling and scenario analysis at their respective watershed level, so as to prepare for a resilient future and achieve global targets, such as Sustainable developments goal 6.0 [4].

As a holistic approach, climate change adaptation actions at the watershed level are needed to reduce risks related to extreme hydro-meteorological weather conditions as well as to manage water quality and freshwater ecosystem [5]. For example, Proctor et al. [6] found that implementing a watershed management plan could be a win–win situation in terms of reducing water scarcity and generating economic revenue. At the national level, [7] found that a resilience-focused national strategy, consisting of broad-based resilience building across financial, human, social, natural and physical capital provides substantial future reduction of costs associated with climate disasters (El Niño).

Many researchers, in theory, argue that climate change adaptation and mitigation policies should be simultaneously developed to maximize their effectiveness, however, there is little understanding for how to do it in practice [8–12]. Holistic approaches to land-use planning and management, as well as incorporating other key drivers and pressures like climate change and population growth, may prove pivotal to bridge the adaptation and mitigation gaps, particularly against the backdrop of rapidly urbanizing developing countries.

The Philippines is a rapidly growing country (GDP over 7%), but the growth has sustained uncoordinated rapid urban expansion, together with inadequate wastewater management infrastructure [13]. As a result, surface water resources remain under severe pressure due to direct discharge of wastewater. A series of extreme weather events, including floods and typhoons, further exacerbates the water quality deterioration. Nevertheless, only a couple of studies actually attempted to understand the status of water resources and their management strategies for the near future. Considering the gap in integrating climate change adaptation and mitigation measures into the land-use planning process at the watershed level, we indulged a multi-stakeholder participatory approach to develop a robust and resilient water resource management strategy. The present work was conducted in the Santa Rosa Sub-watershed, located near Metro Manila and alongside the Laguna de Bay lake (largest lake in the Philippines), and involved local governments as well as national agencies (the Laguna Lake Development Authority). We selected this study site because Santa Rosa City, the largest city within this watershed, is witnessing rapid industrial growth and expected to become the "Silicon Valley" and "Detroit" of the Philippines [14]. This also resulted in swift land use change, population influx and deterioration of water quality. So far, only a few scientific studies have addressed the water quality deterioration in this region using different methods like machine learning and/or time series analysis, and they found that river water was greatly polluted in reference to national guidelines [15,16]. However, the main limitations of the above studies were the limited period of observation and the lack of management guidelines both at short and long time scale. Looking in to the above knowledge gap, we have proposed an integrated approach also called "Participatory Watershed Land-use Management (PWLM) approach" to target water management issues from cradle to grave. This work will contribute to evidence-based policymaking (through giving meaningful inputs to local governments' "Local Climate Change Action Plans" and "Comprehensive Land Use Plans") related to sustainable water management by assessing the current and future water quality situation in the Santa Rosa River. We used a numerical simulation tool, the Water Evaluation and Planning (WEAP), to model river water quality under different scenarios. Finally, based on the hydrological simulation output, necessary adaptation and mitigation countermeasures and there inclusion in to local policy planning for four local government units within one watershed (Silang-Santa Rosa watershed) of the Laguna de Bay lake, Philippines.

2. Study Area

Santa Rosa is one of the major rivers that passes through cities of Biñan, Santa Rosa, Silang City, Cabuyao and very small portion of Tagaytay as shown in Figure 1. Annual rainfall in the study area is 1950 mm [17]. Population in Santa Rosa Watershed area is 641,884 as of year 2013 [18]. Whole area under the watershed is 108.2 Km^2 [5]. A Land Use/land cover map for the year 2014, developed through a supervised classification of a Landsat 8 image, indicated that the watershed is comprised of seven categories, namely: coconut (7%), built up (25%), forest (8%), grassland (25%), industrial (9%), mixed crop (8%), and rice crop (17%) [5]. For the modeling purpose, we have considered only four local government units (LGUs), i.e., Binan, Santa Rosa, Siland and Cabuyao, as the area of Tagaytay falling under the Santa Rosa sub-watershed is extremely small.

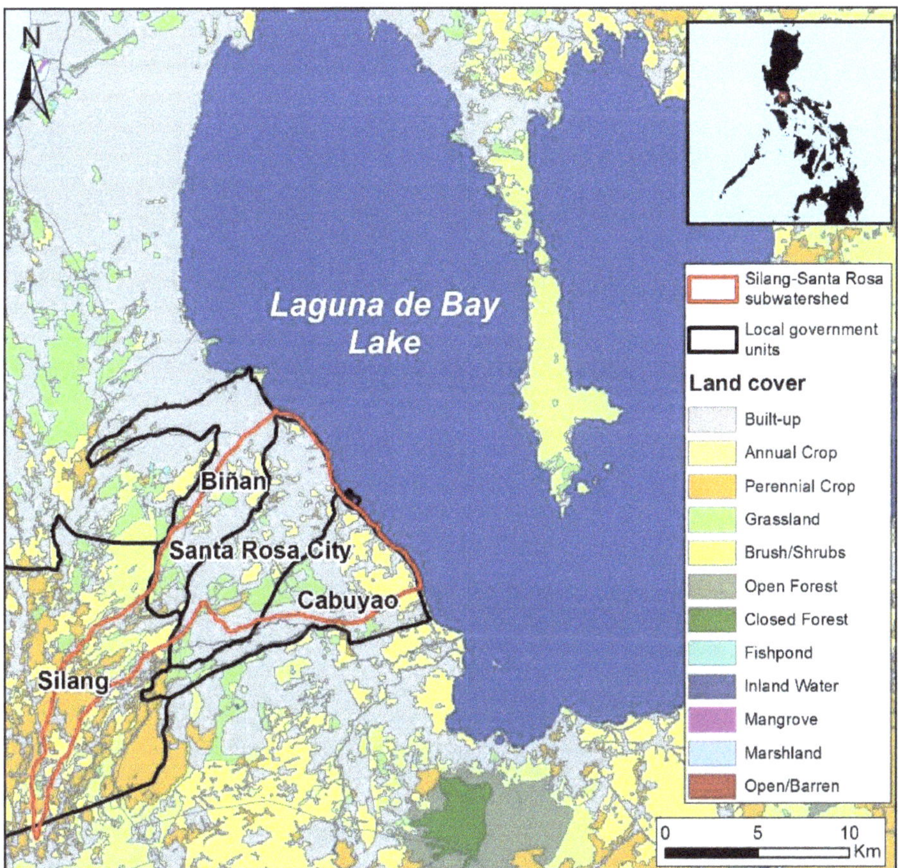

Figure 1. Santa Rosa watershed map.

3. Methodology

3.1. Participatory Watershed Land-Use Management (PWLM) Approach

This approach consists of following four steps: (i) Scenario analysis, (ii) impact assessment, (iii) Climate change adaptation and mitigation (CCA&M) measure development, and (iv) Land-use plan improvement (Figure 2).

1. The first step, scenario analysis, aims at understanding the problems that the local governments face in terms of water quality deterioration, as well as the future land-use and development plans of the city/municipality. For this analysis, participatory rapid appraisal activities including key informant interviews, focus group discussions, and participatory land-use mapping, are conducted with representatives from each LGU. The government officials participating in these discussions typically come from different departments, including urban planning, agriculture, environment, and disaster risk reduction and water resource management. Through key informant interviews and focus group discussion, we identified the main issues pertaining water resource in terms of both quality and quantity, using basic sets of questions. For the participatory mapping, officials produce a map of the land-use changes expected to occur over the next decade (until 2030) by sketching the future land conversions on a (poster-sized) sheet of tracing paper overlaid onto the current land-use map. This information is later digitized and georeferenced using Geographical Information Systems (GIS) software as shown in Figure 2.

2. The second step, impact assessment, involves estimating the impacts of the planned land-use changes, climate change and population growth on water quality. For this, Water Evaluation and Planning (WEAP), an integrated water resource management (IWRM) hydrological model, was used. The WEAP model is capable to estimate both water quality and quantity using scenario analysis. The other reason to select this model was that it is free for developing countries, hence easy for the hydrological officers to replicate such studies by themselves after capacity development.

3. The third step, CCA&M measure development, aims to devise possible climate actions for both adaptation and mitigation in consultation with the local governments, and prioritize these actions according to their feasibility and urgency. Another focus group discussion (FGD) session was organized including participants from Water Quality Management Authority (WQMA) board members, Laguna Lake Development Authority (LLDA) (a prime nodal agency dedicated to water resource management in and around Laguna De Bay area) officials from environment and water divisions, environmental officers in-charge of water management from four LGUs falling within Santa Rosa river watershed. We requested the officials to identify measures based on the needs of each local government. Through this FGD, a set of possible countermeasures like building wastewater treatment plants (WWTPs), river rehabilitation, reduction in industrial effluent discharge without treatment were presented. Further consultation then led to the identification of priority measures.

4. Step four, land-use plan improvement, aims to support local governments to strengthen their land-use and related development plans through dialogue on the recommendations generated from the previous three steps. At present, there is no adaptation and mitigation measures for water quality improvement, mentioned in the LGUs' land-use plans. Therefore, we have evaluated the hypothetical installation of WWTPs of capacity 164MLD and its impact on water quality improvement. Based on the positive simulation result we got, suggestions were provided to local government units (LGUs) policy documents i.e., Comprehensive land use plan (CLUP) [19] or Local climate change action plan (LCCAP).

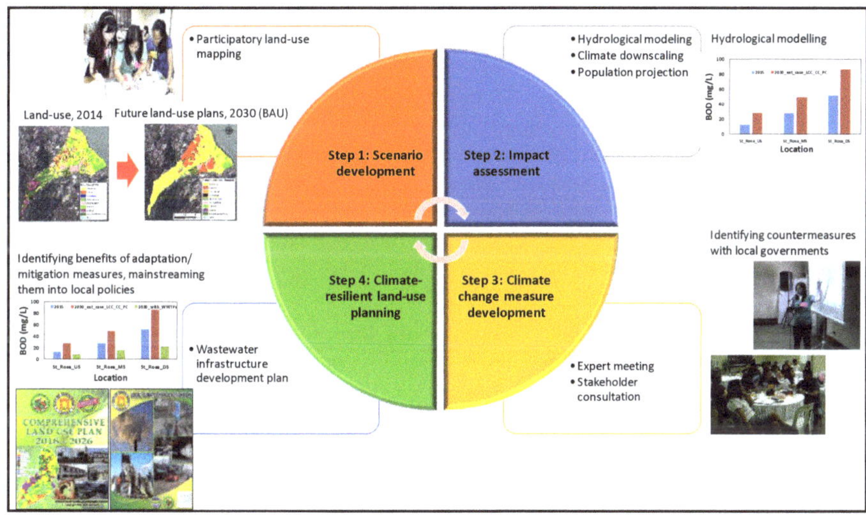

Figure 2. Research methodology.

3.2. Hydrological Model

WEAP model is used to simulate river discharge and water quality variables. Using GIS files for administrative boundary of the study area and drainage network, schematic diagram was developed. Whole study area is divided into smaller catchments based on the topographical, hydrological, confluence points, and climatic characteristics of the river basin Different catchment methods, namely rainfall runoff (simplified coefficient method), irrigation demands only (simplified coefficient method) and rainfall runoff (soil moisture method) that are available in the WEAP platform enable it to simulate different components of hydrological cycle including the catchment's potential evapotranspiration (by using crop coefficients), surface runoff and infiltration of rainfall. Based on the data availability, the rainfall runoff method (simplified coefficient method), is used for the catchment simulation [20].

Streeter–Phelps model within WEAP was used to estimate pollution concentrations in water bodies. Simulation of oxygen balance in a river within this model is governed by two processes i.e., consumption by decaying organic matter and reaeration induced by an oxygen deficit. The removal BOD from water is a function of water temperature, settling velocity, and water depth Equations (1) and (2):

$$BOD_{final} = BOD_{init} \exp^{\frac{-k_{rBOD}L}{U}} \tag{1}$$

$$k_{rBOD} = k_{d20}^{1.047(t-20)} + \frac{v_s}{H} \tag{2}$$

where BOD_{init} = BOD concentration at the top of the reach (mg/L), BOD_{final} = BOD concentration at the end of the reach (mg/L), t = water temperature (in degrees Celsius), H = water depth (m), L= reach length (m), U = water velocity in the reach, v_s = settling velocity (m/s), k_r, k_d and k_a = total removal, decomposition and aeration rate constants (1/time), k_{d20} = decomposition rate at reference temperature (20° Celsius).

Data Requirement for Model Set Up

Scenario based future (year 2030) prediction of the water quality is done to visualize plausible alternative water management policies in the Santa Rosa River. Datasets used for the modeling were domestic wastewater discharge, past River water quality at different monitoring stations, population, rainfall, temperatures, river cross section, river length, river discharge, land use/land cover etc.

For water quality parameters, we have used biochemical oxygen demand (BOD) and total coliforms (which was later assumed as equivalent to Escherichia coli counts) collected at three points on Santa Rosa River. Data for both water quality parameters was analyzed by LLDA following the standard method of water quality analysis from APHA [21]. Principal reasons for selecting sampling locations were the accessibility of water samples, saving cost and to observe effect of urbanization at approximately equidistant throughout the river length passing through the watershed. For hydrological modeling, four catchment areas in the Santa Rosa watershed, which experienced inter-basin transfers were considered. Pollutant transport from a catchment accompanied by rainfall-runoff is enabled by ticking the water quality modeling option During non-rainy days, pollutants accumulate on the catchment surfaces and reach water bodies through surface runoff. Regarding future precipitation data, three different Global Climate Models (GCMs) (MRI-CGCM3, MIROC5, HadGEM-ES) and Representative Concentration Pathway (RCP) (4.5 and 8.5) output were used after downscaling and bias correction., We have evaluated average value of the change in monthly average precipitation for both RCP 4.5 and 8.5 in order to evaluate the climate change on water quality. The whole study area is divided into four demand sites for estimating the effect of population growth and its associated domestic wastewater discharge on river water quality status. Primarily, these demand sites denotes population of different LGUs lying on either side of the Santa Rosa River within our study area. Future population in these demand sites were estimated by ratio method using UNDESA projected growth rate [22]. In the absence of exact information on the total domestic wastewater production, the daily volume of domestic wastewater generation per person considered for this study was 130 liters [23]. Whole simulation process is divided in three phases: (a) Model set-up and data input, (b) calibration and validation, and (c) future simulation using scenario analysis. Schematic diagram for the model set-up is shown in Figure 3.

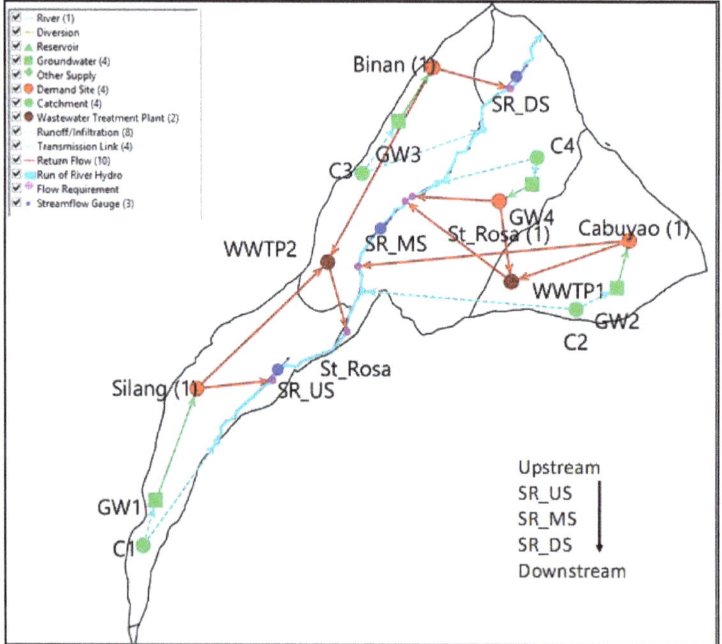

Figure 3. Schematic diagram showing the problem domain for water quality modeling in Santa Rosa River using Water Evaluation and Planning (WEAP) interface. Here, WWTP—Wastewater treatment plant, SG—Stream gauge, C—Catchment, GW—Groundwater.

4. Results and Discussion

4.1. Future Prediction of Key Drivers and Pressures

For this study, three drivers and pressures i.e., population growth, climate change and land use land cover were considered to estimate impact of its change in water quality. Here source for current population (2013) data is Philippine Statistics Authority, while future population is estimated with growth rate provided by UNDESA [22]). UNDESA provides future population of different countries for different administrative levels with population growth rate at every five years. This growth rate is predicted based on the population growth in past twenty years, rate of economic growth, industrial growth and other likely factors. We have also adopted these population growth rate to predict population of the study area and result is show in Figure 4. It is found that total population, which was 638,711 in year 2013, will be projected to grow to 1,283,202 by year 2030.

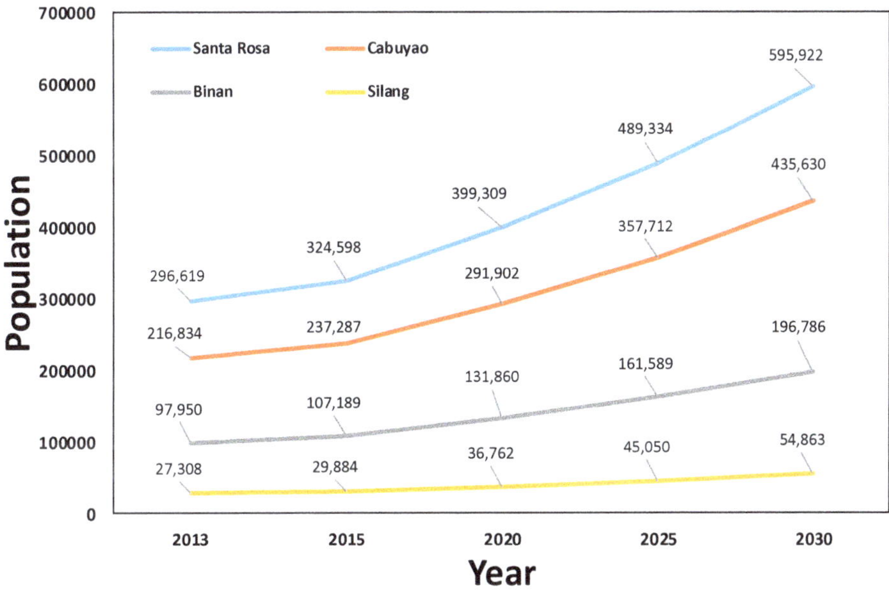

Figure 4. Plot showing trend in population growth in the study area.

A summary of the observed and projected monthly precipitation (mm) values after climate downscaling is presented in Figure 5. For annual precipitation, the observed value for the year 2015 was 2803.2 mm. On the other hand, projected annual precipitation values for the year 2030 using different GCMs and RCPs ranged from 1658.5 to 2806.9 mm. The output from the HadGEM-ES GCM showed a sharp decrease in precipitation, especially during summer month, which was considered as an anomaly and hence not included during our scenario building. Looking at the outputs from other GCMs (MRI-CGCM and MIROC5), it was clear that the annual precipitation projected from GCM downscaling is not much different from the current observed value. However, looking carefully, there is some difference between these values at the monthly scale. Therefore, in this research, we wanted to estimate whether this small changes on precipitation has any significant changes on the water quality or not.

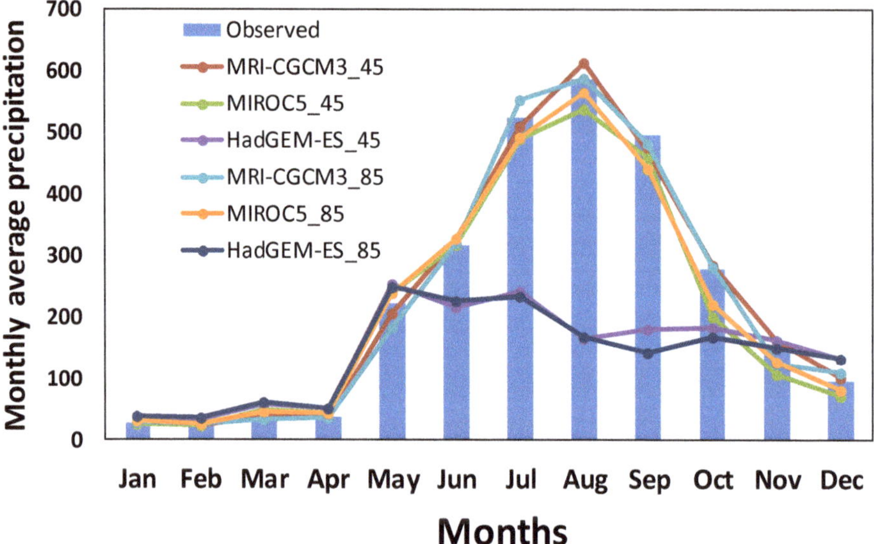

Figure 5. Comparison between observed and downscaled Global Climate Models (GCMs) output for monthly average rainfall. Here three GCMs used are MRI-CGCM, MIROC5 and HadGEM-ES. Suffix 45 and 85 represents RCP 4.5 and 8.5 used here.

Other drivers used to observe water quality deterioration was land use land cover change and the result is shown are shown in Figure 6. From the figure, it is found that among five different land use land cover classes; only built-up is showing positive changes i.e., percentage share of built-up area to the total area is increasing. Also, this change is happening at the expanse of decrease in agriculture, idle land/grassland, rice field and tree. Further, percentage increase in built-up area will be maximum for Silang followed by Cabuyao, Binan and Santa Rosa. In case of loss of agricultural land, highest change was observed for Silang, followed by Cabuyao, Santa Rosa and Binan. For percentage reduction in the idle land/grassland the order was Silang, Cabuyao, Binan and Santa Rosa. Looking into the percentage reduction of rice field, both Binan and Santa Rosa are suffering most followed by Cabuyao, where Silang is least impacted. Looking into the percentage reduction in tree cover the order was Silang, Binan, Cabuyao, and Santa Rosa.

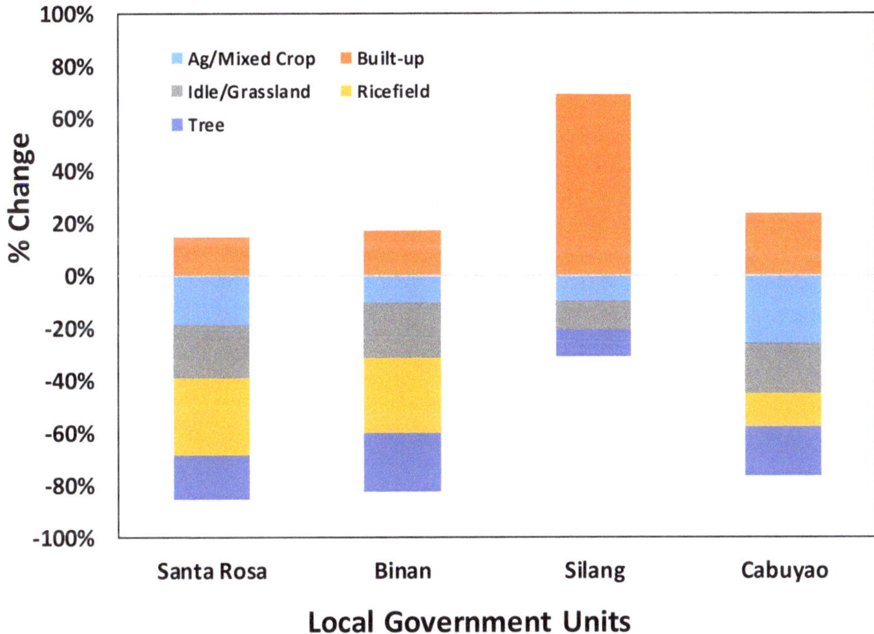

Figure 6. Percentage of land use land cover change among four different local government units (LGUs) coming within Santa Rosa sub-watershed.

4.2. Hydrological Simulation Using WEAP

4.2.1. Model Performance Evaluation

Before doing future simulation, it is important to check the performance credibility of the WEAP model through calibration and validation. For model validation, stream flow and BOD were used for hydrological and water quality components respectively. For calibration, trial and error method was applied during simulation in order to get best-fit results. Mainly effective precipitation and runoff/infiltration were adjusted to reproduce the observed monthly stream flows for the year of 2015 (Table 1). The final best-fit values for both entities was 92% and 55/45 respectively. Validation result for both stream flow and BOD is shown in Figure 7a,b respectively. Figure 7a, shows a relation of monthly simulated and observed stream flows for Santa Rosa midstream station. Significant association was found for most of the months with correlation coefficient (R^2) $\cong$ 0.78, root-mean-square error (RSME) $\cong$ 0.23, and an average error of 13%. For water quality validation, both simulated and observed concentration of BOD from the year 2014 was considered. Here, also results show a strong relation between observed and simulated BOD concentrations (with error of 14%) (Figure 7b) authorizing credibility of the model performance in study area. Selection of year 2015 and 2014 for stream flow and BOD respectively, were made because of the observed data availability.

Table 1. Summary of parameters and steps used for calibration.

Parameter	Initial Value	Step	Final Value
Effective precipitation	100%	±0.5%	92%
Runoff/infiltration ratio	50/50	±5/5	55/45

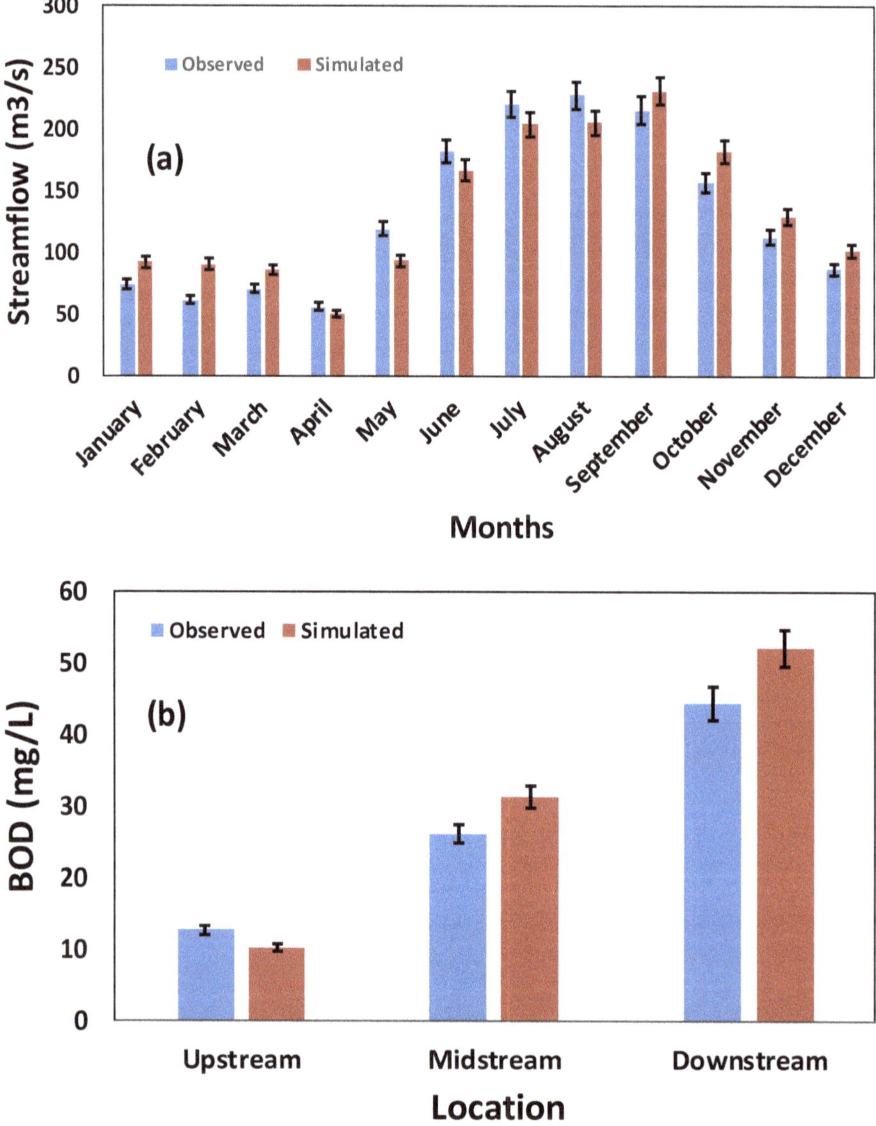

Figure 7. Validation of the model output by comparing simulated and observed (**a**) average monthly river discharge for year 2015 at Santa Rosa midstream; (**b**) average biochemical oxygen demand (BOD) values for different locations for the year 2014. Here small vertical black lines on the top of the bars are showing percentage error.

4.2.2. Scenario Analyses

For water quality, simulation is done using two scenarios namely future scenario without adaptation measures (for the risk assessment i.e., step two of PWLM) and future scenarios with adaptation measures (after the stakeholders for identifying countermeasures i.e., step three).

For impact assessment, we have considered all three drivers and pressures. Five sets of simulated concentration were obtained considering following "what-if" situations without adaptation measures,

(a) with population growth only; (b) with population growth and moderate climate change; (c) with population growth and extreme climate change; (d) with population growth, moderate climate change and land use land cover change; and (e) with population growth, extreme climate change and land use land cover change. Result obtained through hydrological simulation were compared with national guideline for class B i.e., swimmable category (BOD < 3 mg/L and E. coli < 500CFU/100 mL) [24] is shown in Figure 8.

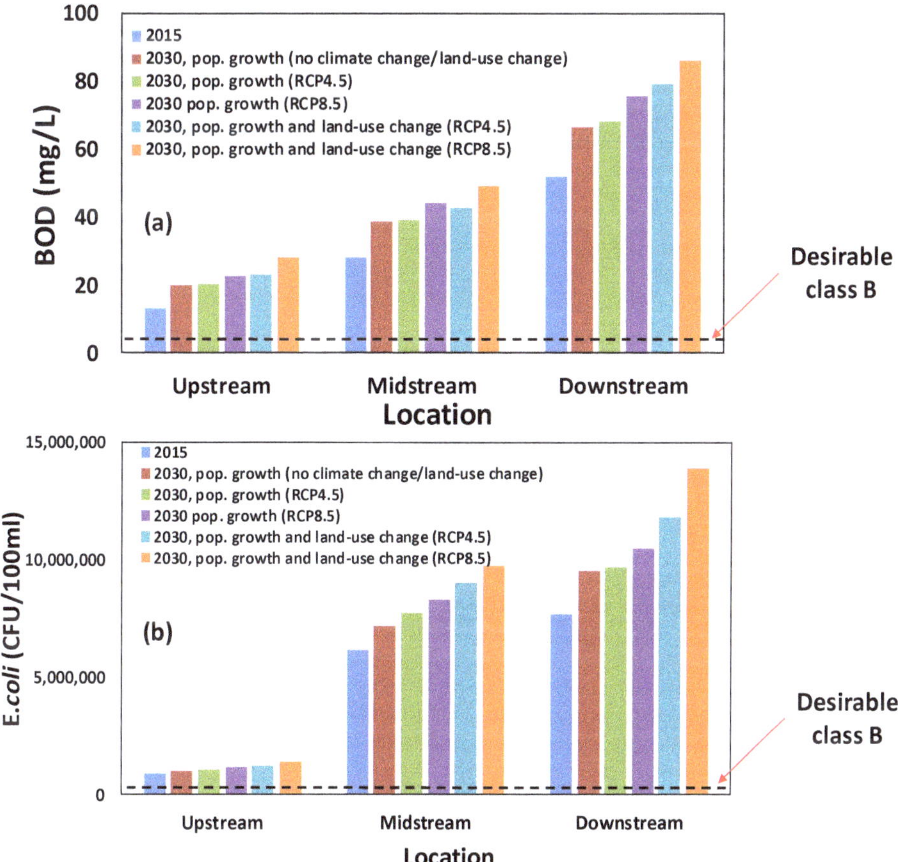

Figure 8. Result shows simulated water quality parameters (**a**) BOD (**b**) E. coli for future scenario without adaptation measures.

Simulated concentration of BOD for year 2015 varies from 13 to 52 mg/L. When compared to the desirable concentration of BOD from Class B, it can be said that all samples throughout the year falls under moderately to extremely polluted category. Average percentage increase in the BOD and E. coli by 2030 is 76% and 104% respectively when compared with situation in 2015. Looking into the result from future scenarios, it is found that the effect of all climate change, land use/land cover change and population changes are prominent in water quality status and water quality will deteriorate further in 2030 when compared to the current situation. We have further analyzed the impact of individual drivers and pressure on water quality deterioration and result is shown in Table 2. It was found that, the order of average deterioration effect of water quality (both for BOD and E. coli) due to population growth, climate change and land use/land cover change were population growth > land use/land cover change > climate change. This result can be explained, as amount of wastewater being generated will grow automatically with population increase without any countermeasures. On the other hand, land use land cover change as well as climate change induced extreme weather condition will trigger the changes in River discharge. This change in River discharge will cause change in water quality parameters due to oxidation-reduction process. This clearly depicts that building any local climate resilient plan must consider these three key drivers and pressures which likely to impact water resources. Although effect of climate change is not huge but it is significant enough to make the local master plan more robust for sustainable future water environment. In addition, based on the simulated water quality, river water likely to pose potential health risks like gastroenteritis if consumed accidently (microbial contamination), and death of aquatic organisms like fish (because of high BOD).

Table 2. Summary for order of contribution of different drivers on water quality deterioration in Santa Rosa sub-watershed.

Parameter	Average % Increase (2015 to 2030)	% Contribution from Population Growth	% Contribution from LULC Change	% Contribution from Climate Change
BOD	76	66	23	11
E. coli	104	71	20	9

4.2.3. Scenario with Adaptation Measures

In step three of PWLM, an experts meeting and stakeholder consultation workshop was organized to identify possible countermeasures to mitigate the expected water quality deterioration. Participants were governmental officials from department of water resources, environment, irrigation, fisheries, infrastructure etc., both at national and provincial level; consultants, NGOs, academia. List of possible countermeasures, which were discussed were: (a) Building Wastewater Treatment Plants (WWTPs) and sewerage connection to each households, (b) cleaning the riverfronts, (c) strict regulation on discharge of untreated/partially treated industrial effluents to the waterbodies, (d) diligent monitoring (sampling and analysis) of water quality, (e) minimizing the human impacts on river catchment areas etc.

In this study, we have quantified the effect of first suggested countermeasure i.e., Building Wastewater Treatment Plants (WWTPs) and sewerage connection to each households. Here the capacity of WWTP was decided based on the projected wastewater discharge by year 2030, which was 166,816,291 liter of wastewater generated per day. Hence, we have considered two WWTPs of combined capacity of 164 MLD estimated as shown in Table 3. All considered WWTPs are of Upflow Anaerobic Sludge Blanket Reactor Coupled with Sequencing Batch Reactor (UASB-SBR) type. These WWTPs are designed with high contaminant removal efficiency as 97% for BOD and 99.69% for fecal coliform [25]. In addition, we have also considered 100% sewerage connection rate.

Table 3. Summary for WWTPs considered for the scenario building.

WWTP	Future Population (2030)	Wastewater Generated in Liter Per Day for the Year 2030@130 liter/capita/day	WWTP Capacity in Million Liter Per Day (MLD) Assumed for Numerical Simulation
WWTP 1 (will serve Santa Rosa and Cabuyao)	1,031,553	134,101,847	134
WWTP 2 (will serve Binan and Silang)	251,650	32,714,443	32

Based on the above assumption, water quality for scenario with countermeasures was simulated and the result is shown in Figure 9. Simulated value of BOD ranged from 8 to 17 mg/L and average BOD reduction of 62 percent when compared with current status. On the other hand, simulated value of E. coli ranging from 35,783 to 112,948 CFU/100 mL with an average reduction of 98 percent when compared to current situation. Although water quality improved a lot which is an encouraging sign but to achieve water quality of Class B, government need to adopt more than one adaptation measures together.

Finally, suggestions for adaptation measures with scientific evidence of their impact were communicated to the Local Government Officials to incorporate this suggestion in their policy documents like i.e., Comprehensive land use plan (CLUP) or Local climate change action plan (LCCAP) for sustainable water resource management [19].

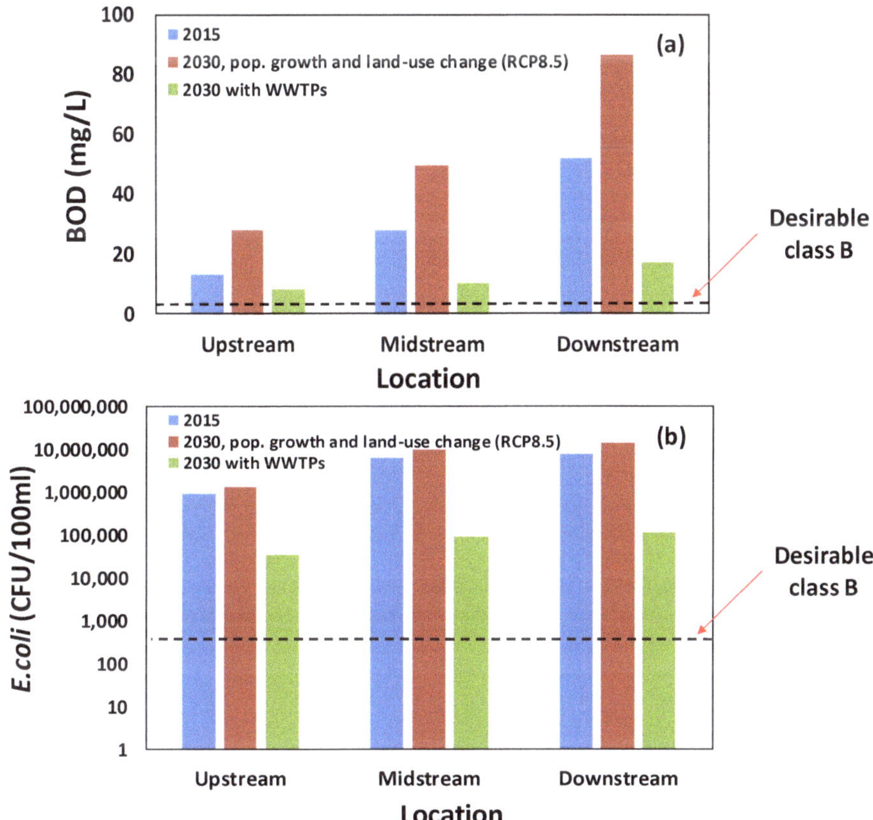

Figure 9. Result shows simulated water quality parameters (**a**) BOD (**b**) E. coli for scenario with adaptation measures.

5. Conclusions

Although there are typically multiple institutions dealing with land-use and water quality issues within a watershed (e.g., local and provincial governments, NGOs, and private companies), disconnects and/or redundancies often occur in their work, which presents a major obstacle in institutional set-up. Therefore, a transdisciplinary and integrated (both bottom-up and top-down) approach should be employed for solving this complex issue of water resource management. Along these lines, this study emphasized that the Participatory Watershed Land-use Management (PWLM) approach can aid the sustainable management of water resources if all the key stakeholders and LGUs involved work in a holistic manner rather in silos. Hydrological simulation presented a clear picture for the current status as well as future prediction of water quality throughout the Santa Rosa River using scenario analysis. The PWLM approach provides an integrated framework for water resource management, where retrofitting models are being developed with regular feedbacks of different stakeholders to meet their need and to feel ownership of the simulated results. Although PWLM has been applied in a previous work to estimate the impacts of land-use change and climate change on urban flooding [5], this is the first study to utilize PWLM for water quality assessment and management. As the number of sampling locations used is relatively few in this study (and in the area around the Laguna de Bay lake in general), more sampling locations along with increased frequency of monitoring (as well as monitoring of additional organic/inorganic water quality parameters) should be a priority to promote for further studies in this critical region.

Author Contributions: Conceptualization, P.K., B.A.J.; methodology, P.K., B.A.J.; investigation, P.K., B.A.J.; writing, P.K., B.A.J.; writing—review and editing, P.K., B.A.J., R.D., R.A., S.C., M.K., D.B.M.-M. All authors have read and agreed to the published version of the manuscript.

Funding: This research received no external funding.

Acknowledgments: This paper is generally based upon outputs produced under the "Adaptation Initiative" project for the fiscal year 2019, commissioned work of the Japanese Ministry of the Environment. We also show our appreciation for the participation of LGU officials from the four units (Santa Rosa, Silang, Cabuyao, Binan), support and data from LLDA, and the support from the Institute of Biological Sciences, University of Philippines, Los Banos.

Conflicts of Interest: The authors declare no conflict of interest.

References

1. Mekonnen, M.M.; Hoekstra, A.Y. Four billion people facing severe water scarcity. *Sci. Adv.* **2016**, *2*, e1500323. [CrossRef] [PubMed]

2. Hosseini, N.; Johnston, J.; Lindenschmidt, K. Impacts of climate change on the water quality of a regulated Prairie River. *Water* **2017**, *9*, 199. [CrossRef]

3. Kumar, P.; Masago, Y.; Mishra, B.K.; Fukushi, K. Evaluating future stress due to combined effect of climate change and rapid urbanization for Pasig-Marikina River, Manila. *Groundw. Sustain. Dev.* **2018**, *6*, 227–234. [CrossRef]

4. Kumar, P. Numerical quantification of current status quo and future prediction of water quality in eight Asian Mega cities: Challenges and opportunities for sustainable water management. *Environ. Monit. Assess.* **2019**, *191*, 319. [CrossRef] [PubMed]

5. Endo, I.; Magcale-macandog, D.B.; Kojima, S.; Johnson, B.A.; Bragais, M.; Beatrice-macandog, P.; Scheyvens, H. Participatory land-use approach for integrating climate change adaptation and mitigation into basin scale local planning. *Particip. Sustain. Cities Soc.* **2017**, *64*, 184–193. [CrossRef]

6. Procter, A.; McDaniels, T.; Vignola, R. Using expert judgments to inform economic evaluation of ecosystem-based adaptation decisions: Watershed management for enhancing water supply for Tegucigalpa, Honduras. *Environ. Syst. Decis.* **2017**, *37*, 410–422. [CrossRef]

7. French, A.; Mechler, R. *Managing El Niño Risks Under Uncertainty in Peru: Learning from the Past for a More Disaster-resilient Future*; International Institute for Applied Systems Analysis: Laxenburg, Austria, 2017.

8. IPCC. *Climate Change 2014: Synthesis Report*; Contribution of Working Groups I, II and III to the Fifth Assessment Report of the Intergovernmental Panel on Climate Change; Core Writing Team, Pachauri, R.K., Meyer, L.A., Eds.; IPCC: Geneva, Switzerland, 2014; p. 151.

9. Jones, R.; Patwardhan, A.; Cohen, S.; Dessai, S.; Lammel, A.; Lempert, R.; Mirza, M.M.Q.; von Storch, H. Foundations for decision making. In *Climate Change 2014: Impacts, Adaptation, and Vulnerability. Part A: Global and Sectoral Aspects*; Contribution of Working Group II to the Fifth Assessment Report of the Intergovernmental Panel on Climate Change; Field, C.B., Barros, V.R., Dokken, D.J., Mach, K.J., Mastrandrea, M.D., Bilir, T.E., Chatterjee, M., Ebi, K.L., Estrada, Y.O., Genova, R.C., et al., Eds.; Cambridge University Press: Cambridge, UK; New York, NY, USA, 2014; pp. 195–228.

10. Duguma, L.A.; Wambugu, S.W.; Minang, P.A.; van Noordwijk, M. A systematic analysis of enabling conditions for synergy between climate change mitigation and adaptation measures in developing countries. *Environ. Sci. Policy* **2014**, *42*, 138–148. [CrossRef]

11. Wilbanks, T.; Sathaye, J. Integrating mitigation and adaptation as responses to climate change: A synthesis. *Mitig. Adapt. Strateg. Glob. Chang.* **2007**, *12*, 957–962. [CrossRef]

12. VenkataRaman, S.; Iniyan, S.; Goic, R. A review of climate change, mitigation and adaptation. *Renew. Sustain. Energy Rev.* **2012**, *16*, 878–897. [CrossRef]

13. Asian Development Bank (ADB); Felipe, J.; Estrada, G.B. *Why has the Philippines's Growth Performance Improved? From Disappointment to Promising Success*; ADB economics working paper series, Number 542; ADB publications: Manila, Philippines, 2018.

14. Webpage for City of Santa Rosa, Province of Laguna. Available online: www.santarosacity.gov.ph/home/ (accessed on 16 August 2018).

15. Zimmer, V.; Bendoricchio, G. Nutrient and suspended solid loads in the Laguna de Bay, Philippines. *Water Sci. Technol.* **2001**, *44*, 77–86. [CrossRef] [PubMed]

16. Victoriano, J.M.; Lacatan, L.L.; Vinluan, A.A. Predicting river pollution using random forest decision tree with GIS model: A case study of MMORS, Philippines. *Int. J. Environ. Sci. Dev.* **2020**, *11*, 1. [CrossRef]

17. Philippine Atmospheric Geophysical Astronomical Services Administration (PAGASA). 2020. Available online: http://bagong.pagasa.dost.gov.ph/information/climate-philippines (accessed on 12 January 2020).

18. Philippine Statistical Authority (PSA). *Philippines in Figures*; Databank and information services division: Quezon City, Philippines, 2015; p. 102.

19. Comprehensive Land Use Plan (CLUP) City of Santa Rosa. *Comprehensive Land Use Plan (2018–2026)*; CLUP: City of Santa Rosa, Philippine, 2018.

20. Seiber, J.; Purkey, D. *WEAP—Water Evaluation and Planning System User Guide for WEAP 2015*; Stockholm Environment Institute: Stockholm, Sweden, 2015.

21. American Public Health Association (APHA). *Standard Methods for the Examination of Water and Wastewater*, 22nd ed.; Rice, E.W., Baird, R.B., Eaton, A.D., Clesceri, L.S., Eds.; American Public Health Association (APHA), American Water Works Association (AWWA) and Water Environment Federation (WEF): Washington, DC, USA, 2012; p. 1496.

22. United Nations, Department of Economic and Social Affairs, Population Division (UN DESA). *World Urbanization Prospects: The 2014 Revision, 2015, (ST/ESA/SER.A/366)*; United Nations Publication: New York, NY, USA, 2015; p. 517.

23. *The United Nations World Water Development Report 2017*; Wastewater: The Untapped Resource; UNESCO: Paris, France, 2017.

24. Department of Environment and Natural Resources (DENR). *Water Quality Guidelines and General Effluent Standards of 2016*; Republic of Philippines: Quezon City, Philippines, 2016; p. 25.

25. Khan, A.A.; Gaur, R.Z.; Diamantis, V.; Lew, B.; Mehrotra, I.; Kazmi, A.A. Continuous fill intermittent decant type sequencing batch reactor application to upgrade the UASB treated sewage. *Bioprocess Biosyst. Eng.* **2013**, *36*, 627–634. [CrossRef] [PubMed]

Article

Impact of Rice Intensification and Urbanization on Surface Water Quality in An Giang Using a Statistical Approach

Huynh Vuong Thu Minh [1], Ram Avtar [2], Pankaj Kumar [3], Kieu Ngoc Le [1,4], Masaaki Kurasaki [2] and Tran Van Ty [5,*]

[1] Department of Water Resources, CENREs, Can Tho University, Can Tho 900000, Vietnam; hvtminh@ctu.edu.vn (H.V.T.M.); knle@uark.edu (K.N.L.)
[2] Faculty of Environmental Earth Science, Hokkaido University, Sapporo 060-0810, Japan; ram@ees.hokudai.ac.jp (R.A.); kura@ees.hokudai.ac.jp (M.K.)
[3] Natural Resources and Ecosystem Services, Institute for Global Environmental Strategies, Hayama 240-0115, Japan; kumar@iges.or.jp
[4] Department of Biological and Agricultural Engineering, University of Arkansas, Fayetteville, AR 72701, USA
[5] Department of Hydraulic Engineering, College of Technology, Can Tho University, Can Tho 900000, Vietnam
* Correspondence: tvty@ctu.edu.vn; Tel.: +84-939501909

Received: 26 May 2020; Accepted: 12 June 2020; Published: 15 June 2020

Abstract: A few studies have evaluated the impact of land use land cover (LULC) change on surface water quality in the Vietnamese Mekong Delta (VMD), one of the most productive agricultural deltas in the world. This study aims to evaluate water quality parameters inside full- and semi-dike systems and outside of the dike system during the wet and dry season in An Giang Province. Multivariable statistical analysis and weighted arithmetic water quality index (WAWQI) were used to analyze 40 water samples in each seasons. The results show that the mean concentrations of conductivity (EC), phosphate (PO_4^{3-}), ammonium (NH_4^+), chemical oxygen demand (COD), and potassium (K^+) failed to meet the World Health Organization (WHO) and Vietnamese standards for both seasons. The NO_2^- concentration inside triple and double rice cropping systems during the dry season exceeds the permissible limit of the Vietnamese standard. The high concentration of COD, NH_4^+ were found in the urban area and the main river (Bassac River). The WAWQI showed that 97.5 and 95.0% of water samples fall into the bad and unsuitable, respectively, for drinking categories. The main reason behind this is direct discharge of untreated wastewater from the rice intensification and urban sewerage lines. The finding of this study is critically important for decision-makers to design different mitigation or adaptation measures for water resource management in lieu of rapid global changes in a timely manner in An Giang and the VMD.

Keywords: triple-rice cropping system; full-dike; surface water quality; WAWQI; An Giang Province; the Vietnamese Mekong Delta

---

1. Introduction

Deltas around the world have played a vital role in food security and economic development. However, the rapid exploitation of natural resources and changes in land use land cover (LULC) have also caused severe environmental degradation, such as water quality deterioration in many deltas in recent years [1–4]. The heavy metal concentrations and high bacterial pathogens due to industrial, agricultural activities, poor sanitation, and hygiene were found in the Middle Nile Delta, Egypt [5]. Several studies have also reported irregulated urban expansion and animal husbandry and its impact on water quality deterioration in Irrawaddy delta, Myanmar [6,7]. Consequently, when this polluted water flows into the city during monsoon, it causes several waterborne diseases such as cholera, gastroenteritis, skin diseases [6,8,9]. Surface water pollution from organic pollutants, microbial contamination, pesticides, metals, etc. is revealed in the Mekong Delta Basin, in both the Cambodian (Phnom Penh) and Vietnamese (Chau Doc, Tan Chau, and Can Tho) part [10–15].

The well-known trans-boundary river of the Mekong River Basin (MRB) in the Asian region has a natural area of 795,000 km^2 and mean annual discharge of 14,500m^3/s [16–18]. The glaciers in the Himalaya mountains is the source of the international Mekong River, which flows to China, Myanmar, Thailand, Laos, Cambodia, Vietnam, and finally to the Pacific Ocean [18]. Therefore, the lower Mekong Delta in Vietnam, located in the downstream of the MRB and accounting for 8% of the entire basin, has dominant diurnal tidal seawater entering twice a day. Changes in water quality and quantity in the upstream region would directly affect the health of proximally 242 million people (2018 data) [19] who live in the lower Mekong river [18,20]. The upper region of the VMD receives from 60% to 80% discharge from outside of the VMD, in which the only location of An Giang Province lies between the two main rivers of Mekong and Bassac. Therefore, the covered lands of An Giang are of fertile soil due to the abundance of water resources and fluvial sedimentation from the Mekong River. Consequently, An Giang has large agricultural areas with dominant rice production [21], but this province has also faced substantial damage by natural flooding phenomena annually from August to November due to the monsoon season in the Asian region [21–23].

The full- and semi-dike systems in An Giang were rapidly built since the 1990s to prevent flooding and to grow rice both for food security and economic development [22,24,25]. The full-dike system and the hydraulic infrastructure were developed to protect the triple-rice cropping system as well as the urban cities [21,25]. Local farmers can grow two or three rice crops per year inside the dike systems instead of single rice crops per year as in the past [21]. Although the dike systems can protect residential areas and increase income for the local farmers, the most critical disadvantage of this system is the surface water quality deterioration [21,22,25]. Water quality degradation may be derived from both natural conditions like rock–water interaction, ion exchange, groundwater–surface water interaction, evapotranspiration, and human activities such as a discharge of untreated wastewater from a point or nonpoint source in natural water bodies [16,21,26].

Water demand for agriculture and aquaculture alone consumes a significant portion of total available water, resulting in high waste discharged from agriculture [27]. Although few studies have reported the impact of land use on stream water quality [21,28,29], studies focusing on different types of dike development for agricultural intensification and its impacts on water quality remain scarce. Henceforth, the objective of this study is to assess the physicochemical properties of the surface water in An Giang Province using the multivariate statistical analysis approach and the weighted arithmetic water quality index (WAWQI). The primary focus of this study is to evaluate the impact of dike development on surface water quality compared to other remaining areas in An Giang. The hypothesis of this study is that the water quality inside the full-dike systems was worse than the outside ones, and water quality in the dry season was worse than that of the wet season.

2. Methodology

2.1. Study Area

An Giang Province (10°12′ N to 10°57′ N and 104°46′ to 105°35′) is located in the most upper part of the VMD and borders with Cambodia in the northwest (104 km long). An Giang is a home to over 2.4 million people (2019) [30], and the total area of 3536 km^2, 70% of which is for agricultural production. There are two distinct seasons: dry and wet (monsoon) in the region. The wet season occurs between May and November annually in which the high rainfall usually occurs at the end of the wet season from October to November (Figure 1). Although total annual rainfall in An Giang is low compared with the average rainfall of the VMD, the rainfall occurs nearly at the same time with the flooding season leading risk at deep inundation. Thus, An Giang has to build a large area of the dike systems (Figure 2) to increase agricultural production and to protect crops during the flooding season

(July to November). Multi-dike protection systems have been built to protect residential areas from flooding, and have mainly supported agricultural intensification since the early 1990s. In addition, hydropower plants were built along the Mekong River, and its branches have led to a change in the water regime (Figure 1). During 1991 and 2015, the average discharge was decreased in the wet season and increased in the dry season. The primary soil type is alluvial soil, accounting for 44.5% of all 37 different soil types present in the province. About 72% of the area is alluvial soil or land receiving huge sediment supply and is suitable for many kinds of crops. The dike systems and hydropower plants have reduced the amount of alluvial soil to be added to the region annually [31,32].

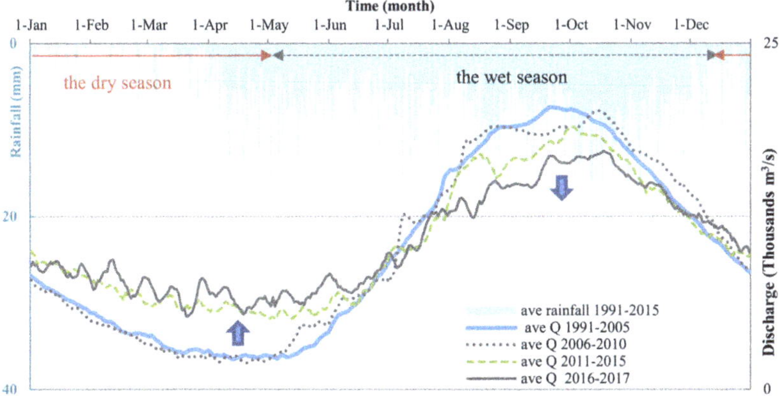

Figure 1. Average hourly discharge (Q) from 2006 to 2017 and average daily rainfall from 1991 to 2015 at Tan Chau Station in An Giang. The discharge imposes a decreasing trend in the wet season and an increasing trend in the dry season. All data were collected from the Southern Regional Hydro-meteorological Center (SRHMC) in Vietnam [33].

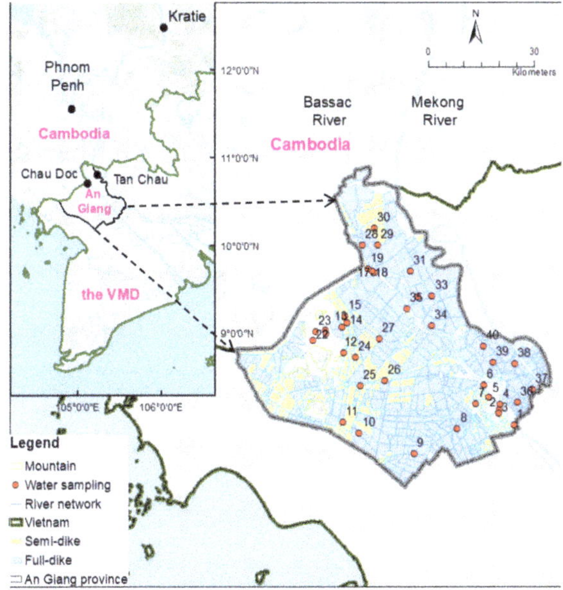

Figure 2. Study area and water sampling sites in An Giang, the Mekong Delta in Vietnam.

2.2. Collection of Water Samples and Analytical Methods

Surface water quality samples were collected and analyzed in the wet and the dry seasons inside the full- and semi-dike systems and outside of the dike system (on the main river and single rice cropping system), as shown in Figure 3. Analyzed data were processed using statistical tools and used to calculate water quality indicators. Finally, the obtained result is discussed to observe spatio-temporal water quality classification and the impact of the dike system on water quality parameters.

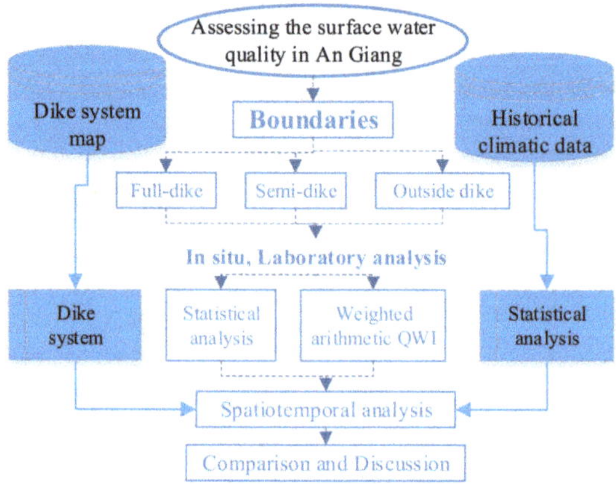

Figure 3. Flowchart for study methodology.

Each season, 40 surface water samples were taken from inside the full- and semi-dike systems, and outside the dike system in An Giang (Figure 3). Sampling was done both for the dry season (22–28 April 2018) and the wet season (6–13 October 2018). Water sample locations were taken by geotagged photos, which were marked in the global positioning system (GPS). The stratified random sampling technique was conducted to select the sampling sites: Cluster 1 includes ten samples outside of the dike system (6 in the main rivers and 4 in single-rice cropping system), Cluster 2 includes ten samples inside the semi-dike system (3 in the forest and 7 in the double-rice cropping system), and Cluster 3 includes 20 samples inside the full-dike system (6 in the urban area and 14 in triple-rice cropping system). After collection, water samples were brought to the laboratory in an ice chest and stored below 4 °C. The collected samples were analyzed for twelve water quality parameters: pH, EC, chloride (Cl^-), nitrite (NO_2^-), nitrate (NO_3^-), NH_4^+, COD, PO_4^{3-}, sodium (Na^+), calcium (Ca^{2+}), magnesium (Mg^{2+}), and K^+. The HORIBA multi-parameter meter (Kyoto, Japan) with a precision of 1% and a handheld meter (Oaklom; Tokyo, Japan) was used for in situ analysis of the physical parameters such as pH, Cl^-, EC, and some chemical parameters of NO_2^-, NO_3^-, NH_4^+, COD and PO_4^3 were measured using pack test⁻. Anions were analyzed by DIONEX ICS-90 ion chromatography with an error percentage of <2%, while cations were analyzed by a Shimadzu mass spectrometer with a precision of <1% using duplicates. The historical meteorological data were collected from the Southern Regional Hydro-meteorological Center (SRHMC) [33].

2.3. Statistical Analyses

2.3.1. Multivariate Statistical Analysis

Multivariate statistical analysis was completed to obtain a better understanding of the processes governing water quality [34–40]. First, we conducted correlation and discriminant analysis (DA) [41] to find out the significant relationship among parameters and discriminant among clusters in terms of water quality characteristics. Second, we used box plots to show differences among different clusters in the dry and wet seasons. Finally, we used the WAWQI method to classify the water quality for human use. XLSTAT Software version 2018 (Addinosoft SARL, Paris, France) and the inverse distance weighting (IDW) interpolation were used to make different plots and display the results [42–45].

We conducted Spearman rank–order to evaluate the relationship among parameters at each season since most of the dataset had a non-normal distribution. Spearman rank–order consumption does not require any distribution test, such as a person correlation with a normal distribution [46,47]. Moreover, Spearman rank–order is used to identify the correlation between related parameters by producing the significance of the data, as reported in previous studies [45,48].

In this study, we use the DA technique to determine the most significant parameters among 40 samples sites as well as between the dry and wet seasons. The DA was also found in various studies [48,49]. The standard DA, forward stepwise, and backward stepwise were applied, which was previously documented [21,48,50]. The forward stepwise adds a parameter in each step, starting from the most significant fit improvement until no change was found. In the case of backward stepwise, each parameter is excluded step-by-step, starting from the least significant fit improvement until no significant changes [51,52]. After standard DA, the backward stepwise model helped to clarify which parameters are the most important. In this standard model, step-by-step, variables were removed from the beginning of the less significant until no significant changes in removal criteria are achieved [48,51].

2.3.2. Weighted Arithmetic Water Quality Index (WAWQI) Model

The WAWQI is an index number that represents the overall quality of water and is a standard tool for the classification of water pollution (Figure 4). The WAWQI can be identified as a reflection of the composite influence of multivariable quality parameters [53]. Thus, WAWQI becomes an important indicator for the assessment and management of water resources. Here, all the selected water quality parameters are aggregated into an overall index, which is the most effective tool to express water quality [54].

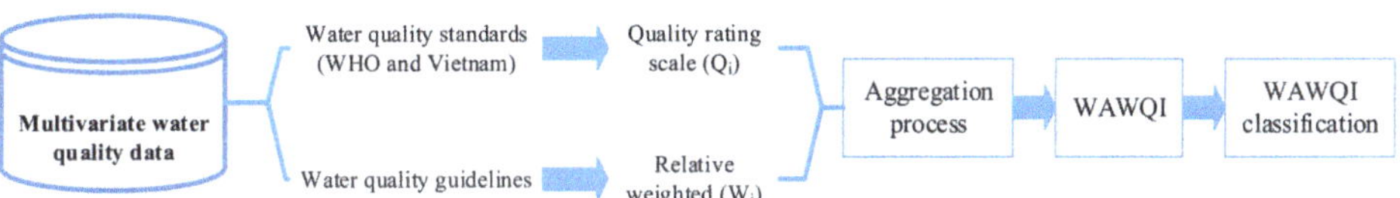

Figure 4. Flowchart of the weighted arithmetic water quality index (WAWQI) model.

In this study, we chose the Horton method to calculate the WAWQI [21,35,54]. The standard for the drinking water was based on the permissible standard for drinking water set by WHO guidelines [55]. These all variables were turned into sub-indices such as quality rating (q$_i$) and unit weights (W$_i$). The sub-indices were expressed on a single scale, and water quality was classified. The WAWQI was estimated using Equation (1) [56]:

$$\text{WAWQI}_i = \frac{\sum_{i=1}^{n} Q_i \times W_i}{\sum_{i=1}^{n} W_i} \tag{1}$$

where,

WAWQI is weighted arithmetic water quality index;

Q_i is a quality rating of nth parameters, $Q_i = [(V_i - V_{di})/(S_i - V_{di})] \times 100$ in which V_i is estimated value of nth parameters based on sample location, V_d is ideal value in pure water for nth parameters (pH = 7.0 and other parameters is 0); S_i is permissible limits of nth parameters;

W_i is the unit weight of nth parameters, $W_i = K/S_i$, in which K is proportionality constant, $K = 1/\sum_{i=1}^{n}(1/S_i)$.

Based on the ranges of WAWQI value, the corresponding status of water quality and their possible drinking use are summarized in Table 1.

Table 1. Water quality classification for human consumption using the weighted arithmetic water quality index (WAWQI) [55].

WAWQI Range	Water Quality Classification
<25	Excellent
26–50	Good
51–75	Bad
76–100	Very bad
>100	Unsuitable for drinking

3. Results

3.1. Statistical Assessment Using Correlation

The results of correlations matrices among 12 water quality parameters in the dry and wet season are shown in Tables 2 and 3, respectively. The parameters showing weak correlation coefficients with others in both seasons in An Giang have been affected by multiple sources such as agriculture, urbanization, and industry [13,21,57]. In the dry season, COD had a strong statistically significant correlation with Mg^{2+} (0.61) and EC (0.61) and a moderately positive relation with PO_4^{3-} (0.49) and NH_4^+ (0.461). In contrast, in the rainy season, COD had no correlation with PO_4^{3-} and Mg^{2+} parameters, excluding EC, pH, and NH_4^+, with which it showed weak correlations. PO_4^{3-} had a weak correlation with EC and NH_4^+ in both seasons and had a very weak relationship with the only NO_2^- in the wet season. On the other hand, NO_3^- had a strong correlation with NO_2^-, while NO_3^- did not correlate to others in both seasons. During flooding, a large amount of water flowing from the upper Mekong River discharges into An Giang with high COD concentration, supported by previous observation [13].

Interestingly, the characteristics of physical parameters in the dry season are strongly correlated than those in the wet season. Physical parameters such as EC and pH had a negative correlation in the wet season and had almost no correlation in the dry season. In the dry season, EC correlated with COD, NH_4^+, and PO_4^{3-} while pH only correlated with NO_2^-. In the wet season, pH and EC had a moderate correlation with COD and NH_4^+. Besides, EC correlated with PO_4^{3-} and pH correlated with Mg^{2+} in the wet season. The EC parameter qualitatively reflects the status of inorganic pollution [58]. The significantly high relation between EC and NH_4^+ for both seasons signifies the excess of breakdown/decomposition of organic matters, animal, and human waste. Nitrogen fixation is an indicator of anthropogenic input, excess of fertilizer application in the agricultural fields. During the wet season, pH and EC are negatively correlated, indicating a lower prevalence of cations and anions when water becomes alkaline. The strong correlation between EC and COD for both seasons indicates high organic pollutants, while the moderate association with PO_4^{3-} implies anthropogenic input. A strong association between NO_2^- and NO_3^- suggest the same source of origin, likely an agricultural runoff with high fertilizer input.

3.2. Spatial Assessment of Water Quality Using DA

The analysis technique of DA method was used to determine how many discriminant water quality parameters between the two seasons. The DA result shows a temporal comparison of the three discriminant significant parameters: pH, Cl^-, and Ca^{2+} between the dry and wet seasons (Figure 5). The pH, Cl^-, and Ca^{2+} showed different behaviors between the two seasons. The pH measures acidity in water or represents the negative logarithm of the hydrogen-ion activity [59,60]. The pH value beyond 6.5 to 8.5 range represents its contamination or pollution [61]. On the other hand, pH has a significant association with dissolved oxygen (DO) in freshwater. Therefore, the breakdown of organic matter exceeds synthesis activities caused oxygen consumption to increase. In this study, the pH 7.42 ± 0.63 (dry season) and 6.97 ± 1.06 (wet season) were neither highly alkaline nor highly acidic. In the dry season, the water is slightly alkaline, while the water is slightly acidic in the wet season. This result also confirms that the fluctuations in the value of water quality parameters in the dry season are greater than those in the wet season. On the other hand, the concentrations of Cl^- and Ca^{2+} were also relatively higher for the dry season than that of the wet season. Relatively low river discharge and higher evapotranspiration cause this seasonal difference in the concentration. Even though Cl^- occurs naturally in water, the larger value of Cl^- level can increase the corrosiveness of water, and in combination with sodium, it creates a salty taste.

Table 2. Correlation matrices in the dry season using Spearman rank–order.

Variables	PH	EC	Cl^-	NO_2^-	NO_3^-	NH_4^+	COD	PO_4^{3-}	Na^+	Ca^{2+}	Mg^{2+}	K^+
PH	1											
EC	0.050	1										
Cl^-	0.022	0.109	1									
NO_2^-	**0.413**	−0.079	−0.040	1								
NO_3^-	0.250	−0.163	−0.241	**0.775**	1							
NH_4^+	0.022	**0.570**	**0.319**	0.075	−0.006	1						
COD	−0.161	**0.605**	0.306	−0.114	−0.278	**0.461**	1					
PO_4^{3-}	0.200	**0.475**	**0.387**	−0.228	−0.296	**0.478**	**0.488**	1				
Na^+	0.120	0.229	0.032	0.213	−0.009	0.275	0.303	0.014	1			
Ca^{2+}	0.131	0.086	−0.290	0.171	0.029	0.046	−0.049	−0.119	**0.336**	1		
Mg^{2+}	−0.279	**0.380**	0.119	−0.146	−0.133	**0.317**	**0.607**	0.250	**0.394**	0.155	1	
K^+	−0.042	0.194	−0.012	−0.080	−0.075	0.275	0.311	0.111	**0.477**	−0.049	**0.562**	1

Values in bold are different from 0 with a significance level at alpha = 0.05. Concentrations of conductivity (EC), phosphate (PO_4^{3-}), ammonium (NH_4^+), chemical oxygen demand (COD), nitrite (NO_2^-); nitrate (NO_3^-).

Table 3. Correlation matrices in the wet season using Spearman rank–order.

Variables	PH	EC	Cl^-	NO_2^-	NO_3^-	NH_4^+	COD	PO_4^{3-}	Na^+	Ca^{2+}	Mg^{2+}	K^+
PH	1											
EC	**−0.509**	1										
Cl^-	0.176	−0.115	1									
NO_2^-	0.127	−0.042	0.005	1								
NO_3^-	0.143	−0.162	0.216	**0.627**	1							
NH_4^+	**−0.424**	**0.710**	0.077	0.116	0.003	1						
COD	**−0.444**	**0.490**	−0.107	0.005	−0.248	**0.427**	1					
PO_4^{3-}	0.125	**0.404**	0.134	**0.351**	0.223	**0.314**	0.067	1				
Na^+	−0.150	0.179	−0.146	−0.106	0.014	0.038	0.110	0.078	1			
Ca^{2+}	−0.307	0.085	0.133	−0.037	0.094	0.204	0.012	−0.109	0.094	1		
Mg^{2+}	**−0.313**	0.308	−0.263	−0.119	−0.130	0.095	0.178	−0.013	**0.338**	**0.434**	1	
K^+	−0.157	0.178	−0.028	0.180	0.107	0.224	0.111	0.111	**0.410**	0.301	**0.381**	1

Values in bold are different from 0 with a significance level at alpha = 0.05.

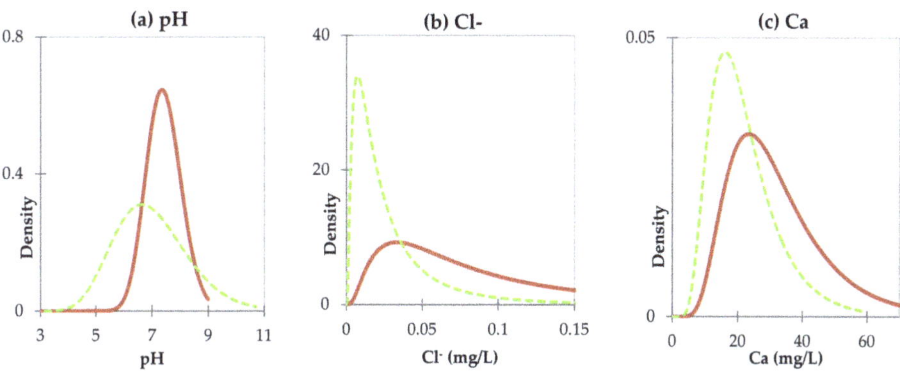

Figure 5. Log-normal probability distribution of (**a**) pH, (**b**) Cl^-, and (**c**) Ca^{2+} during the dry (red line) and wet seasons (green line).

The DA approach was also applied to identify the contribution of the most important parameters of water quality seasonal variations, especially concerning the contribution of the variables in discriminating in space. Therefore, the DA approach is used to determine the discriminant among clusters in the dry and wet seasons (Tables 4 and 5). The significant parameters among clusters are the concentrations of NO_2^-, NO_3^-, and pH in the dry season and Cl^- and Mg^{2+} in the wet season.

Table 4. Unidimensional lambda test of the quality of water parameter equality in the dry season.

Variable	Backward Model			Forward Model		
	Lambda	F	*p*-Value	Lambda	F	*p*-Value
NO_2^-	0.600 ***	8.002	0.000	0.600 ***	8.002	0.000
NO_3^-	0.817 **	2.688	0.006	0.817 **	2.688	0.006
NH_4^+	0.748	4.037	0.014			
COD						
Cl^-						
PO_4^{3-}						
PH	0.609 ***	7.720	0.000	0.609 ***	7.720	0.000
EC						
Na^+	0.805	2.913	0.048			
Ca^{2+}						
Mg^{2+}						
K^+						

Note: Significance levels are denoted as follows: ** $p < 0.01$, *** $p < 0.001$.

Table 5. Unidimensional lambda test of the quality of water parameter equality in the wet season.

Variable	Backward Model			Forward Model		
	Lambda	F	*p*-Value	Lambda	F	*p*-Value
NO_2^-	0.729 **	4.468	0.009			
NO_3^-				0.712 **	4.858	0.006
NH_4^+						
COD						
Cl^-	0.699 **	5.159	0.005	0.699 **	5.159	0.005
PO_4^{3-}						
PH						
EC						
Na^+						
Ca^{2+}						
Mg^{2+}	0.768 **	3.626	0.002	0.768 **	3.626	0.002
K^+						

Note: Significance levels are denoted as follows: ** $p < 0.01$.

The discriminant of water pollutant level among different clusters (Cluster 3: inside the full-dike system, Cluster 2: inside the semi-dike system, and Cluster 1: outside of the dike system) was evaluated. The discriminant among clusters for selected parameters in both seasons was displayed by using box and whisker plots (Figures 6 and 7). For the dry season, concentrations of pH, NO_3^-, NO_2^- were high in Cluster 3 in comparison with Clusters 1 and 2. Meanwhile, in the wet season, the highest concentration of Mg^{2+} was found in Cluster 2, followed by Cluster 3 and Cluster 1. The concentration of Cl^- was found higher in Cluster 3 than that in Clusters 1 and 2 in the wet season.

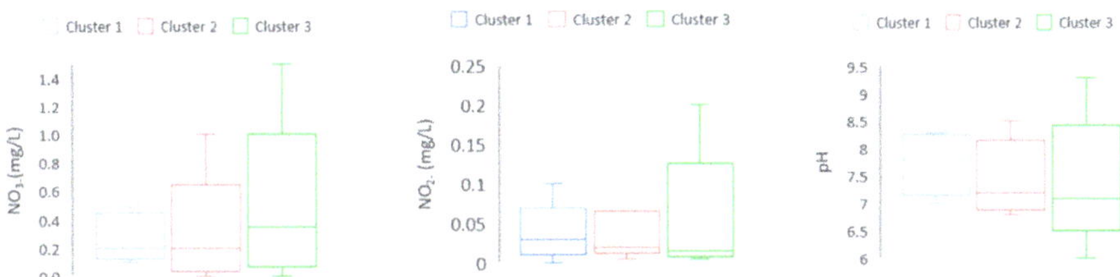

Figure 6. Water quality variables among three Clusters in the dry season. NO_3^-, NO_2^-, and pH were found higher in Cluster 3 than those in Clusters 1 and 2.

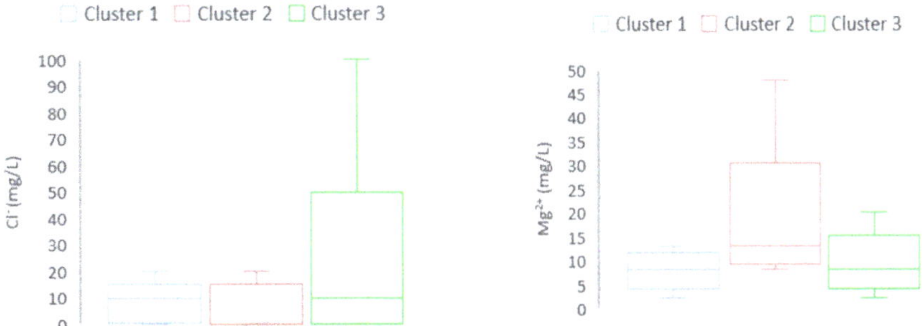

Figure 7. Water quality variables among the three clusters in the wet season. Mg^{2+} was high in Cluster 2 while Cl^- was high in Cluster 3.

3.3. Water Quality Classification Using WAWQI

Table 6 shows the range, mean, and standard deviation values of parameters, some of which were found to exceed the permissible standard for drinking water set by WHO and Vietnam national standard for both seasons. The higher values of these water quality parameters would lead to an increase in WAWQI. Overall, EC, NO_2^-, NH_4^+, COD, PO_4^{3-}, and K^+ were above the permissible standard set by WHO and Vietnamese standards. The EC is a measure of current carrying capacity due to the electrical current being carried by ions in a solution [62]; thus, as the concentration of dissolved salts increases, conductivity value also increases. On the other hand, EC is also used to determine the suitability of water for irrigation and firefighting [61]. Both NO_3^- and NO_2^- are nitrogen-containing compounds that generally indicate contamination from a pasture, decomposed vegetation, agricultural fertilizers, sewage, and rock–water interaction. NO_3^- is the essential nutrients in an ecosystem. Generally, water polluted by organic matter exhibits higher values of nitrate. In this study, the mean concentration of nitrate was 0.34 mg/L in the dry season and 0.5 mg/L in the wet season. Nitrate in all sample sites was below permissible standards.

The Cl^- mean values are 90 mg/L in the dry season and 20 mg/L in the wet season. The concentration of Cl^- in surface water may come from human activities, namely, agricultural runoff and wastewater sources [61,63]. In this study, the high concentration of Cl^- is also considered to be an indication of pollution due to the high organic waste from irrigation drainage, septic tank effluent, animal feed, and landfill leachates [59,60]. This also indicates poor governance and infrastructure to manage wastewater coming from both agricultural fields and urbanized areas.

The WAWQI of the present investigation from 40 sampling sites in both seasons were calculated. The WAWQI calculated from sampling Number 2 in the dry season is shown in Table 7 as an example.

Table 6. Standards for drinking water and relative weight of parameters.

Parameters	Unit	Dry Season			Wet Season			S_i	V_{di}	$(1/S_i)$	K	W_i
		Range	Mean	SD	Range	Mean	SD					
PH		6–9.3	7.37	0.63	3.1–9.3	6.92	1.06	8.5 *	7	0.118	0.0292	0.0034
EC	$S.cm^{-1}$	150–990	359	186.68	90–1160	340	221	300 *	0	0.003	0.0292	0.0001
Cl^-	mg/L	0–200	90	60	0–100	20	30	200 *	0	0.005	0.0292	0.0001
NO_2^-	mg/L	0–0.9	0.06	0.15	0–0.25	0.04	0.05	0.05 **	0	20	0.0292	0.5848
NO_3^-	mg/L	0–0.5	0.34	0.48	0–2	0.50	0.52	2 **	0	0.5	0.0292	0.0146
NH_4^+	mg/L	0.1–10	1.53	2.91	0.05–12	1.10	2.27	0.3 **	0	3.33	0.0292	0.0975
COD	mg/L	4–100	23.75	18.89	7–100	25.53	19.34	10 **	0	0.1	0.0292	0.0029
PO_4^{3-}	mg/L	0.05–5	0.64	1.09	0.08–4	0.59	0.87	0.1 **	0	10	0.0292	0.2924
Na^+	mg/L	6.1–1610	71.80	251.6	0.56–55.5	17.48	13.56	200 *	0	0.005	0.0292	0.0001
Ca^{2+}	mg/L	5.6–65.4	32.21	13.13	5.6–467.4	22.37	10.43	75 *	0	0.013	0.0292	0.0004
Mg^{2+}	mg/L	1.5–34.7	13.11	5.35	2.2–47.7	10.55	7.08	50 *	0	0.02	0.0292	0.0006
K^+	mg/L	1.4–43.6	15.84	11.4	2.5–129	13.39	22	10 *	0	0.1	0.0292	0.0029

Permissible limits for drinking * WHO and ** Vietnamese standard. Measured values (V_i), standard values of water quality parameters (S_i), corresponding ideal values (V_{di}), Q_i is a quality rating of n-th parameters, and unit weights (**W_i**) for sampling.

Table 7. Weighted arithmetic water quality index (WAWQI) calculation for sampling Number 2 as an example in the dry season.

Parameters	Unit	V_i	S_i	V_{di}	Q_i	$(1/S_i)$	K	W_i	$Q_i \times W_i$
PH		8.3	8.5	7	86.7	0.12	0.0292	0.0034	0.30
EC	$S.cm^{-1}$	240	300	0	80	0.00	0.0292	0.0001	0.01
Cl^-	mg/L	0.1	200	0	0.05	0.01	0.0292	0.0001	0.00
NO_2^-	mg/L	0.04	0.05	0	80	20	0.0292	0.5848	46.79
NO_3^-	mg/L	0.4	2	0	20	0.50	0.0292	0.0146	0.29
NH_4^+	mg/L	0.1	0.3	0	33.3	3.33	0.0292	0.0975	3.25
COD	mg/L	4	10	0	40	0.10	0.0292	0.0029	0.12
PO_4^{3-}	mg/L	0.05	0.1	0	50	10	0.0292	0.2924	14.62
Na^+	mg/L	12	200	0	5.8	0.01	0.0292	0.0001	0.00
Ca^{2+}	mg/L	29	75	0	38.2	0.01	0.0292	0.0004	0.01
Mg^{2+}	mg/L	10	50	0	20.8	0.02	0.0292	0.0006	0.01
K^+	mg/L	6	10	0	59.4	0.10	0.0292	0.0029	0.17
	Sum					34.20		1	66

Measured values (V_i), standard values of water quality parameters (S_i), corresponding ideal values (V_{di}), Q_i is a quality rating of n-th parameters, and unit weights (**W_i**) for sampling.

The WAWQI is commonly used for the detection and evaluation of overall water pollution since it can reflect the influence of different quality parameters on the quality of water. The application of WAWQI is a useful method in assessing the suitability of water for various beneficial uses. The WAWQI was analyzed for two seasons, as shown in Appendix A. From the WAWQI of the dry season samples, 70% of the total water samples was unsuitable for drinking, 10% was very bad, 17.7% was bad, and only 2.5% was good. The water quality of the wet season showed that 60% of the total water samples was unsuitable for drinking, 10% was very bad, 20% was bad, and 10% was good. In general, the surface water quality was better in the wet season than in the dry season.

Besides, the WAWQI of both the wet and dry seasons was mapped to show the spatial distribution of WAWQI using the IDW method (Figure 8). The bad conditions of water quality (high values of

WAWQI) were located in the rice intensification areas. Some bad water quality could be found at tributaries of the Bassac River. It might be caused by water discharged from intensive rice crop areas, tourism and urban areas. In the area surrounded by the Mekong and Bassac Rivers in the northeast, the water quality is found to be better. It may be because the proper operation of the sluice-gates system and the alternatives of intensive rice crops (instead of 3 crops/year, it had shifted to 8 crops for every 3-years, and 5 crops for every 2-years by now). Being surrounded by the two large rivers is also advantageous in that the exchange of inside and outside dike systems may lead to a reduction in pollution by dilution.

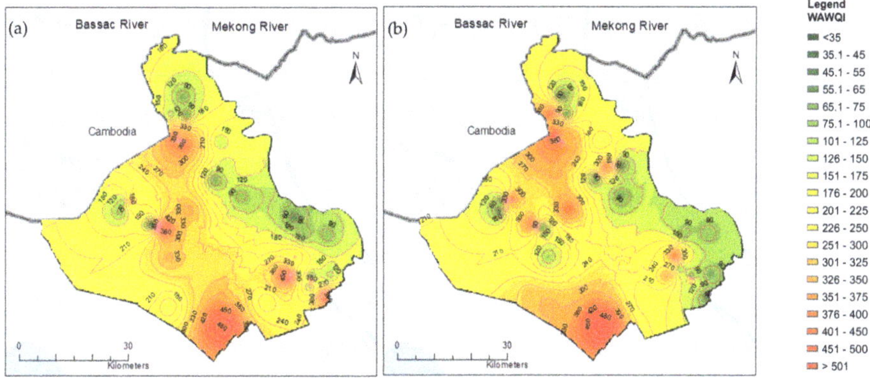

Figure 8. Spatial distribution of weighted arithmetic water quality index (WAWQI) in the (**a**) dry and (**b**) wet seasons using inverse distance weighting (IDW) interpolation.

Overall, the WAWQI values in the wet season are more scattered among the different sites compared to that of the dry season. For example, extreme high WAWQI were found in the northwest and the southwest of An Giang, while the southeast of An Giang was found with good water quality. Regions with high WAWQI were mainly found in the triple-rice system, and the urban area inside the full-dike system was linked with high concentrations of EC, NH_4^+, COD, NO_2^-, and PO_4^{3-}. Contrastingly, locations with low WAWQI mainly represent orchards located inside the full-dike system. The heavy rain in the wet season can dilute pollutant concentrations. Therefore, water quality in this region in the wet season is better than the dry season. The "hotspot" of water quality in the south most of An Giang province is found in both dry and wet seasons. This can be explained by the full triple rice cropping system inside the full-dike system in this location being linked with high concentrations of EC, COD, NO_2^-, and PO_4^{3-}.

4. Discussion

Water is a precious resource for various activities in An Giang. However, due to a rapid rate of increase in rice intensification, urbanization, and tourist area, the water quality has decreased dramatically. This issue was found in various studies in the VMD in recent years [15,21]. The clarification of the seasonal change in water quality was important to evaluate the temporal variations of surface water pollution.

The results show that the concentration of NH_4^+, COD, PO_4^{3-}, and K^+ was relatively higher compared to the World Health Organization (WHO) and the Vietnamese standard for both seasons. Figures 9 and 10 show the concentrations of COD and PO_4^{3-} at the stations of Tan Chau and Chau Doc, respectively, which is close to the Cambodian border. The concentrations of COD showed an increasing trend from 1985 to 2011 at Tan Chau and in 2013 at Chau Doc station. Although COD concentration from 1996 to 2010 in Cambodia was higher than those in Vietnam, most of the COD values were below the permissible standard of Vietnam. From 2015 to 2017, COD has exceeded the Vietnamese standard for domestic use. Linear progress analysis shows the R^2 values at 0.47 and 0.39 for Tan Chau and Chau Doc stations, respectively. The PO_4^{3-} concentrations from 1995 to 2005 (Figure 10) in the Cambodia side were below the standard of Vietnam, while those concentrations in the Vietnam part fluctuated seasonally and were higher than the permissible standard of Vietnam for several years.

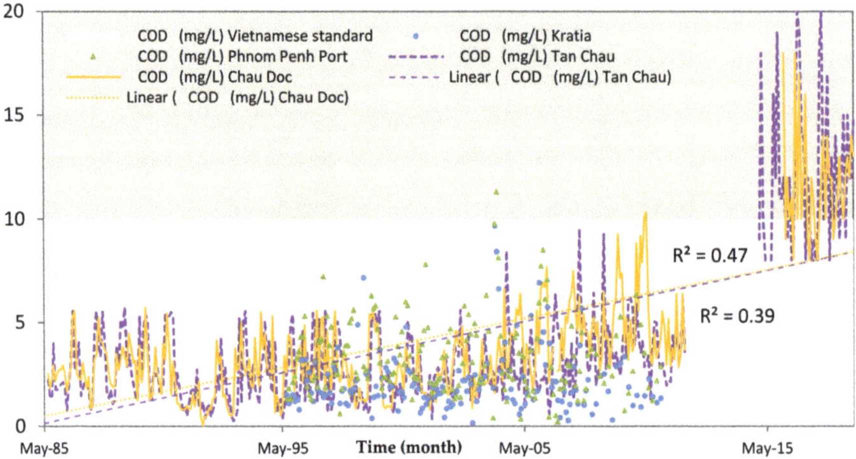

Figure 9. Temporal concentrations of chemical oxygen demand (COD) in the Vietnamese side (Tan Chau and Chau Doc stations) from 1985 to 2017 and in the Cambodia side (Phnom Penh Port and Kratie) from 1995 to 2010.

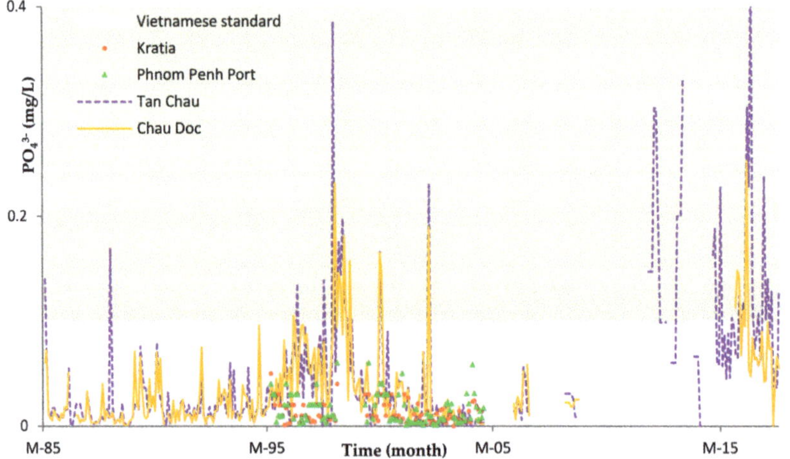

Figure 10. Temporal concentrations of PO_4^{3-} in the Vietnamese side (Tan Chau and Chau Doc stations) from 1985 to 2017 and in the Cambodia side (Phnom Penh Port and Kratie) from 1995 to 2005.

The results of this study show that pH, Cl^-, and calcium were significant discriminant parameters between the two seasons. Cl^- was chosen as an important indicator parameter since its values represent the degree of organic pollution, as mentioned above. The concentration of Cl^- in the dry season was found extremely higher than that in the wets season.

The classification of water quality in this study clearly shows that the status of water bodies in the study area is eutrophic, and it is unsuitable for drinking. It is also observed that most of the pollution loads relatively high in the dry season compared to those in the wet season except NH_4^+ and COD.

Anthropogenic pollutant load is relatively high, as indicated by a higher concentration of PO_4^{3-}, NO_2^- and NO_3^-. These results support the hypothesis that considered water quality deterioration in the dry season.

Furthermore, high concentrations of NO_3^-, NO_2^- and pH in water samples of Cluster 3 inside the full-dike system in the dry season were detected. Meanwhile, high Cl^- and Mg^{2+} were found in water samples of Cluster 3 and Cluster 2, respectively. Minh et al. [21] also found high nitrite and nitrate inside the full-dike system where the triple rice cropping system was dominant in An Giang. The high mean concentration of 90 mg/L in the dry season for Cl^- inside the full-dike system was identified for the influence of wastewater surrounding the urban area and rice fields. Rivers typically have concentrations of Cl^- less than 50 mg/L [64]. The high level of Cl^- may have a negative impact on an ecosystem [64]. This may be an indicator of sewage pollution, which may be from a water softener or sewage contamination discharge from city, located inside the full-dike system. In summary, it also supports the hypothesis that water quality inside the full-dike system is worse than that of outside ones.

The WAWQI for 40 samples ranges from 34 to 1847 in the dry season and from 40 to 1584 in the wet season. Although the range of WAWQI, as well as the minimum values in the dry season, was lower than those in the wet season, the good water quality index of 10% of the location in the wet season was higher than 2.5% of the location in the dry season. The high value of WAWQI at these stations has been found to be mainly due to the higher levels of EC, NH_4^+, and COD. Spatial distribution of water quality using WAWQI values helped to identify factors and processes responsible for water quality evolution.

5. Conclusions

Overall, this study provides an approach for assessing surface water pollutant levels. Water quality in An Giang in the dry and wet seasons has deteriorated tremendously due to urban wastewater discharge and rice intensification in the past 30 years. During the flood season, water from the Upper Mekong River carries high concentrations of pollutants into An Giang. We found high NO_3^-, NO_2^-, Cl^- concentrations inside the full-dike system, while high concentrations of COD and NH_4^+ were found in the urban area and the main river (Bassac River). Most of the water quality samples in both dry and wet seasons were bad or unsuitable for drinking. Thus, the water in An Giang Province should be treated before supplying for drinking water or domestic use. Water quality observation stations along the border should be strengthened to provide a better understanding of the primary pollutant sources that have influenced the surface water quality during the flood season in An Giang as well as the entire VMD.

Author Contributions: Conceptualization—H.V.T.M., R.A., M.K. and T.V.T.; methodology—H.V.T.M., R.A., M.K., P.K., K.N.L. and T.V.T.; writing—original draft preparation, H.V.T.M., R.A., M.K., P.K., K.N.L. and T.V.T.; writing—review and editing, H.V.T.M., R.A., M.K., P.K., K.N.L. and T.V.T. All authors have read and agreed to the published version of the manuscript.

Funding: This research received no external funding.

Acknowledgments: The authors thank the Vietnamese Ministry of Education and Training, Can Tho University, and Hokkaido University for supporting us to complete this research.

Conflicts of Interest: The authors declare no conflict of interest.

Appendix A

Table A1. The weighted arithmetic water quality index (WAWQI) at 40 sampling sites in the dry and wet season in 2018.

Sampling Site	Dry Season		Wet Season	
	WAWQI	Water Classification	WAWQI	Water Classification
1	443	Unsuitable for drinking	51	Bad
2	66	Bad	65	Bad
3	241	Unsuitable for drinking	361	Unsuitable for drinking
4	123	Unsuitable for drinking	208	Unsuitable for drinking
5	64	Bad	66	Bad
6	222	Unsuitable for drinking	205	Unsuitable for drinking
7	101	Unsuitable for drinking	124	Unsuitable for drinking
8	198	Unsuitable for drinking	98	Very Bad
9	1847	Unsuitable for drinking	1489	Unsuitable for drinking
10	448	Unsuitable for drinking	239	Unsuitable for drinking
11	1813	Unsuitable for drinking	1584	Unsuitable for drinking
12	187	Unsuitable for drinking	198	Unsuitable for drinking
13	561	Unsuitable for drinking	487	Unsuitable for drinking
14	159	Unsuitable for drinking	312	Unsuitable for drinking
15	339	Unsuitable for drinking	361	Unsuitable for drinking
16	111	Unsuitable for drinking	368	Unsuitable for drinking
17	131	Unsuitable for drinking	151	Unsuitable for drinking
18	131	Unsuitable for drinking	349	Unsuitable for drinking
19	161	Unsuitable for drinking	43	Good
20	81	Very Bad	63	Bad
21	111	Unsuitable for drinking	61	Bad
22	63	Bad	143	Unsuitable for drinking
23	76	Very Bad	75	Bad
24	73	Bad	76	Very Bad
25	52	Bad	107	Unsuitable for drinking
26	595	Unsuitable for drinking	307	Unsuitable for drinking
27	217	Unsuitable for drinking	281	Unsuitable for drinking
28	34	Good	369	Unsuitable for drinking
29	736	Unsuitable for drinking	66	Bad
30	334	Unsuitable for drinking	97	Very Bad
31	334	Unsuitable for drinking	310	Unsuitable for drinking
32	61	Bad	62	Bad
33	67	Bad	48	Good
34	94	Very Bad	77	Very Bad
35	319	Unsuitable for drinking	139	Unsuitable for drinking

Table A1. *Cont.*

Sampling Site	Dry Season		Wet Season	
	WAWQI	**Water Classification**	**WAWQI**	**Water Classification**
36	1075	Unsuitable for drinking	177	Unsuitable for drinking
37	277	Unsuitable for drinking	369	Unsuitable for drinking
38	185	Unsuitable for drinking	274	Unsuitable for drinking
39	78	Very Bad	40	Good
40	102	Unsuitable for drinking	49	Good

References

1. Venkatramanan, S.; Chung, S.; Ramkumar, T.; Rajesh, R.; Gnanachandrasamy, G. Assessment of groundwater quality using GIS and CCME WQI techniques: A case study of Thiruthuraipoondi city in Cauvery deltaic region, Tamil Nadu, India. *Desalin. Water Treat.* **2016**, *57*, 12058–12073. [CrossRef]
2. Van Stokkom, H.; Witter, J. Implementing integrated flood risk and land-use management strategies in developed deltaic regions, exemplified by The Netherlands. *Int. J. River Basin Manag.* **2008**, *6*, 331–338. [CrossRef]
3. Minh, H.V.T.; Ngoc, D.T.H.; Ngan, H.Y.; Men, H.V.; Van, T.N.; Ty, T.V. Assessment of groundwater level and quality: A case study in O Mon and Binh Thuy districts, Can Tho City, Vietnam. *Fac. Eng. Naresuan Univ.* **2016**, *11*, 25–33.
4. Minh, H.V.T.; Avtar, R.; Kumar, P.; Tran, D.Q.; Ty, T.V.; Behera, H.C.; Kurasaki, M. Groundwater quality assessment using fuzzy-AHP in an Giang Province of Vietnam. *Geosciences* **2019**, *9*, 330. [CrossRef]
5. El-Kowrany, S.I.; El-Zamarany, E.A.; El-Nouby, K.A.; El-Mehy, D.A.; Abo Ali, E.A.; Othman, A.A.; Salah, W.; El-Ebiary, A.A. Water pollution in the Middle Nile Delta, Egypt: An environmental study. *J. Adv. Res.* **2016**, *7*, 781–794. [CrossRef]
6. Akaishi, F.; Satake, M.; Otaki, M.; Tominaga, N. Surface water quality and information about the environment surrounding Inle Lake in Myanmar. *Limnology* **2006**, *7*, 57–62. [CrossRef]
7. Bowles, J. *The Ayeyarwady River Endangered*; Myanmar Development Research Institute (MDRI): Yangon, Myanmar, 2013; p. 42.
8. Charron, D.F.; Thomas, M.K.; Waltner-Toews, D.; Aramini, J.J.; Edge, T.; Kent, R.A.; Maarouf, A.R.; Wilson, J. Vulnerability of waterborne diseases to climate change in Canada: A review. *J. Toxicol. Env. Health Part A* **2004**, *67*, 1667–1677. [CrossRef]
9. Phung, D.; Huang, C.; Rutherford, S.; Chu, C.; Wang, X.; Nguyen, M. Climate change, water quality, and water-related diseases in the Mekong Delta Basin: A systematic review. *Asia Pac. J. Public Health* **2015**, *27*, 265–276. [CrossRef]
10. Dat, T.Q.; Kanchit, L.; Thares, S.; Trung, N.H. Modeling the influence of river discharge and sea level rise on salinity intrusion in Mekong Delta. In Proceedings of the 1st Environment Asia International Conference, Bangkok, Thailand, 23–26 March 2011; Volume 35, pp. 685–701.
11. Wilbers, G.-J.; Becker, M.; Nga, L.T.; Sebesvari, Z.; Renaud, F.G. Spatial and temporal variability of surface water pollution in the Mekong Delta, Vietnam. *Sci. Total Environ.* **2014**, *485*, 653–665. [CrossRef]
12. Ikemoto, T.; Tu, N.P.C.; Watanabe, M.X.; Okuda, N.; Omori, K.; Tanabe, S.; Tuyen, B.C.; Takeuchi, I. Analysis of biomagnification of persistent organic pollutants in the aquatic food web of the Mekong Delta, South Vietnam using stable carbon and nitrogen isotopes. *Chemosphere* **2008**, *72*, 104–114. [CrossRef]
13. Chea, R.; Grenouillet, G.; Lek, S. Evidence of water quality degradation in lower Mekong Basin revealed by self-organizing map. *PLoS ONE* **2016**, *11*, e0145527. [CrossRef] [PubMed]
14. Berg, M.; Stengel, C.; Trang, P.T.K.; Viet, P.H.; Sampson, M.L.; Leng, M.; Samreth, S.; Fredericks, D. Magnitude of arsenic pollution in the Mekong and Red River Deltas—Cambodia and Vietnam. *Sci. Total Environ.* **2007**, *372*, 413–425. [CrossRef] [PubMed]
15. Chau, N.D.G.; Sebesvari, Z.; Amelung, W.; Renaud, F.G. Pesticide pollution of multiple drinking water sources in the Mekong Delta, Vietnam: Evidence from two provinces. *Environ. Sci. Pollut. Res.* **2015**, *22*, 9042–9058. [CrossRef] [PubMed]

16. Bartram, J.; Ballance, R. *Water Quality Monitoring: A Practical Guide to the Design and Implementation of Freshwater Quality Studies and Monitoring Programmes*; CRC Press: London, UK, 1996; p. 383.

17. Mekong River Commission. *Overview of the Hydrology of the Mekong Basin*; PDR: Vientiane, Laos, 2005; p. 82.

18. Wang, W.; Lu, H.; Ruby Leung, L.; Li, H.; Zhao, J.; Tian, F.; Yang, K.; Sothea, K. Dam construction in Lancang-Mekong River Basin could mitigate future flood risk from warming-induced intensified rainfall. *Geophys. Res. Lett.* **2017**, *44*, 10–378. [CrossRef]

19. Open Development Mekong Population and Censuses. Available online: https://opendevelopmentmekong.net/topics/population-and-censuses (accessed on 26 May 2020).

20. Mekong River Commission (MRC). Mekong Basin. Available online: http://www.mrcmekong.org/mekong-basin (accessed on 18 April 2020).

21. Minh, H.V.T.; Kurasaki, M.; Ty, T.V.; Tran, D.Q.; Le, K.N.; Avtar, R.; Rahman, M.M.; Osaki, M. Effects of multi-dike protection systems on surface water quality in the Vietnamese Mekong Delta. *Water* **2019**, *11*. [CrossRef]

22. Minh, H.V.T.; Avtar, R.; Mohan, G.; Misra, P.; Kurasaki, M. Monitoring and mapping of rice cropping pattern in flooding area in the Vietnamese Mekong Delta using Sentinel-1A data: A case of An Giang Province. *ISPRS Int. J. Geo Inf.* **2019**, *8*. [CrossRef]

23. Tuan, L.A.; Minh, H.V.T.; Tuan, D.D.A.; Thao, N.T.P. *Baseline Study for Community Based Water Management Project*; Mekong Water Governance Program Vietnam: Hanoi, Vietnam, 2015.

24. Fujii, H.; Fujihara, Y.; Hoshikawa, K. Expansion of full-dyke system and its impact in flood-prone rice area in the Mekong Delta. *Trans. Jpn. Soc. Irrig. Drain. Rural Eng.* **2013**, *81*, 271–278. [CrossRef]

25. Nguyen, V.K.T.; Nguyen, V.D.; Fujii, H.; Kummu, M.; Merz, B.; Apel, H. Has dyke development in the Vietnamese Mekong Delta shifted flood hazard downstream? *Hydrol. Earth Syst. Sci.* **2017**, *21*, 3991–4010. [CrossRef]

26. Ty, T.V. Scenario-based impact assessment of land use/cover and climate changes on water resources and demand: A case study in the Srepok River Basin, Vietnam—Cambodia. *Water Resour. Manag.* **2012**, *26*, 1387–1407. [CrossRef]

27. DONRE Water Resource Distribution in an Giang. Available online: http://sotainguyenmt.angiang.gov.vn/TongQuan_TNN1.aspx (accessed on 2 September 2019).

28. Allan, J.D. Landscapes and riverscapes: The influence of land use on stream ecosystems. *Annu. Rev. Ecol. Evol. Syst.* **2004**, *35*, 257–284. [CrossRef]

29. Johnson, L.; Gage, S. Landscape approaches to the analysis of aquatic ecosystems. *Freshw. Biol.* **1997**, *37*, 113–132. [CrossRef]

30. Wikipedia. An Giang Provice. Available online: https://en.wikipedia.org/wiki/An_Giang_Province (accessed on 10 May 2020).

31. Yoshida, Y.; Lee, H.S.; Trung, B.H.; Tran, H.-D.; Lall, M.K.; Kakar, K.; Xuan, T.D. Impacts of mainstream hydropower Dams on fisheries and agriculture in lower Mekong Basin. *Sustainability* **2020**, *12*, 2408. [CrossRef]

32. Tran, D.; Weger, J. Barriers to implementing irrigation and drainage policies in An Giang Province, Mekong Delta, Vietnam. *Irrig. Drain.* **2018**, *67*, 81–95. [CrossRef]

33. The Southern Regional Hydro-meteorological Center. Supply and Exploit Data. Available online: http://www.kttv-nb.org.vn/index.php/dich-vu-kttv/cung-cap-khai-thac-so-lieu (accessed on 6 June 2020).

34. Helena, B.; Pardo, R.; Vega, M.; Barrado, E.; Fernandez, J.M.; Fernandez, L. Temporal evolution of groundwater composition in an alluvial aquifer (Pisuerga River, Spain) by principal component analysis. *Water Res.* **2000**, *34*, 807–816. [CrossRef]

35. Avtar, R.; Kumar, P.; Singh, C.K.; Sahu, N.; Verma, R.L.; Thakur, J.K.; Mukherjee, S. Hydrogeochemical assessment of groundwater quality of Bundelkhand, India using statistical approach. *Water Qual. Expo. Health* **2013**, *5*, 105–115. [CrossRef]

36. Kido, M.; Yustiawati; Syawal, M.S.; Sulastri; Hosokawa, T.; Tanaka, S.; Saito, T.; Iwakuma, T.; Kurasaki, M. Comparison of general water quality of rivers in Indonesia and Japan. *Environ. Monit. Assess.* **2008**, *156*, 317. [CrossRef] [PubMed]

37. Shammi, M.; Rahman, M.M.; Islam, M.A.; Bodrud-Doza, M.; Zahid, A.; Akter, Y.; Quaiyum, S.; Kurasaki, M. Spatio-temporal assessment and trend analysis of surface water salinity in the coastal region of Bangladesh. *Environ. Sci. Pollut. Res.* **2017**, *24*, 14273–14290. [CrossRef]

38. Avtar, R.; Kumar, P.; Singh, C.; Mukherjee, S. A comparative study on hydrogeochemistry of Ken and Betwa Rivers of Bundelkhand using statistical approach. *Water Qual. Expo. Health* **2011**, *2*, 169–179. [CrossRef]

39. Kumar, P.; Ram, A. Chapter 4: Integrating major ion chemistry with statistical analysis for geochemical assessment of groundwater quality in coastal aquifer of Saijo plain, Ehime prefecture, Japan. In *Water Quality: Indicators, Human Impact and Environmental Health*; Nova Publication: Haryana, India, 2013; pp. 99–108.

40. Avtar, R.; Kumar, P.; Oono, A.; Saraswat, C.; Dorji, S.; Hlaing, Z. Potential application of remote sensing in monitoring ecosystem services of forests, mangroves and urban areas. *Geocarto. Int.* **2017**, *32*, 874–885. [CrossRef]

41. Johnson, R.A.; Wichern, D.W. *Applied Multivariate Statistical Analysis*, 3rd ed.; Prentice-Hall: Englewood Cliffs, NJ, USA, 1992; p. 642.

42. Ogbozige, F.J.; Adie, D.B.; Abubakar, U.A. Water quality assessment and mapping using inverse distance weighted interpolation: A case of River Kaduna, Nigeria. *Niger. J. Technol.* **2018**, *37*, 249–261. [CrossRef]

43. Ke, W.; Cheng, H.P.; Yan, D.; Lin, C. The application of cluster analysis and inverse distance-weighted interpolation to appraising the water quality of Three Forks Lake. *Procedia Environ. Sci.* **2011**, *10*, 2511–2517. [CrossRef]

44. Mirzaei, R.; Sakizadeh, M. Comparison of interpolation methods for the estimation of groundwater contamination in Andimeshk-Shush Plain, Southwest of Iran. *Environ. Sci. Pollut. Res.* **2016**, *23*, 2758–2769. [CrossRef] [PubMed]

45. Avtar, R.; Kumar, P.; Surjan, A.; Gupta, L.; Roychowdhury, K. Geochemical processes regulating groundwater chemistry with special reference to nitrate and fluoride enrichment in Chhatarpur area, Madhya Pradesh, India. *Environ. Earth Sci.* **2013**, *70*, 1699–1708. [CrossRef]

46. Algina, J.; Keselman, H. Comparing squared multiple correlation coefficients: Examination of a confidence interval and a test significance. *Psychol. Methods* **1999**, *4*, 76. [CrossRef]

47. Govindarajulu, Z. Rank correlation methods. *Technometrics* **1992**, *34*, 108. [CrossRef]

48. Shrestha, S.; Kazama, F. Assessment of surface water quality using multivariate statistical techniques: A case study of the Fuji river basin, Japan. *Environ. Model. Softw.* **2007**, *22*, 464–475. [CrossRef]

49. Molekoa, M.D.; Avtar, R.; Kumar, P.; Minh, H.V.T.; Kurniawan, T.A. Hydrogeochemical assessment of groundwater quality of Mokopane area, Limpopo, South Africa using statistical approach. *Water* **2019**, *11*, 1891. [CrossRef]

50. Duan, W.; He, B.; Nover, D.; Yang, G.; Chen, W.; Meng, H.; Zou, S.; Liu, C. Water quality assessment and pollution source identification of the Eastern Poyang Lake Basin using multivariate statistical methods. *Sustainability* **2016**, 8. [CrossRef]

51. Wunderlin, D.A.; del Pilar Díaz, M.; Amé, M.V.; Pesce, S.F.; Hued, A.C.; de Los Ángeles Bistoni, M. Pattern recognition techniques for the evaluation of spatial and temporal variations in water quality. A case study: Suquía River Basin (Córdoba–Argentina). *Water Res.* **2001**, *35*, 2881–2894. [CrossRef]

52. Poulsen, J.; French, A. Discriminant Function Analysis 2008. Available online: http://userwww.sfsu.edu/~{}efc/classes/biol710/discrim/discrim (accessed on 10 May 2020).

53. Horton, R.K. An index number system for rating water quality. *J. Water Pollu. Cont. Fed.* **1965**, *37*, 300–305.

54. Kachroud, M.; Trolard, F.; Kefi, M.; Jebari, S.; Bourrié, G. Water quality indices: Challenges and application limits in the literature. *Water* **2019**, *11*, 361. [CrossRef]

55. World Health Organization. *Guidelines for Drinking-Water Quality*, 4th ed.; World Health Organization: Geneva, Switzerland, 2017; p. 631.

56. Brown, R.M.; McClelland, N.I.; Deininger, R.A.; O'Connor, M.F. A water quality index—Crashing the psychological barrier. In *Proceedings of the Indicators of Environmental Quality*; Thomas, W.A., Ed.; Plenum Press: New York, NY, USA, 1972; pp. 173–182.

57. Thuy, P.T.; Van Geluwe, S.; Nguyen, V.-A.; Van der Bruggen, B. Current pesticide practices and environmental issues in Vietnam: Management challenges for sustainable use of pesticides for tropical crops in (South-East) Asia to avoid environmental pollution. *J. Mater. Cycles Waste Manag.* **2012**, *14*, 379–387. [CrossRef]

58. McCutcheon, S.; Martin, J.; Barnwell, T., Jr. *Water Quality, Handbook of Hydrology*; Maidment, D.R., Ed.; McGraw-Hill Inc.: New York, NY, USA, 1993.

59. Edzwald, J.K. *Water Quality and Treatment A Handbook on Drinking Water*, 6th ed.; McGrawHill Education: New York, NY, USA; American Water Works Association: Denver, CO, USA, 2010.

60. Spellman, F.R. *The Drinking Water Handbook*; CRC Press: Boca Raton, FL, USA, 2017; p. 388.

61. Omer, N.H. Water quality parameters. In *Water Quality-Science, Assessments and Policy*; IntechOpen: London, UK, 2019; p. 18.

62. APHA; AWWA; WEF. Standard methods for the examination of water and wastewater 21st Edition Method 5310 B. High temperature combustion method. In *Standard Methods for the Examination of Water and Wastewater*; American Public Health Association: Washington, DC, USA, 2005; pp. 5–21.

63. Sharma, S.; Bhattacharya, A. Drinking water contamination and treatment techniques. *Appl. Water Sci.* **2017**, *7*, 1043–1067. [CrossRef]

64. Brandt, M.J.; Johnson, K.M.; Elphinston, A.J.; Ratnayaka, D.D. Chemistry, microbiology and biology of water. In *Twort's Water Supply*, 7th ed.; Brandt, M.J., Johnson, K.M., Elphinston, A.J., Ratnayaka, D.D., Eds.; Butterworth-Heinemann: Boston, MA, USA, 2017; pp. 235–321.

Article

Spatio-Temporal Analysis of Surface Water Quality in Mokopane Area, Limpopo, South Africa

Mmasabata Dolly Molekoa [1], Ram Avtar [1,2,*], Pankaj Kumar [3], Huynh Vuong Thu Minh [4], Rajarshi Dasgupta [3], Brian Alan Johnson [3], Netrananda Sahu [5], Ram Lal Verma [6] and Ali P. Yunus [7,8]

1   Graduate School of Environmental Science, Hokkaido University, Sapporo 060-0810, Japan; sabimolekoa@gmail.com
2   Faculty of Environmental Earth Science, Hokkaido University, Sapporo 060-0810, Japan
3   Natural Resources and Ecosystem Services, Institute for Global Environmental Strategies, Hayama 240-0115, Japan; kumar@iges.or.jp (P.K.); dasgupta@iges.or.jp (R.D.); johnson@iges.or.jp (B.A.J.)
4   Department of Water Resources, College of Environment and Natural Resources, Cantho University, Cantho City 900000, Vietnam; hvtminh@ctu.edu.vn
5   Department of Geography, Delhi School of Economics, University of Delhi, Delhi 110007, India; nsahu@geography.du.ac.in
6   Regional Resource Centre for Asia and the Pacific, Asian Institute of Technology, Pathum Thani 12120, Thailand; verma@rrcap.ait.ac.th
7   State Key Laboratory for Geohazard Prevention and Geoenvironment Protection, Chengdu University of Technology, Chengdu 610059, Sichuan, China; pulpadan.yunusali@nies.go.jp
8   Center for Climate Change Adaptation, National Institute for Environmental Studies, Tsukuba, Ibaraki 305-8056, Japan
*   Correspondence: ram@ees.hokudai.ac.jp; Tel.: +81-011-706-2261

Citation: Molekoa, M.D.; Avtar, R.; Kumar, P.; Thu Minh, H.V.; Dasgupta, R.; Johnson, B.A.; Sahu, N.; Verma, R.L.; Yunus, A.P. Spatio-Temporal Analysis of Surface Water Quality in Mokopane Area, Limpopo, South Africa. *Water* **2021**, *13*, 220. https://doi.org/10.3390/w13020220

Received: 23 December 2020
Accepted: 14 January 2021
Published: 18 January 2021

Publisher's Note: MDPI stays neutral with regard to jurisdictional claims in published maps and institutional affiliations.

Abstract: Considering the well-documented impacts of land-use change on water resources and the rapid land-use conversions occurring throughout Africa, in this study, we conducted a spatiotemporal analysis of surface water quality and its relation with the land use and land cover (LULC) pattern in Mokopane, Limpopo province of South Africa. Various physico-chemical parameters were analyzed for surface water samples collected from five sampling locations from 2016 to 2020. Time-series analysis of key surface water quality parameters was performed to identify the essential hydrological processes governing water quality. The analyzed water quality data were also used to calculate the heavy metal pollution index (HPI), heavy metal evaluation index (HEI) and weighted water quality index (WQI). Also, the spatial trend of water quality is compared with LULC changes from 2015 to 2020. Results revealed that the concentration of most of the physico-chemical parameters in the water samples was beyond the World Health Organization (WHO) adopted permissible limit, except for a few parameters in some locations. Based on the calculated values of HPI and HEI, water quality samples were categorized as low to moderately polluted water bodies, whereas all water samples fell under the poor category (>100) and beyond based on the calculated WQI. Looking precisely at the water quality's temporal trend, it is found that most of the sampling shows a deteriorating trend from 2016 to 2019. However, the year 2020 shows a slightly improving trend on water quality, which can be justified by lowering human activities during the lockdown period imposed by COVID-19. Land use has a significant relationship with surface water quality, and it was evident that built-up land had a more significant negative impact on water quality than the other land use classes. Both natural processes (rock weathering) and anthropogenic activities (wastewater discharge, industrial activities etc.) were found to be playing a vital role in water quality evolution. This study suggests that continuous assessment and monitoring of the spatial and temporal variability of water quality in Limpopo is important to control pollution and health safety in the future.

Keywords: surface water quality; WQI; HPI; HEI

1. Introduction

Water is an essential resource to sustain life on the Earth. Different key drivers of global change *viz.* urbanization, population growth and extreme weather conditions induced by climate change are severely affecting this finite resource, both in terms of quantity and quality [1]. Indiscriminate exploitation of groundwater, resulting in the depletion of groundwater levels and consequently greater dependency on surface water resources, is occurring in many regions around the world. The diligent monitoring and analysis of surface water quality are essential for sustainable management and use of surface water resources [2,3]. It is also useful for assessing processes that govern hydro-geochemical evolution of water resources [4].

South Africa has a population of over 51 million people, and out of that, 60% live in urban environments [5]. Because of the uneven distribution of water resources, approximately 77% of South African people are dependent on surface water resources [5]. Approximately 40% of African people lack improved water supply and more than 60% have no access to improved sanitation facilities [6,7]. There are valuable surface water resources such as rivers, dams and streams that are priceless assets, irreplaceable and provide important habitat for recreations, economic growth and nature conversation [8]. Preserving and ensuring the sustainable use of surface water resources can contribute towards the implementation of Sustainable Development Goals (SDGs 6) [9]. The increasing population, economic growth, and change in lifestyle cause an increase in the requirement of fresh water, which amplifies the pressure on limited water resources [10]. The surface water resources are at risk of contamination because of rapid industrialization, urbanization, extensive agriculture activities, mining and population growth [5,11].

Among different contaminants in water resources, heavy metal pollution is one of the most serious, and it poses threats to human life even at minor concentrations [12,13]. The major sources of heavy metal pollution in water are both natural (such as chemical weathering of minerals and soil leaching) and anthropogenic (such as industrial and domestic effluents, landfill leachate, water runoff, urban storm, mining activities, etc.). Several studies [14–16] have shown that heavy metal pollution of water can lead to various diseases such as tumors, head congestion, muscular edema etc. To evaluate the pollution load in water bodies, calculating the heavy metal pollution index is one of the most common approaches, as it can decipher the source of heavy metals [17–19]. A study on the distribution of heavy metals conducted by [18] showed how human activities could have impacts on aquatic ecosystems as a result of discharged wastes. According to [20,21], poorly planned industrialization and urbanization still exist in many developing countries and that deteriorates the situation on environmental pollution. Untreated waste disposal from refineries and various industries worsen the water quality. Therefore, monitoring heavy metals in surface water is an essential need in order to ensure the safety of both animal and human health. Villanueva et al., 2013 [22] reported that increased effluent from industrial, urban and agricultural areas elevates heavy metal pollution in surface water bodies. With the above background, in the absence of any significant work on surface water quality and factors playing key roles in determining this quality in the Mokopane area, Limpopo, South Africa, this study strives to quantify the spatio-temporal trend of different physico-chemical parameters and their relationship with the land use and land cover (LULC) pattern. In particular, the focus of this study is to quantify heavy metal pollution in the study area because of nearby mining activities as well as the absence of heavy metal pollution information in the Mokopane area.

2. Study Area

2.1. Site Description

The study area is located in Mokopane, Limpopo province of South Africa, approximately 250 km from Johannesburg city and is situated at the latitude 24.1944° S and longitude 29.0097° E (Figure 1). The total population of Mokopane is approximately 328,905 in 2016 [23]. It is one of the richest agricultural areas, producing wheat, cotton, maize, citrus fruits, etc. Recently, there have been mining industries introduced in the area. The mean annual maximum and minimum temperature ranges from 23.4 °C and 13 °C, respectively (Figure 2). It is a steppe climate with a mean annual precipitation of 490 mm that normally occurs from December to April and less rainfall during the winter season from June to September (Figure 2) [24]. The region is served with water mainly by four rivers, the Dithokeng, Mogalakwena Deep pool (Ngwaditse), Rooisloot and the Dorps Rivers, which supply water for various domestic and irrigation purposes [25]. Sahu et al. [26] studied the impact of climatic variability on the streamflow of river; therefore, the study area's climatic data were analyzed to see the rainfall and temperature patterns.

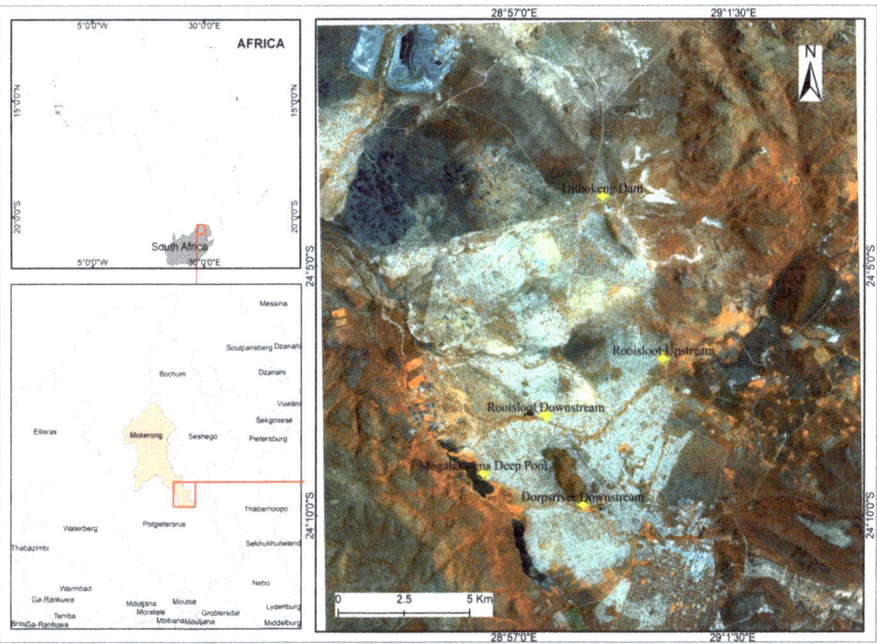

Figure 1. Study area map with sampling location. The sampling sites are indicated through upstream to downstream.

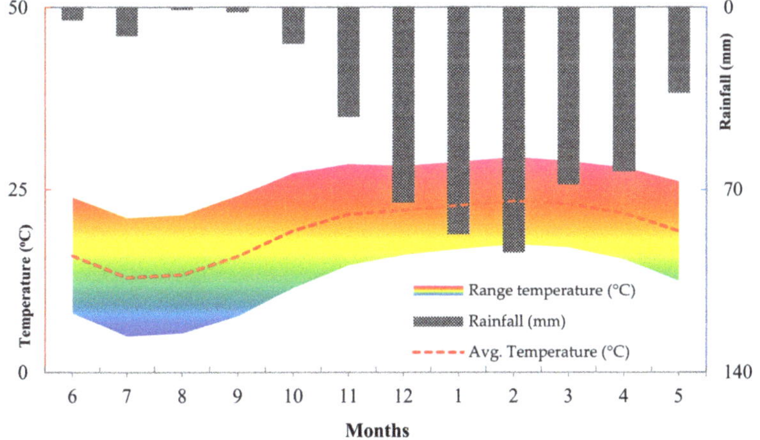

Figure 2. Monthly average rainfall and temperature pattern of the study area [23].

There are five sampling sites selected for this study—Mogalakwena Deep pool, Rooisloot downstream and upstream, Dithokeng dam and Dorpsrivier, as shown in Figure 1. Water samples for physico-chemical analysis were collected mid-stream directly into clean polyethylene bottles. The most socially and economically important site is the Mogalakwena Deep pool, because all of the other streams are flowing into it. Many local people rely on the Mogalakwena deep pool for their primary source of water, as well as for fishing. It was found that there is no water in Rooisloot upstream during the dry seasons, hence there is no water sampling done during that period.

2.2. LULC Classification

Land use/land cover (LULC) classification involves the extraction of thematic information about various landscape features from satellite data. Landsat-8 OLI data were acquired on 6th May, 2019 from the USGS Earth Explorer [27] in order to produce a LULC map of the study area. Figure 3 illustrates the LULC map of the study area. LULC information is useful for the management and planning of land resources [28]. Various classification algorithms have been developed to classify satellite data. However, in this study, the most common Maximum likelihood classification algorithm was performed using ENVI 5.2 software. We have noticed some misclassification in the built-up area using the MLC algorithm; therefore, the built-up area was manually digitized to improve the accuracy of the LULC map. The study area was classified into five classes, namely—agriculture, bare land, built-up, mountain/vegetation and water bodies (Figure 3). Results showed that most of the study area is covered by mountain/vegetation with 48.7%, followed by agriculture (29.6%), built-up (19.8%), bare land (1.5%) and water body (0.36%), respectively. The study area is one of the richest agricultural areas, producing wheat, cotton, maize, citrus, etc., with the supply of water from the surrounding river system.

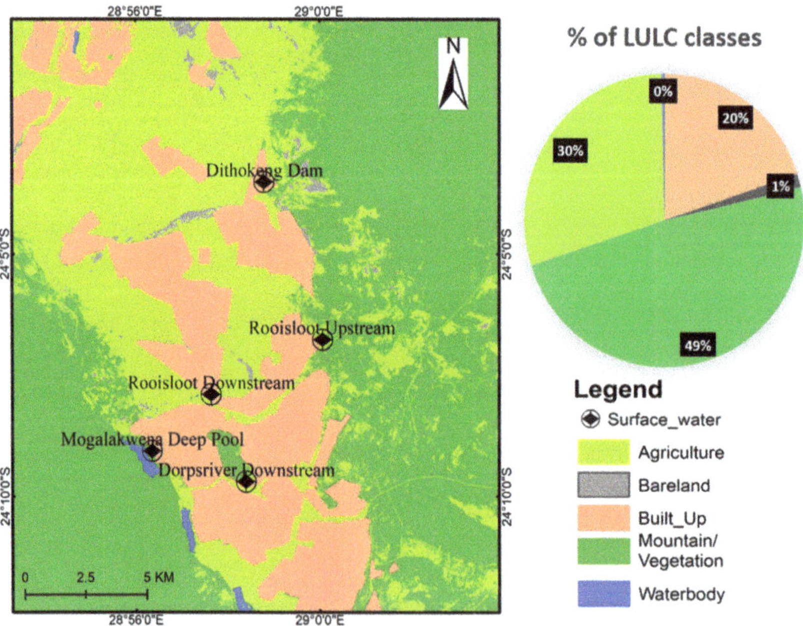

Figure 3. Land use/land cover (LULC) map of the study area and pie chart shows percentage of various classes.

3. Methodology

To get the insight of surface water quality, water samples were collected from five monitoring sites—Dithokeng River, Rooisloot upstream, Rooisloot downstream, Mogalakwena deep pool and Dorps River. Sampling locations were selected in such a way that they represent a significant stretch of rivers from upstream to downstream, as well as distance from Mogalakwena and Ivanplats platinum mines. To analyze spatio-temporal variation in river water quality, water samples were collected and analyzed four times a year (except for year 2020) from March 2016 until November 2020 by the Environmental Department of Ivanplats mine [29] in South Africa. Twenty samples were collected from each monitoring point except Dithokeng Upstream, Rooisloot upstream regions because of non-accessibility and non-availability of water, respectively, during some sampling periods. Field measurements for pH, EC and temperature were done using an Orion Model Number, 01915. After in situ analysis, water samples were filtered by 0.20 μm Millipore filter paper and then collected in pre-rinsed uncontaminated polyethylene bottles. To prevent any fluctuation in the concentration of trace metals, the collected samples for major cation and trace metal analysis were acidified by 1% HNO_3 at pH ~2. The concentration of HCO_3^- was analyzed by acid titration (using Metrohm Multi-Dosimat); while other anions Cl^-, NO_3^-, SO_4^{2-}, and PO_4^{3-} were analyzed by DIONEX ICS-90 ion chromatograph. Inductively coupled plasma-mass spectrometry (ICP-MS) was used to evaluate major cations and trace metals. The summary of different techniques used for water quality parameter analysis is also shown in Table 1. After obtaining all the analyzed water quality data for the aforementioned period from the Environmental Department of Ivanplats mine, different techniques and software were used to deduce the factors responsible for spatio-temporal variation in the water quality. Heavy metal pollution index (HPI) and Heavy metal evaluation index (HEI) were calculated to provide overall quality of the water with regard to heavy metals. In this study, the permission limits are taken from WHO, 2009 [29].

Table 1. Statistical summary for observed water quality parameters.

Parameters	Condition of Sampled Water	Method of Analysis	Precision Level	Method of Validation
pH, EC, Temperature	Natural condition	Multi parameter probe (Orion Model Number, 01915)	<5%	Repetition after each five analysis
HCO_3^-	Natural condition	Acid titration using Metrohm Multi-Dosimat	<5%	Repetition after each five analysis
Cl^-, SO_4^{2-}, F^-, NO_3^-, PO_4^{3-}	Natural condition	DIONEX ICS-90 ion chromatography with a detection limit of 10μg/L.	<2%	Repetition after each five analysis
Ca^{2+}, Mg^{2+}, Na^+, K^+, Al, Cr, Cu, Fe, Mn, Sr, Ti, Zn, Si	Acidic condition (by addition of 1% HNO_3)	Agilent 7500 Series Inductively coupled plasma-mass spectrometry (ICP-MS)	<2%	Repetition after each five analysis

3.1. Heavy Metal Pollution Index (HPI) Calculation

Metal pollution is one of the most significant problems in water bodies, causing serious health hazards to human beings. The HPI, based on the weighted arithmetic sum of water quality parameters, is a powerful technique for the assessment of water quality based on the heavy metal concentration and effect of individual trace metals on human health [30,31].

The HPI model has been proposed in Equation (1) given by Mohan et al., 1996 [30]. Heavy metal concentrations were compared with the drinking water standards set by the WHO.

$$\text{HPI} = \frac{\sum_{i=1}^{n} Q_i \times W_i}{\sum_{i=1}^{n} W_i} \tag{1}$$

where, n and i are the number of parameters considered and denote ith parameter;

- W_i is the unit weight of ith parameter, $W_i = K/S_i$ in which K is constant of proportionality, $K = 1/\sum_{i=1}^{n}(1/S_i)$;

- Q_i is the sub-index of the ith parameter, $\sum_{i=1}^{n} Q_i = \frac{(M_i - I_i)}{(S_i - I_i)} \times 100$ in which M_i is monitored values; and

- I_i is the ideal value, S_i is suggested permissible values.

3.2. Heavy Metal Evaluation Index (HEI) Calculation

We also conducted HEI to interpret the water quality in response to heavy metals and trace elements present in water, as proposed in Equation (2).

$$\text{HEI} = \sum_{i=1}^{n} \frac{M_i}{S_i} \tag{2}$$

where, M_i—monitored value of ith and S_i—standard value of ith parameter.

The classifications of the HEI index is as follows—low heavy metal (less than 10), moderate-heavy metal (between 10 and 20), and high heavy metal (more than 20).

3.3. Water Quality Index

Water quality index is one of the effective methods which has been applied in various studies for both surface and groundwater [18,25,31–35].

Water quality index (WQI) is used in this study, which has been considered as one of the most reliable tools for classifying water pollution levels for both groundwater and surface water [24,32,33]. The following steps were taken in order to calculate WQI:

1. Calculating relative weight: It was calculated using Equation (3).

$$Wi = \frac{wi}{\sum_{i}^{n} wi} \tag{3}$$

where W_i represents the relative weight of each parameter sampled, wi represents the weight of each parameter, and n represents the total number of parameters.

2. Calculating Q value: It was calculated using Equation (4).

$$Q_i = \frac{C_i \times 100}{S_i} \tag{4}$$

where Q_i = quality rating, C_i = Concentration of each parameter (mg/L), and S_i is derived from the WHO water quality standard.

3. Finally, the Water quality Index (WQI) was calculated using Equation (5).

$$\text{WQI} = \sum W_i \times Q_i \tag{5}$$

Water Quality assessment in terms of the WQI is shown in Table 2.

Table 2. Water quality classification based on Water quality Index (WQI) values [24].

Ranking	Water Quality
<50	Excellent
50–100	Good water
100–200	Poor water
200–300	Very poor water
>300	Likely not suitable for drinking

4. Results and Discussion

4.1. General Water Chemistry

A statistical summary of the analyzed river water quality is shown in Table 3. The pH values of the water samples varied from 6.63 to 9.43, with an average value of 8.12, depicting the alkaline nature of the water due to high soil–water interaction during the flow course of the drainage system [35]. The electrical conductivity values varied from 91.19–2686.6 μS/cm, with an average value of 1022.17μS/cm, indicating high ionic activity in the area. Furthermore, the arid/semiarid climate, with relatively low rainfall and high evaporation, supports high mineral concentration in the water bodies. Looking into the ionic abundance, $Na^+ > Mg^{2+} > Ca^{2+} > K^+$ was the order among cations, whereas the order among anions was $HCO_3^- > Cl^- > SO_4^{2-} > PO_4^{3-} > NO_3^- > F^-$. For cations, $Na^+ > Mg^{2+} > Ca^{2+} > K^+$ and the average milli-equivalent ratio of $Mg^{2+} + Ca^{2+}/Na^+ + K^+$ was found to be 1.23, indicating the ascendency of carbonaceous weathering in the study area. The dominance of Na^+ in the water sample might be because of its conservative nature. Excess of both Mg^{2+} and Ca^{2+} can be explained by the presence of a common source of minerals like dolomite. The highest average concentration of HCO_3^- among the anions is due to the weathering of the carbonaceous sandstones in the watershed and the weathering of the carbonaceous minerals through runoff. Higher Cl^- and SO_4^{2-} concentration in the river water witnessed the anthropogenic inputs coming along surface runoff in the watershed area. In particular, higher concentrations of SO_4^{2-} can be due to leaching of organic matter and agricultural runoff carrying unused SO_4^{2-}. This organic matter can range from landfills area with piles of organic wastes or leaching from organic matter-rich sediment present in the study area like peat or clay. The concentration of PO_4^{3-}, NO_3^- and F- are not a concern as they are well below the permissible limits of WHO for all surface water samples. The time series value for EC and Ti is shown in Figure 4a,b, respectively. Here, it is found that the EC value has an increasing tendency towards downstream. It can be supported by higher values of major cations and anions, a strong indicator of inputs from both anthropogenic (runoff carrying pollutants) and natural sources (mineral weathering). Among different trace metals, the concentration of Ti is of major concern for this study area, especially in Rooisloot Upstream as compared to Downstream from 2016–2017. A lot of animals such as cows and pigs were observed during field surveying and grazing-led sedimentation can exaggerate the water quality deterioration. Ti is among the most abundant chemical elements on the earth's crust, ranking ninth of all the elements and among transition metals, it follows second after iron [36–38]. Human activities are among the factors that cause Ti to enter water, especially in its nanoparticle form and this affects aquatic life. The migration mobility of Ti is generally low. To analyze the spatio-temporal variation of water quality, time series evaluation of key water quality parameters is plotted and shown in Figure 4. Looking at the spatial trend, EC displays an increasing trend when moving from the upstream region towards the downstream region. This can be justified because of the transportation and continuous accumulation of contaminants from different point and non-point sources throughout the stretch of the river. On the other hand, the spatial trend of Ti shows some different patterns. Here, the result shows a higher concentration in the upstream region, which is decreasing when going towards the downstream region. Hence, after their release in the river body through surface runoff or leachate, the concentration gradually decreases because of the dilution effect. Looking at the temporal variation, in general, the concentration of water quality parameters shows higher concentration during dry periods compared to the wet periods. The possible reason behind this is that because of the reduction in river discharge, these parameters attenuate and hence the concentration increased. Looking at the year 2018, Mokopane received an increase of rainfall (Figure 5), which might have caused a sudden decrease in Ti found in water. The high concentration of EC in Rooisloot Downstream could result from domestic effluents and affected by Rooisloot Upstream as it recharges this stream. To further support the increasing causes of water quality deterioration, the land use land cover map was prepared for the years 2015 and 2020 as shown in Figure 6. Here, it is found that built-up areas are significantly increased, especially in the upstream region, at the expense of bare land and water bodies. This increase in built-up areas represents the source of both point and non-point sources of water pollution. Based on the above findings, a conceptual diagram is developed as shown in Figure 7, which is depicting the processes governing water quality evolution in the study area.

Table 3. Statistical summary for observed water quality parameters.

Parameters	Minimum	Maximum	Average	St. Dev.
pH	6.6300	9.4300	8.1246	0.6962
EC (μs/cm)	91.9000	2686.0000	1094.1306	389.5242
HCO_3^- (mg/L)	37.5000	738.8000	316.1714	190.3866
Cl^- (mg/L)	1.4300	609.4300	109.2615	147.8633
SO_4^{2-} (mg/L)	0.2500	467.3300	38.0170	78.9900
F^- (mg/L)	0.1500	19.4000	1.2675	2.6694
NO_3^- (mg/L)	0.0600	53.4400	3.5649	8.5746
PO_4^{3-} (mg/L)	0.0900	15.1500	3.8751	4.1173
Ca^{2+} (mg/L)	4.0400	112.7200	35.9860	20.1310
Mg^{2+} (mg/L)	1.3560	143.0800	57.7536	38.6605
K^+ (mg/L)	0.4200	13.6410	4.0060	2.5411
Na^+ (mg/L)	1.5270	477.1500	113.0088	98.2543
Al (mg/L)	0.0040	1.1770	0.1310	0.2708
Cr (mg/L)	0.0054	0.0120	0.0081	0.0016
Cu (mg/L)	0.0050	0.0323	0.0121	0.0091
Fe (mg/L)	0.0069	1.8700	0.1501	0.3563
Mn (mg/L)	0.0060	1.0655	0.3187	0.2675
Sr (mg/L)	0.0210	0.2480	0.1179	0.0521
Ti (mg/L)	0.0010	0.0800	0.0165	0.0190
Zn (mg/L)	0.0050	0.0320	0.0133	0.0101
Si (mg/L)	0.0320	12.1400	5.7580	3.9085

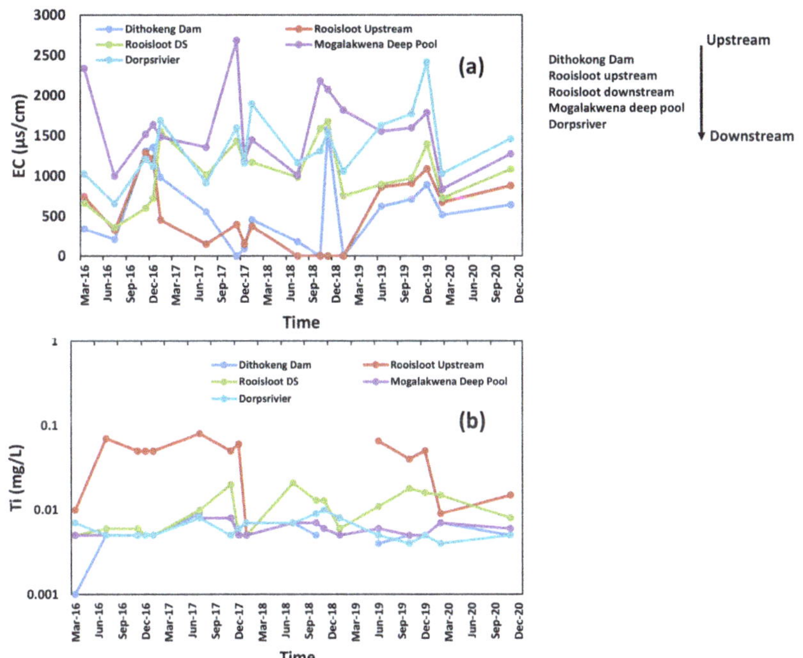

Figure 4. Time series concentration values for (**a**) EC and (**b**) Ti for water samples at five sampling locations.

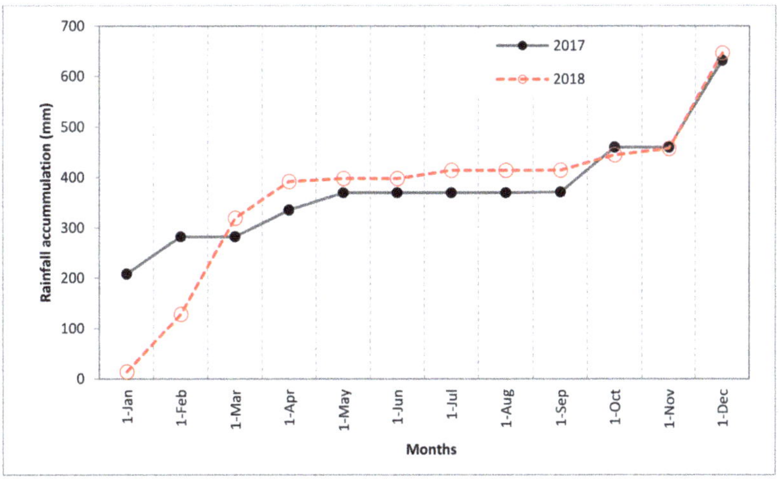

Figure 5. Comparison of rainfall accumulation between 2017 and 2018.

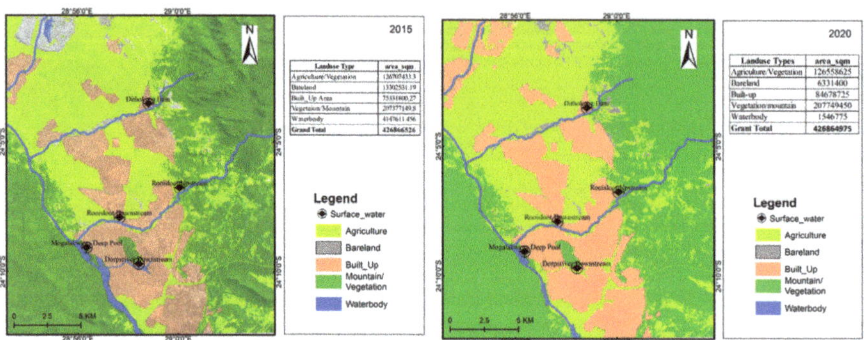

Figure 6. Land-use map of the study area for year 2015 and 2020.

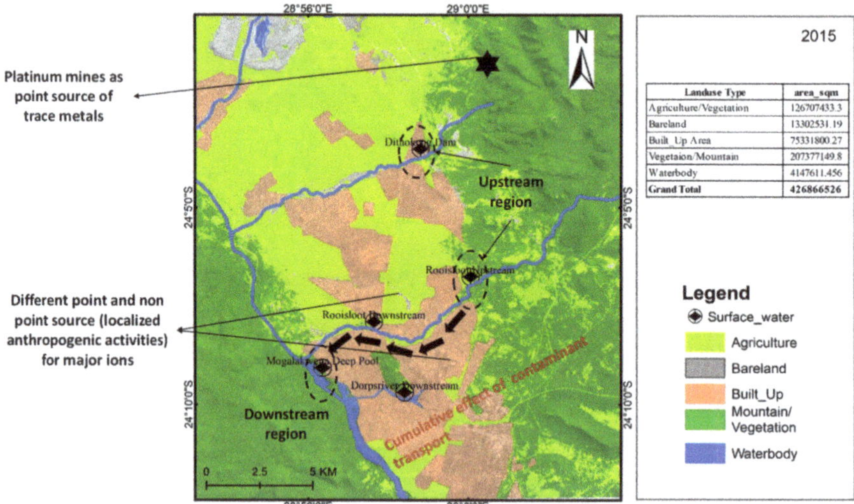

Figure 7. Conceptual diagram showing processes involved in water quality evolution.

4.2. Heavy Metal Evaluation Index (HEI)

To calculate the heavy metal evaluation index, first of all the unit weight for different metals at the individual levels was calculated, which was used further as an input to calculate the heavy metal pollution index and the heavy metal evaluation index for different water samples at a different time period. Results for heavy metal unit weight are shown in Table 4.

The results for HPI and HEI are shown in Tables 5 and 6, respectively. These values represent the cumulative value of different heavy metals. It was found that sampling locations in the upstream region, namely the Dithokeng upstream and Rooisloot upstream locations, have low concentrations of heavy metals. On the other hand, both the Rooisloot downstream and Mogalakwena sites had moderate heavy metal content. Finally, Dorpsriver has low-to-moderate heavy metal content, which can be explained by a dilution effect on heavy metal concentration by river discharge. Here, the main attribute for heavy metal contamination in the water samples can be related to the land use pattern as shown in Figure 3. In this area, the spatial distribution of built-up areas that are dominant in the southern side of the study area is significantly correlated with the heavy metal composition of the water samples. Built-up areas may act as a non-point source of heavy metal due to different activities like small-scale industries (leather, textile, etc.), human settlements;

where wastewater and effluent discharge bring different heavy metals like Fe, Zn, Mn, etc. into the river water bodies. On the other hand, mining sand, natural factors such as rock weathering and other domestic effluents near Rooisloot upstream region also exaggerate the concentration of heavy metals like Ti, Cu, Cr, Ni [25] etc. An uncontrolled flow of sewage into Dorps River (the downstream sampling locations) was also observed during the field survey. Looking at the result, it is found that both HPI and HEI are showing lower values, especially for the year 2020. This can be justified with lower anthropogenic activities like mining, industrial activities during COVID-19-induced lockdown.

Table 4. Unit weight calculation of the heavy metal evaluation index (HEI).

Parameters	WHO Standards (mg/L)-S_i	The Ideal Values	$1/S_i$	K	W_i
Ag	0.05	0.005	20	0.0015	0.029
Al	0.9	0.2	1	0.0015	0.002
As	0.01	0.001	100	0.0015	0.146
B	2.4	0.01	0	0.0015	0.001
Ba	0.7	0.002	1	0.0015	0.002
Cd	0.003	0.0001	33	0.0015	0.049
Co	0.05	-	20	0.0015	0.029
Cr	0.05	0.0002	20	0.0015	0.029
Cu	2	0.0005	167	0.0015	0.243
Fe	0.3	-	33	0.0015	0.049
Hg	0.01	0.006	20	0.0015	0.029
Mn	0.1	-	10	0.0015	0.015
Mo	0.01	-	100	0.0015	0.146
Ni	0.07	0.001	14	0.0015	0.021
Pb	0.01	-	20	0.0015	0.029
Se	0.04	0.0005	25	0.0015	0.036
Ti	0.03	0.007	30	0.0015	0.037
Zn	3	0.01	100	0.0015	0.146

Table 5. Heavy metal pollution index (HPI) calculation.

Year	Classification (HPI) from Upstream to Downstream Location				
	Dithokeng Dam	Rooisloot Upstream	Rooisloot Downstream	Mogalakwena Deep Pool	Dorps River
2016	Good (49)	Bad (51)	Bad (51)	Bad (51)	Good (49)
2017	Good (50)	Bad (53)	Good (50)	Bad (50)	Bad (69)
2018	Good (49)	Good (50)	Bad (60)	Bad (65)	Bad (51)
2019	Good (50)	Bad (54)	Bad (51)	Bad (56)	Good (49)
2020	Good (49)	Good (50)	Good (48)	Good (48)	Good (49)

Table 6. Heavy metal evaluation index (HEI) calculation.

Year	Classification (HEI) from Upstream to Downstream Location				
	Dithokeng Dam	Rooisloot Upstream	Rooisloot Downstream	Mogalakwena Deep Pool	Dorps River
2016	Low heavy metal (9)	Low heavy metal (10)	Low heavy metal (10)	Low heavy metal (10)	Low heavy metal (9)
2017	Low heavy metal (9)	Low heavy metal (11)	Moderate heavy metal (17)	Moderate heavy metal (14)	Low heavy metal (10)
2018	Low heavy metal (10)	Low heavy metal (10)	Moderate heavy metal (15)	Moderate heavy metal (16)	Low heavy metal (9)
2019	Low heavy metal (10)	Low heavy metal (10)	Low heavy metal (10)	Low heavy metal (9)	Moderate heavy metal (13)
2020	Low heavy metal (10)	Low heavy metal (10)	Low heavy metal (9)	Low heavy metal (9)	Low heavy metal (9)

4.3. Water Quality Index (WQI)

The result for the water quality index calculated for the four-year time period is shown in Table 7. Calculated WQI values ranged from 120.71 to 4643.71, which indicates that the water in all of these locations falls under the "very poor water" and "likely not suitable for drinking purposes" categories. The highest values were mainly found near the downstream, i.e., Dorpsriver, which shows the accumulative effects of different contaminants along with the river flow course. One of the major concerns regarding poor WQI is heavy metal contamination. Values for the year 2018 were relatively on the higher end because of high rainfall, which results in high sedimentation and ionic activities. Lower values for the year 2019 are only because of the giving input of incomplete datasets for the year 2019. The temporal variations of WQI showed that surface water quality at five sampling sites has not improved much over the 2016–2019 period. All sampling sites were considered as "poor water quality" to "likely not suitable for drinking". The reason behind this was inefficient water resource management practices during that time. In this regard, it shows that the water quality did not improve in the period from 2016–2019. However, for the year 2020, water quality is relatively improved because of lower environmental perturbances due to COVID-19-induced lockdown period, as discussed earlier.

Table 7. Water quality index (WQI) results for the period 2016–2019.

Year	Classification (WQI) from Upstream to Downstream Location				
	Dithokeng Dam	Rooisloot Upstream	Rooisloot Downstream	Mogalakwena Deep Pool	Dorps River
2016	Likely not suitable for drinking (318.33)	Likely not suitable for drinking (975.02)	Likely not suitable for drinking (1536.39)	Likely not suitable for drinking (1829.67)	Likely not suitable for drinking (1930.49)
2017	Very poor water (245.4)	Very poor water (265.46)	Likely not suitable for drinking (1219.11)	Likely not suitable for drinking (1032.61)	Likely not suitable for drinking (1082.92)
2018	Likely not suitable for drinking (318.55)	Likely not suitable for drinking (786.67)	Likely not suitable for drinking (1110.04)	Likely not suitable for drinking (1920.07)	Likely not suitable for drinking (4643.71)

Table 7. Cont.

Year	Classification (WQI) from Upstream to Downstream Location				
	Dithokeng Dam	Rooisloot Upstream	Rooisloot Downstream	Mogalakwena Deep Pool	Dorps River
2019	Likely not suitable for drinking (1566.46)	Likely not suitable for drinking (950.5)	Poor water (120.71)	Likely not suitable for drinking (751.13)	Likely not suitable for drinking (681.05)
2020	Likely not suitable for drinking (408.12)	Very poor water (223.09)	Poor water (175.57)	Likely not suitable for drinking (440.24)	Very poor water (205.85)

5. Conclusions and Recommendation

This study strived to quantify spatio-temporal water quality in the Mokopane area of South Africa and identify the processes which governed water quality changes. The results indicated that the concentration of most of physico-chemical species in the water samples was within permissible limits, except for a few parameters and a few locations. There was a trend showing water quality deterioration towards the downstream, as contaminants accumulated with the river flow. The water quality for the streams was found to be worsened or unchanged over the four-year period. For example, the Dorps River and Dithokeng dam showed no significant change from the 2016–2019 periods as the water quality fell under "likely not suitable for drinking". However, for the year 2020, water quality shows an improvement in terms of WQI, HPI, HEI owing to the suspension of different human activities like mining, industrial, agricultural, etc., due to the lockdown imposed by COVID-19. This means without proper management that ensures good water quality in these areas, the water will continue to not being fit for humans, animals and plants for their survival. Focusing on spatio-temporal variation, water quality concentration showed an increasing trend from upstream to downstream as pollutants get accumulated. Also, temporally, rainfall has a significant impact on water quality parameters by dilution and attenuation during wet and dry seasons, respectively. Land use has a significant relation with water quality, and we found that built-up areas had a bad impact on water quality in the study site. Looking into the processes, both natural processes (rock weathering) and anthropogenic activities (household wastewater discharge, industrial especially mining activities etc.) were playing a major role in governing water quality. In the absence of any previous credible scientific study or reports, this study sheds light on issues regarding water resource management. The sampling location and number of water samples in this study are less due to lack of financial support. Detailed coverage of the river stretch with the inclusion of more sampling locations for the time-series analysis of water quality data along with analyzing both point and non-point sources of pollutants is recommended as a future study. A participatory approach for watershed management and making land use climate-resilient might be investigated in the future to plan the best suitable adaptation and mitigation measures for water resource management. A comparative study of surface water quality in the study site and nearby Doorndraai dam is necessary considering the impacts of LULC change.

Author Contributions: Conceptualization, M.D.M., R.A., P.K., H.V.T.M.; methodology, M.D.M., R.A., P.K., H.V.T.M.; investigation, M.D.M., R.A.; resources, M.D.M., R.A., R.L.V.; data curation, M.D.M., R.A., R.L.V., N.S.; writing, M.D.M., R.A., P.K., R.D., B.A.J., N.S., R.L.V., A.P.Y.; writing—review and editing, M.D.M., R.A., P.K., R.D., B.A.J., N.S., R.L.V., A.P.Y.; supervision, R.A., P.K.; funding acquisition, M.D.M., R.A. All authors have read and agreed to the published version of the manuscript.

Funding: The publication fee was supported by JICA-ABE research grant.

Acknowledgments: Mmasabata Dolly Molekoa extends gratitude to the JICA-ABE Initiative to provide a scholarship to study master course at Hokkaido University, Japan. The authors are thankful to the Environmental Department of Ivanplats mine for providing data for this research and the Graduate School of Environmental Science (Hokkaido University) for facilities, and anonymous reviewers for their comments.

Conflicts of Interest: The authors declare no conflict of interest.

References

1. Avtar, R.; Kumar, P.; Singh, C.; Mukherjee, S. A Comparative Study on Hydrogeochemistry of Ken and Betwa Rivers of Bundelkhand Using Statistical Approach. *Water Qual. Exposure Health* **2011**, *2*, 169–179. [CrossRef]
2. Kazi, T.; Arain, M.; Jamali, M.K.; Jalbani, N.; Afridi, H.; Sarfraz, R.; Baig, J.; Shah, A.Q. Assessment of Water Quality of Polluted Lake Using Multivariate Statistical Techniques: A Case Study. *Ecotoxicol. Environ. Saf.* **2009**, *72*, 301–309. [CrossRef] [PubMed]
3. Kumar, P. Numerical Quantification of Current Status Quo and Future Prediction of Water Quality in Eight Asian Megacities: Challenges and Opportunities for Sustainable Water Management. *Environ. Monit. Assess.* **2019**, *191*, 319. [CrossRef] [PubMed]
4. De Souza, A.; Tundisi, J. Hidrogeochemical Comparative Study of the Jaú and Jacaré-Guaçu River Watersheds, São Paulo, Brazil. *Rev. Bras. Biol.* **2000**, *60*, 563–570. [CrossRef]
5. UN Water. Water a Shared Responsibility. 2006. Available online: http://Unesdoc.Unesco.Org/Images/0014/001454/145405e.Pdf#page=519 (accessed on 21 April 2019).
6. Eneh, O.C.; Eneh, A. Potable Water Access and Management and Management in Africa: Implecations for Poverty, Hunger and Health. *Technol. Sci. Rev.* **2014**, *5*, 8036–9694.
7. Sila, O.N. Physico-Chemical and Bacteriological Quality of Water Sources in Rural Settings, a Case Study of Kenya, Africa. *Sci. Afr.* **2019**, *2*, E00018. [CrossRef]
8. Bello, M.; Tolaba, M.; Suarez, C. Factors Affecting Water Uptake of Rice Grain during Soaking. *LWT—Food Sci. Technol.* **2004**, *37*, 811–816. [CrossRef]
9. Avtar, R.; Aggarwal, R.; Kharrazi, A.; Kumar, P.; Kurniawan, T.A. Utilizing Geospatial Information to Implement SDGs and Monitor Their Progress. *Environ. Monit. Assess.* **2020**, *192*. [CrossRef]
10. Shrivastava, A.; Tandon, S.A.; Kumar, R. Water Quality Management Plan for Patalganga River for Drinking Purpose and Human Health Safety. *Int. J. Sci. Res. Environ. Sci.* **2015**, *3*, 71–87. [CrossRef]
11. Lukubye, B.; Andama, M. Bacterial Analysis of Selected Drinking Water Sources in Mbarara Municipality, Uganda. *J. Water Resour. Prot.* **2017**, *9*, 999–1013. [CrossRef]
12. Prasad, B.; Kumari, P.; Bano, S.; Kumari, S. Ground Water Quality Evaluation near Mining Area and Development of Heavy Metal Pollution Index. *Appl. Water Sci.* **2014**, *4*, 11–17. [CrossRef]
13. Prasanna, M.V.; Chidambaram, S.; Hameed, A.S.; Srinivasamoorthy, K. Hydrogeochemical Analysis and Evaluation of Groundwater Quality in the Gadilam River Basin, Tamil Nadu, India. *J. Earth Syst. Sci.* **2011**, *120*, 85–98. [CrossRef]
14. Johnson, F. The Genetic Effects of Environmental Lead. *Mutation Res./Rev. Mutation Res.* **1998**, *410*, 123–140. [CrossRef]
15. Tsuji, L.; Karagatzides, J. Chronic Lead Exposure, Body Condition, and Testis Mass in Wild Mallard Ducks. *Bull. Environ. Contam. Toxicol.* **2001**, *67*, 489–495. [CrossRef] [PubMed]
16. Abbasi, S.; Abbasi, N.S.R. *Heavy Metal in the Environment*, 1st ed.; Mital Publication: New Delhi, India, 1998.
17. Amadi, A.; Okoye, N.; Alabi, A.; Tukur, A.; Angwa, E. Quality Assessment of Soil and Groundwater near Kaduna Refinery and Petrochemical Company, Northwest Nigeria. *J. Sci. Res. Rep.* **2014**, *3*, 884–893. [CrossRef]
18. Tsai, L.-J.; Yu, K.-C.; Chen, S.-F.; Kung, P.-Y. Effect of Temperature on Removal of Heavy Metals from Contaminated River Sediments via Bioleaching. *Water Res.* **2003**, *37*, 2449–2457. [CrossRef]
19. Aliyu, J.; Saleh, Y.; Kabiru, S. Heavy Metals Pollution on Surface Water Sources in Kaduna Metropolis, Nigeria. *Sci. World J.* **2015**, *10*, 1–5.
20. Bhagure, G.R.; Mirgane, S. Heavy Metal Concentrations in Groundwaters and Soils of Thane Region of Maharashtra, India. *Environ. Monit. Assess.* **2011**, *173*, 643–652. [CrossRef]
21. Varalakshmi, L.; Ganeshamurthy, A. Heavy Metal Contamination of Water Bodies, Soils and Vegetables in Peri Urban Areas of Bangalore City of India. In Proceedings of the 19th World Congress of Soil Science, Soil Solutions for a Changing World, Brisbane, Australia, 1–6 August 2010.
22. Villanueva, J.; Le Coustumer, P.; Huneau, F.; Motelica-Heino, M.; Perez, T.; Materum, R.; Espaldon, M.; Stoll, S. Assessment of Trace Metals during Episodic Events Using DGT Passive Sampler: A Proposal for Water Management Enhancement. *Water Resour. Manag.* **2013**, *27*, 4163–4181. [CrossRef]
23. Statistics South Africa 2016. Available online: https://www.gov.za/sites/default/files/gcis_document/201609/40312gen628.pdf (accessed on 29 May 2019).
24. Climate Data Climate Mokopane. Available online: https://En.Climate-Data.Org/Africa/South-Africa/Limpopo/Mokopane-953/ (accessed on 1 July 2019).
25. Platreef Resources (PTY) Ltd. *Platreef Project Pre-Feasibility Study: Draft Hydrology Report for Draft EIA*; Platreef Resources (PTY) Ltd.: Sandton, South Africa, 2013; p. 85.

26. Sahu, N.; Panda, A.; Nayak, S.; Saini, A.; Mishra, M.; Sayama, T.; Sahu, L.; Duan, W.; Avtar, R.; Behera, S. Impact of Indo-Pacific Climate Variability on High Streamflow Events in Mahanadi River Basin, India. *Water* **2020**, *12*, 1952. [CrossRef]
27. USGS Earth Explorer. Available online: https://Earthexplorer.Usgs.Gov/ (accessed on 1 May 2019).
28. Avtar, R.; Herath, S.; Saito, O.; Gera, W.; Singh, G.; Mishra, B.; Takeuchi, K. Application of Remote Sensing Techniques toward the Role of Traditional Water Bodies with Respect to Vegetation Conditions. *Environ. Dev. Sustain.* **2014**, *16*, 995–1011. [CrossRef]
29. Available online: https://Www.Ivanhoemines.Com/Projects/Platreef-Project/ (accessed on 22 December 2019).
30. Panigrahy, B.; Singh, P.; Tiwari, A.; Kumar, B.; Kumar, A. Assessment of Heavy Metal Pollution Index for Groundwater around Jharia Coalfield Region, India. *J. Biodivers. Environ. Sci.* **2015**, *6*, 33–39.
31. Avtar, R.; Kumar, P.; Singh, C.K.; Sahu, N.; Verma, R.L.; Thakur, J.K.; Mukherjee, S. Hydrogeochemical Assessment of Groundwater Quality of Bundelkhand, India Using Statistical Approach. *Water Qual. Exposure Health* **2013**, *5*, 105–115. [CrossRef]
32. Kumar, P.; Kumar, A.; Singh, C.K.; Saraswat, C.; Avtar, R.; Ramanathan, A.; Herath, S. Hydrogeochemical Evolution and Appraisal of Groundwater Quality in Panna District, Central India. *Expo. Health* **2016**, *8*, 19–30. [CrossRef]
33. Minh, H.V.T.; Kurasaki, M.; Van Ty, T.; Tran, D.Q.; Le, K.N.; Avtar, R.; Rahman, M.; Osaki, M. Effects of Multi-Dike Protection Systems on Surface Water Quality in the Vietnamese Mekong Delta. *Water* **2019**, *11*, 1010. [CrossRef]
34. Minh, H.V.T.; Avtar, R.; Kumar, P.; Tran, D.Q.; Ty, T.V.; Behera, H.C.; Kurasaki, M. Groundwater Quality Assessment Using Fuzzy-AHP in An Giang Province of Vietnam. *Geosciences* **2019**, *9*, 330. [CrossRef]
35. Subramanian, V.; Saxena, K. *Hydrogeochemistry of Groundwater in the Delhi Region of India*; IAHS-AISH Publication: Wallingford, UK, 1983.
36. Skrabal, S.A.; Terry, C.M. Distributions of Dissolved Titanium in Porewaters of Estuarine and Coastal Marine Sediments. *Mar. Chem.* **2002**, *77*, 109–122. [CrossRef]
37. Neal, C.; Jarvie, H.; Rowland, P.; Lawler, A.; Sleep, D.; Scholefield, P. Titanium in UK Rural, Agricultural and Urban/Industrial Rivers: Geogenic and Anthropogenic Colloidal/Sub-Colloidal Sources and the Significance of within-River Retention. *Sci. Total Environ.* **2011**, *409*, 1843–1853. [CrossRef]
38. Voitkevich, G.; Kokin, A.; Miroshnikov, A.; Prokhorov, V. *Spravochnik Po Geokhimii. (Geochemistry Handbook)*; Nedra Publishers: Moscow, Russia, 1990.

water

Article

Exploring Artificial Intelligence Techniques for Groundwater Quality Assessment

Purushottam Agrawal [1], Alok Sinha [1], Satish Kumar [2], Ankit Agarwal [3,4], Ashes Banerjee [5], Vasanta Govind Kumar Villuri [2], Chandra Sekhara Rao Annavarapu [6], Rajesh Dwivedi [7], Vijaya Vardhan Reddy Dera [8], Jitendra Sinha [9] and Srinivas Pasupuleti [10,*]

[1] Department of Environmental Science and Engineering, Indian Institute of Technology (Indian School of Mines) Dhanbad, Dhanbad 826004, Jharkhand, India; agrawal.2015dr1174@ese.iitism.ac.in (P.A.); alok@iitism.ac.in (A.S.)

[2] Department of Mining Engineering, Indian Institute of Technology (Indian School of Mines) Dhanbad, Dhanbad 826004, Jharkhand, India; satish.19pr0049@me.iitism.ac.in (S.K.); vgkvilluri@iitism.ac.in (V.G.K.V.)

[3] Department of Hydrology, Indian Institute of Technology Roorkee, Roorkee 247667, Uttarakhand, India; ankit.agarwal@hy.iitr.ac.in or agarwal@pik-potsdam.de

[4] Section Hydrology, GFZ German Research Centre for Geosciences, Telegrafenberg, 14473 Potsdam, Germany

[5] Department of Civil Engineering, Indian Institute of Technology Guwahati, Guwahati 781039, Assam, India; ashes@iitg.ac.in

[6] Department of Computer Science and Engineering, Indian Institute of Technology (Indian School of Mines) Dhanbad, Dhanbad 826004, Jharkhand, India; acsrao@iitism.ac.in

[7] KIET Group of Institutions, Department of Computer Science and Engineering, Ghaziabad 201206, Delhi-NCR, India; anubhav.dwivedi8@gmail.com

[8] SMS India Ltd., Khetri, Rajasthan 333503, India; deravijay@gmail.com

[9] Soil and Water Engineering, SVCAETRS, Indira Gandhi Krishi Vishwavidyalaya, Raipur 492012, Chhattisgarh, India; jsvenusmars@gmail.com

[10] Department of Civil Engineering, Indian Institute of Technology (Indian School of Mines) Dhanbad, Dhanbad 826004, Jharkhand, India

* Correspondence: srinivas@iitism.ac.in

Citation: Agrawal, P.; Sinha, A.; Kumar, S.; Agarwal, A.; Banerjee, A.; Villuri, V.G.K.; Annavarapu, C.S.R.; Dwivedi, R.; Dera, V.V.R.; Sinha, J.; et al. Exploring Artificial Intelligence Techniques for Groundwater Quality Assessment. *Water* **2021**, *13*, 1172. https://doi.org/10.3390/w13091172

Academic Editors: Kwok-wing Chau and Pankaj Kumar

Received: 8 January 2021
Accepted: 20 April 2021
Published: 23 April 2021

Publisher's Note: MDPI stays neutral with regard to jurisdictional claims in published maps and institutional affiliations.

Abstract: Freshwater quality and quantity are some of the fundamental requirements for sustaining human life and civilization. The Water Quality Index is the most extensively used parameter for determining water quality worldwide. However, the traditional approach for the calculation of the WQI is often complex and time consuming since it requires handling large data sets and involves the calculation of several subindices. We investigated the performance of artificial intelligence techniques, including particle swarm optimization (PSO), a naive Bayes classifier (NBC), and a support vector machine (SVM), for predicting the water quality index. We used an SVM and NBC for prediction, in conjunction with PSO for optimization. To validate the obtained results, groundwater water quality parameters and their corresponding water quality indices were found for water collected from the Pindrawan tank area in Chhattisgarh, India. Our results show that PSO–NBC provided a 92.8% prediction accuracy of the WQI indices, whereas the PSO–SVM accuracy was 77.60%. The study's outcomes further suggest that ensemble machine learning (ML) algorithms can be used to estimate and predict the Water Quality Index with significant accuracy. Thus, the proposed framework can be directly used for the prediction of the WQI using the measured field parameters while saving significant time and effort.

Keywords: WQI; Pindrawan tank area; drinking water quality; artificial intelligence; particle swarm optimization; support vector machine; naive Bayes classifier

1. Introduction

A high enough quantity and appropriate quality of freshwater are some of the fundamental requirements for sustaining human life and civilization. Indeed, the tremendous population growth and miraculous achievements in science and technology have increased groundwater utilization for domestic, industrial, and irrigation purposes multiple folds throughout the world over the last few decades. Rapid urbanization, overexploitation, and unscientific waste disposal have also influenced the accessibility and quality of groundwater. Excessive population growth and rapid urbanization have forced the use of chemicals and pesticides for agricultural purposes, which often results in leaching and mixing into the groundwater. As indicated by the World Health Organization (WHO), inappropriate or polluted water causes around 80% of all diseases in human beings. Furthermore, contaminated groundwater quality cannot be improved or re-established by preventing contamination from the source. Therefore, understanding and determining water quality is imperative in the study of water resources and environmental engineering.

Water quality essentially determines the usability of water from a source in terms of the nature and concentration of the impurities present in the sample [1]. As a combined effect of the continuous deterioration in water quality and quantity, approximately one billion people worldwide face a shortage of adequate and safe water supply. These statistics' increasing nature makes it essential to monitor water quality for its efficient management and supply [2,3].

The most efficient method for classifying water quality is using the Water Quality Index (WQI). Water quality is often estimated based on water quality indices [4,5]. It is a tool that has been extensively utilized to assess the performance of water quality management approaches [6]. The approach and methodology used for calculating and interpreting water quality indices have evolved over the years [7–11]. The estimated values of water quality indices have been used to indicate water samples' suitability for day-to-day use. They can be utilized effectively in the execution of water quality overhauling programs.

The WQI's variables comprise biological oxygen demand (BOD), temperature, dissolved oxygen (DO), total suspended solids (TSSs), ammoniacal nitrogen (AN), chemical oxygen demand (COD), and pH [12]. Groundwater quality indices (GQIs) are usually forecasted by measuring the standard variables, such as magnesium (Mg^{2+}), calcium (Ca^{2+}), and nitrate (NO^{-3}) [13–15]. The value provided by the WQI is significant enough to help decision makers. However, estimating the WQI is not that simple because subindex calculations are done in the WQI equations themselves. Several methods are available in the literature for the computation of the WQI worldwide, e.g., United States National Sanitation Foundation Water Quality Index (NSFWQI), the British Columbia Water Quality Index (BCWQI), and the Canadian Water Quality Index (CWQI).

The WQI aims to convert the complicated water quality information into straightforward data that is readily useable by researchers and conveyable to people in general. The calculation process in the case of some approaches applied in several countries, including India [6,16], can be exceptionally intricate and time consuming. As a result, the process always contains the risk of attracting unintended miscalculations [17]. Thus, the limitations for the calculations of WQI are the following: (a) time consuming, (b) lengthy process, (c) complicated process, and (d) different equations are used for WQI calculations, hence there are inconsistencies. It may be obvious from the above discussion that no standard method is available for the WQI.

To conquer the above problems, a few scientists have proposed a nonphysical approach that can successfully predict WQI using machine learning (ML) and artificial intelligence (AI) [18–20]. After satisfactory training, an AI-based model can promptly produce a WQI value by eliminating the sub-index calculations. Awareness of AI algorithms is increasing due to benefits that include nonlinear structures, the capability to calculate complicated trends, the capability to manage huge datasets consisting of different data scales, and insensitivity to absent data. The forecasting capability of ML–AI algorithms greatly relies on the procedures and exactness of the data collection and analysis. The continuous evolution of computational ability has allowed researchers to use diverse arrangements of ML–AI models. Approaches such as artificial neural networks (ANNs) [17,21–26] adaptive neuro-fuzzy inference systems [27–31], and support vector machines (SVMs) [32] have been effectively applied to predict the quality of water worldwide. Abba et al. (2020) [33]

describe in detail the ML–AI techniques that are used for WQI measurement. Most of these ML–AI algorithms can perform with a certain degree of accuracy and it is challenging to compare them based on their performance [25,34].

The AI techniques used in the present study, sometimes include complex manual implementation to reduce its actual effectiveness for water quality management personnel. Practitioners have a great interest in learning the codes such that the codes can be used for solving complex models like the one discussed above. A comprehensive comparison of such models' applications with required software packages must be carried out to improve the accuracy of predictions and the suitability of the AI-based models. However, various data mining programs do not involve vast manipulation of several AI models; instead, the majority of them just support fundamental methods without optimization.

Our study also aimed to develop a user-friendly interface in MATLAB for practitioners that do not have a programming background. The recommended interface is based on a nature-inspired metaheuristic classification system that integrates particle swarm optimization (PSO), along with an SVM and NBC. The water quality was forecasted using fundamental AI techniques, which involved a particle swarm optimization (PSO) algorithm combined with support vector machines (SVMs) for prediction. The classification and predictive AI system investigated in the study was developed using four AI models (single), hybrid metaheuristic regression, and four ensembles (i.e., stacking, voting, bagging, and tiering). The baseline models encompassed single models by using two AI techniques: SVM and NBC, respectively. Subsequently, the ensemble models integrated the registered single models and utilized voting, bagging, tiering, and stacking methods. The goal of the present work was to propose a framework for flexible water quality modeling. The analytical technique had similar goals: the models' predictive accuracy and applicability. The framework will empower administrators and hydrologists to choose the best analytical tools for water management using AI techniques.

These models should be selected based on specific requirements. However, sometimes applying an ensemble model can significantly enhance the model accuracy and reduce the computational cost. In the present study, the combination of the PSO algorithm's applicability with an SVM and NBC was exploited. A framework was proposed for predicting the WQI in the Pindrawan tank area, Raipur region, Chhattisgarh, India.

2. Study Area

The Pindrawan tank command area was the area under study (Figure 1); it is situated within 81°45′–81°50′ E and 21°20′–21°25′ N in the upper Mahanadi River valley (southeastern part) and Raipur district of Chhattisgarh, India. A total of nine villages, namely, Pauni, Amlitalab, Khauna, Deogaon, Bangoli, Dhansuli, Kurra, Baraonda, and Nilja, come under the study area, which has a tropical wet and dry climate. The temperature in this part of India remains moderate throughout the year. The highest temperatures in the year are observed from March to June.

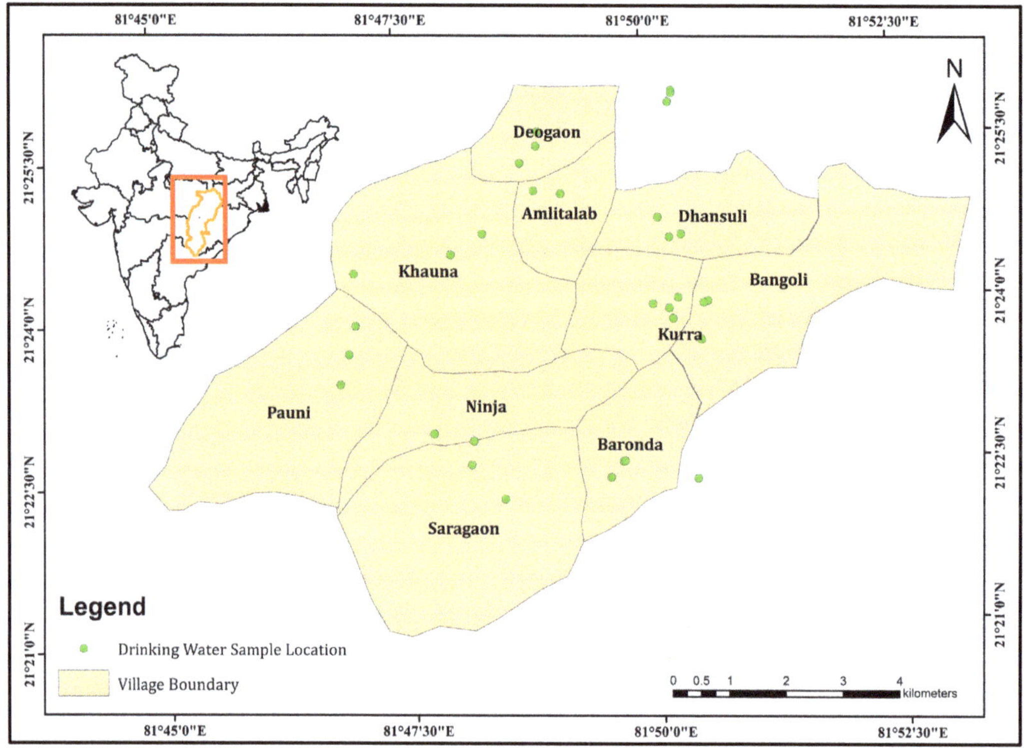

Figure 1. Map of the study area showing the Pindrawan tank command area's geographical location in Chhattisgarh State, India. The figure shows the location of the study area at the country and state levels, as well as the village boundaries that are under the Pindrawan tank command area with drinking water sample locations (green color points).

3. Methodology

3.1. Data Collection and Water Quality Estimation

The groundwater samples were collected in 2018 during the pre-monsoon period from hand pumps and bore wells (37 sites), which are extensively utilized for drinking in the Pindrawan tank area. The identification of the sampling points was performed using topographic sheets and GPS, and the maps were prepared using ArcGIS 10.1 (ESRI, California USA). Topographic sheets were utilized to prepare the base map and recognize the general features of the area. GPS techniques were used to identify the geographic position of each sampling point. The collected groundwater samples were investigated for the concentration of different parameters, namely, electrical conductivity (EC), pH, total dissolved solids (TDSs), total hardness (TH), alkalinity, bicarbonate (HCO_3^-), chloride (Cl^-), sulfate (SO_4^{2-}), nitrate (NO_3^-), fluoride (F^-), calcium (Ca^{2+}), magnesium (Mg^{2+}), sodium (Na^+), potassium (K^+), iron (F^-), and chromium (Cr^{2+}), per the specification of the Federation and American Public Health Association (2005). The EC and pH of the collected samples were measured using an EC and pH meter on the field. Fluoride concentrations were analyzed based on the selective electrode method. TH, chloride, and alkalinity were measured using titrimetric methods. Heavy metals were measured using an atomic absorption spectrum and prescribed safety measures were considered to avoid contamination.

The locations of the sampling stations are presented in Figure 1. The concentrations of the parameters were compared with the acceptable limits prescribed by BIS (2012) [35]. The permissible limits of potassium, bicarbonate, and sodium are reported in [36,37].

The WQI of the collected samples was calculated using the weighted arithmetic Water Quality Index (WQI) method [38–40]. The weights (W_i) that were assigned to each parameter according to their impact on the water quality are shown in Table 1.

Based on the corresponding WQI values, the quality of the groundwater for drinking purposes can be classified into five categories, as presented in Table 2.

Table 1. Water quality parameters used when calculating the WQI.

Parameters	Indian Standards	Weight (W_i)	Unit Weight (W_i)	Parameters	Indian Standards	Weight (W_i)	Unit Weight (W_i)
EC	300	1	0.024	Alkalinity	200	3	0.073
PH	6.5–8.5	2	0.049	TH	300	2	0.049
TDS	500	3	0.073	Fluoride	1	4	0.098
Calcium	75	2	0.049	Iron	0.3	4	0.098
Magnesium	30	2	0.049	Chromium	0.05	4	0.073
Potassium	12	2	0.049	Chloride	250	2	0.049
Sodium	200	1	0.022	Bicarbonate	250	3	0.073
Sulfate	200	3	0.073	Total		41	1
Nitrate	45	3	0.073				

Table 2. WQI classification based on the same WQI used by Ramakrishnaiah et al., 2009 [41].

WQI	Class
0–50	Excellent water quality
50–100	Good water quality
100–200	Poor water quality
200–300	Very poor water quality
>300	Unfit for drinking

3.2. Utilization of AI for the Prediction of the WQI

The present study utilized two powerful machine learning approaches for the estimation of the WQI classes by considering the parameter (variables) values as inputs. All 16 variables resembled a variable vector. The analysis was carried out using 1250 variable vectors (250 for each class), which were generated using PSO to contain the whole array of every class. Calibration was conducted using 1250 variable vectors (250 from each class) by applying tenfold cross-validation, and the assessment was done using 250 variable vectors (50 from every class).

3.2.1. Classification and Prediction Using a PSO–SVM Approach Based on the Water Quality Index

The PSO approach is an extremely powerful algorithm that can optimize different model parameters depending on a population's behavior. The approach was proposed by Eberhart and Kennedy in 1995 [42]. The PSO approach has been efficiently used to solve a multitude of nonlinear problems in diversified fields, such as geology [43,44], landslide analysis [45,46], forest fire mapping [47], and flood modeling [48,49]. The algorithm is initialized with a population of arbitrarily selected solutions between the maximum and minimum range of the parameters. Several advantages of the PSO approach, including the ease of implementation and convergence, fewer parameters, and the use of parallel computing, makes this approach a more comfortable choice compared to other available optimization techniques. The algorithm was developed based on the conduct of a group of fish or birds selecting the smallest path to a food source [50]. The algorithm can improve the exchange of information between samples in a population through an interactive learning process that helps the population arrive at a consistent solution. Each solution is considered as "bird", also known as "particle", in the solution space. Such interactions between members of the population allow this algorithm to demonstrate a robust search proficiency and advanced adaptability to various problems. In PSO, particles (solutions) will be collected randomly, and then the best particles will be found by renewing the generation. In each generation, each particle is modified using the next two "best" parameters. The first is the best value based on fitness that has been obtained by it until now (fitness parameters are also stored). This value is called individual best value (pbest). Pbest is the best value of thepartile among all the values obtained so far. The other "best" parameter, which comes from the particle swarm analyzer, is the best value that ha been obtained by any particle in the current population. This highest value is called global best (gbest). The movement of the particles is controlled by these optimal values of pbest and gbest. After finding an improved position, they will continue to control the movement of the flock. In the solution space [51], a particle is comprised primarily of two vectors, namely, velocity (Vi) and position (Xi) [52], by using Equations (1) and (2) respectively. Figure 2 gives the PSO algorithm that is used for the particle optimization. The optimization of these two vectors in the dth dimension is performed through the following equations:

$$v_{id}^{t+1} = wv_{id}^{t} + c_1 r_{1d}\left(pbest_{id}^{t} - x_{id}^{t}\right) + c_2 r_{2d}\left(gbest_{id}^{t} - x_{id}^{t}\right) \tag{1}$$

$$x_{id}^{t+1} = x_{id}^{t} + v_{id}^{t+1} \tag{2}$$

where, w is known as the inertia weight. The value of these parameters specifies the number of particles following the current velocity. The parameters c_1 (cognitive coefficient) and c_2 (social coefficient) are known as the acceleration factors. The parameters c_1 and c_2 represent the self-reasoning capability and the ability to acquire information from any particle's contemporary global optimal solution, respectively. r_1 and r_2 are two independent arbitrary parameter numbers in the range [0, 1] [53]. $pbest_{id}^{t}$ and $gbest_{id}^{t}$ are known as the local optimum (best-known position value of any particle i) and the global optimum (optimal value obtained by the swarm of all particles).

The coordinate attained by every individual particle in the solution space is recorded by the algorithm. These coordinates are representations of the best solution (fitness value) that has been attained by the particle and is called the local optimum (pbest), whereas the best solution attained by any particle in the vicinity of a specific particle is known as the global optimum (gbest). Although, the particles in the PSO approach tend to move arbitrarily, the best achieved position of the particles (pbest) and the group's best position (gbest) have significant influence over their movement.

Presently, the PSO approach was utilized to produce the optimized values of the WQI, along with all of the 16 water quality variables by considering the variables' lower and upper limits, as presented in Table 3. Based on the corresponding WQI values, the groundwater quality for drinking purposes was classified into five categories (Table 2). To achieve the optimized values of the WQI and water quality variables corresponding to the different classes of water quality, the WQI parameter was considered as the fitness function. The algorithm was set up with an initial population of 50 and processed up to a maximum generation of 500; therefore, a total of 50 × 500 = 25,000 optimized values were generated. The ranges of values for each variable used in the WQI function are presented in Table 3.

The procedure for generating the optimal variables' values was as follows:

Step 1—The fitness function was explained using the WQI function, initializing "50 as population" and "500 as the maximum generation."

Step 2—Each variable's maximum and minimum limits were set while using the WQI function according to Table 3.

Step 3—Every particle's movements were recorded in every generation in the vector form comprising the value of the WQI, together with the subsequent values of the 16 variables.

Step 4—The category (class) of each variable vector was obtained by considering its corresponding WQI, as presented in Table 2.

Step 5—A total of 250 variables vectors were selected from each category in such a manner that the entire range of the particular category should be covered, as given in Table 2.

In every generation, the populace shifted from the initial position to a new appropriate place and produced new fitness values. Every particle's movement in every generation was recorded in the vector form containing the WQI value along with the subsequent variables' values. Every random particle updated its fitness value (WQI) in each generation, which was stored in the database and related variables. In PSO, the population's values (swarm) and max iteration (generation) depend on the user. The flowchart for this work is shown in Figure 3. The classification of the WQI values was performed using a support vector machine and a naive Bayes classifier. Before proceeding with the classification, the dataset was normalized between 0 and 1 to enhance the accuracy. The variables' values in vector format were treated as a feature vector in the normalized dataset.

Table 3. Comparison of chemical parameters with prescribed standards.

Parameter	Experimentally Obtained Range of Concentration in the Collected Samples	Permissible Limits	Percentage of Samples Exceeding Permissible Limits	Undesirable Effect
pH	7.26–8.59	6.5 to 8.5	2.70	Irritation in eyes, skin, and mucous membranes; skin disorders
EC	152–1998	300	89.19	Cardiac dysrhythmias
TDS (mg/L)	98.8–1199	500	21.62	Gastrointestinal irritation
Alkalinity (mg/L)	60–335	200	29.73	Unpleasant and harmful to aquatic life and humans
Chloride (mg/L)	20–330	250	8.11	Salty taste
Calcium (mg/L)	4–60.5	75	0	Scale formation
Magnesium (mg/L)	4–20.2	30	0	Cerebrovascular disease (Yang, 1998)
Potassium (mg/L)	0–30.9	12	16.20	Bitter taste
Sodium(mg/L)	1.2–18.3	200	0	High blood pressure
Nitrate (mg/L)	3.4–8.2	45	0	Methemoglobinemia
Sulfate (mg/L)	25–50	200	0	Laxative effect
Bicarbonate (mg/L)	2.5–6.5	250	0	Vomiting, dehydration, chronic obstructive pulmonary disease
Fluoride (mg/L)	0.25–0.84	1	0	Mottling of teeth, deformation of bones
Iron (mg/L)	0.015–0.785	0.3	5.41	Diabetes, hemochromatosis, stomach problems, nausea, and vomiting
Chromium (mg/L)	0.007–0.737	0.05	56.76	Hearing loss, blood disorders, hypertension, death at high levels
TH (as mg/L)	138–320	200	43.24	Scale formation in pipes anencephaly, urolithiasis, parental mortality

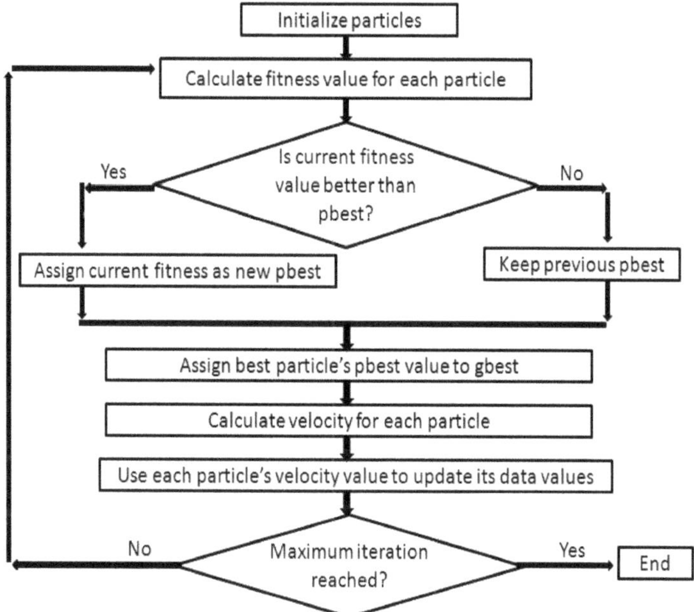

Figure 2. Flowchart for the optimization of the particles.

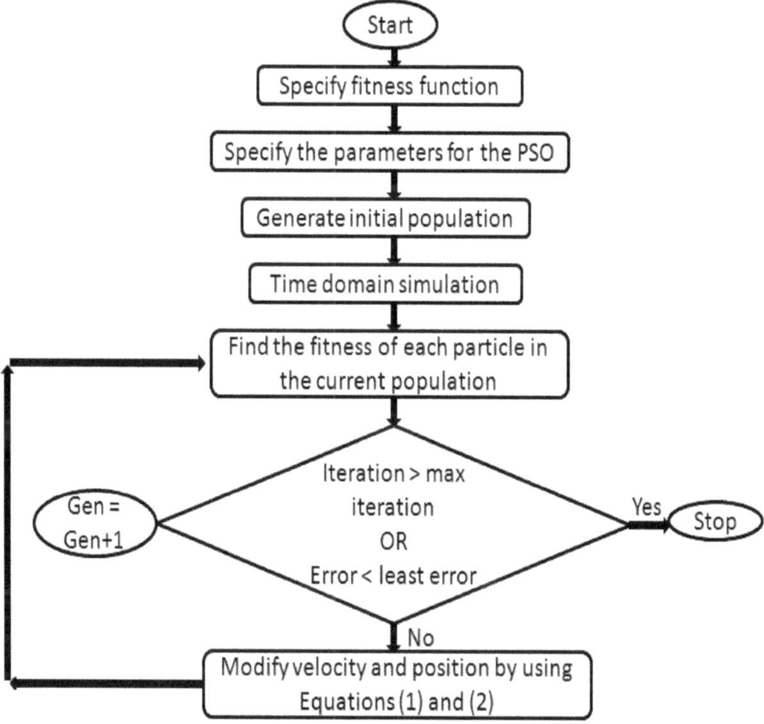

Figure 3. Flowchart describing the workings of the PSO.

3.2.2. Classification Using a Support Vector Machine

The SVM classifier [54] plays an essential and comprehensive role in classification due to its high accuracy and ability to deal with high-dimensional data. The simple form of the classification is the binary used for separating two types of objects belonging to positive (+1) and negative (−1) classes. A support vector machine uses two kinds of concepts to distinguish between two classes: (1) separation from the margin and (2) the kernel function.

The simple two-dimensional data can be classified by using a straight line. The points that fall above the line belong to one class, and the points that fall below the line belong to another class. The high-dimensional data can be classified by using the hyperplanes. However, in a binary classification, multiple planes can be drawn such that they separate the data into two classes. As such, which plane will be selected for the classification? In this case, the hyperplane that gives the maximum margin will be selected for classification. Therefore, we choose the hyperplane such that the distance from it to the nearest data point on each side is maximized. The classification of the data with the best margin hyperplane is shown in Figure 4.

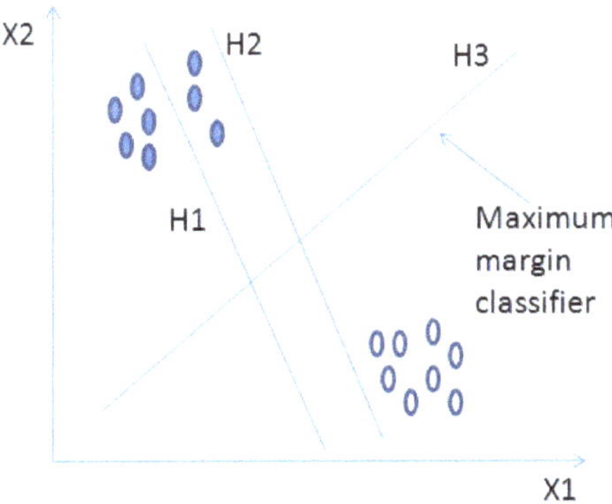

Figure 4. Classifications of data using various hyperplanes.

In Figure 4, there are two types of data points: filled and unfilled dots. Three planes exist, which are named H1, H2, and H3. H1 does not successfully classify the data points. Planes H2 and H3 are both capable of classifying data points, but H2 gives a smaller margin than plane H3.

This is why plane H3 is selected for the classification. Sometimes the data is not classified by hyperplanes because of its distribution in a vast space. In that case, we use a nonlinear separation for the classification. The SVM classifier can efficiently perform this nonlinear classification by using kernel functions. The nonlinear classification is presented in Figure 5. In Figure 5, there are two types of objects, as identified by the solid and hollow dots. The objects represented in this figure cannot be separated using a linear hyperplane; the support vector machine performs this task using kernel functions. The kernel function separates the data in the feature space by using a linear hyperplane.

In this work, the SVM classifier separates the individual water quality classes with hyperplanes by using the radial basis kernel (Gaussian) function [55–58]. The distance of a feature vector from the hyperplanes determines its probability of featuring in a specific class. The normalized dataset and the class labels were used as inputs in the present study. The dataset was randomly divided 80:20, where 80% of the dataset was used for training purposes using tenfold cross-validation. In the tenfold cross-validation, the entire dataset was divided randomly into ten equal-sized subsamples. A single subsample was used for testing purposes, and nine subsamples were used for training purposes on ten subsamples. This process was repeated ten times until each of the 10 subsamples were used exactly once for testing purposes. The remaining 20% of the dataset was used for testing and validation purposes.

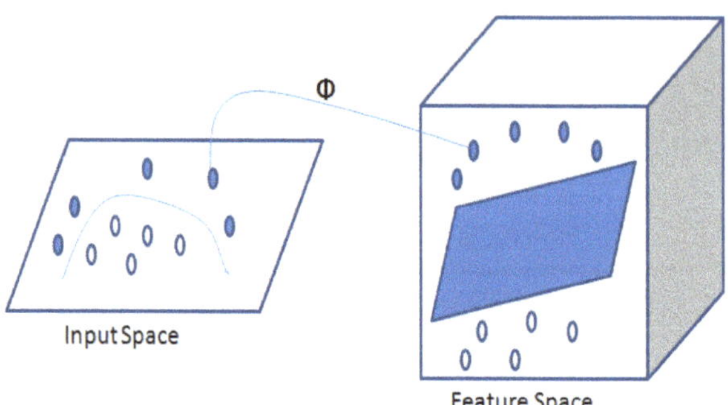

Figure 5. Use of the kernel function in an SVM.

3.2.3. Classification Using Naive Bayes Classifier

Naive Bayes classifiers are based on Bayes Theorem with a family of algorithms with the same principle, i.e., each pair of features being categorized is independent of every other. The fundamental naive Bayes assumption is that every feature makes an unbiased and identical contribution to the outcome. A naive Bayes classifier is a probabilistic machine learning model that is used for a classification task. The crux of the classifier is based on Bayes' theorem:

$$P(A|B) = \frac{P(B|A)P(A)}{P(B)} \tag{3}$$

By using Equation (3), the probability of event A happening can be measured by considering that event B has occurred. Here A is the hypothesis and B is the evidence. One assumption that is considered here is that all features are independent/autonomous, which means the presence of one particular feature does not affect the other. Hence it is called naive. Before the PSO–NBC analysis, the dataset was normalized to enhance the performance of the model. A total of 80% of the dataset was used to train the algorithm, whereas 20% of the dataset was used to study the algorithm's prediction accuracy. In this work, continuous values that were associated with each feature were assumed to be distributed according to a Gaussian/normal distribution.

4. Results and Discussion

4.1. Water Quality Index (WQI) Analysis of the Field-Based Samples

The concentration, distribution, and impact of different physicochemical parameters observed from water samples collected from the Pindarwan tank area are discussed in this section. The ranges of concentrations observed for various parameters and the percentages of total samples exceeding the prescribed limit are presented in Table 3, along with their undesirable effect on groundwater quality and human physiology. This section provides an overview of the spatial distribution of the physicochemical parameters that were measured in the Pindarwan tank area; a more detailed description is provided in Figures A1–A15 in the Appendix A.

Out of 37 samples, 32.43% of the samples had excellent water quality, 43.24% of the samples had good water quality, 21.62% of the samples had poor water quality, and 2.71% of the samples had very poor water quality. This may be due to the heavy concentrations of metals, such as Pb and Cr, due to nearby industries, which involve mining activities, thermal power plants, etc. The areas corresponding to these WQI values are presented in Figure 6.

Figure 7 represents a correlation plot between the WQI and the parameters observed from the study area's water samples. The correlation between the independent parameters can be neglected in the plot since these plots are mostly empirically based on specific values. In decreasing order, the influence of different parameters can be presented as chromium, sodium, fluoride, potassium, chloride, conductivity, total dissolved solids, alkalinity, bicarbonate, and pH. Contributions from the rest of the parameters on the overall water quality were much less compared to these parameters. Through observing Figure 7, it can be concluded that water quality for drinking was susceptible to heavy metal concentrations, such as chromium.

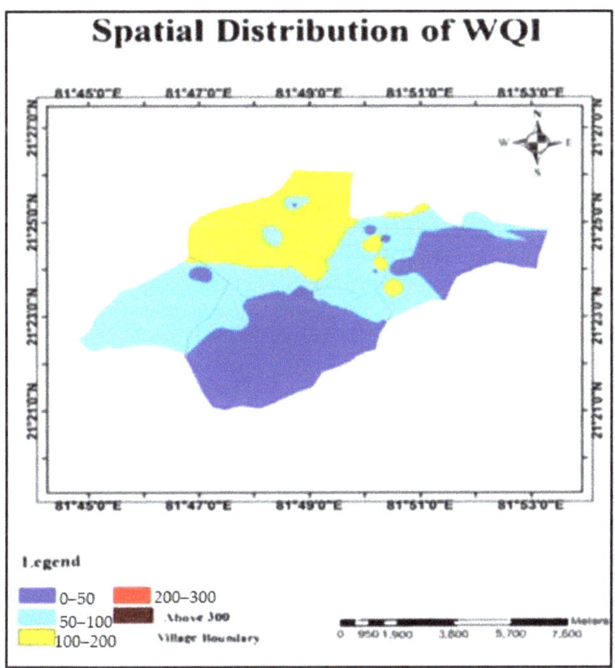

Figure 6. Spatial distribution of the WQI.

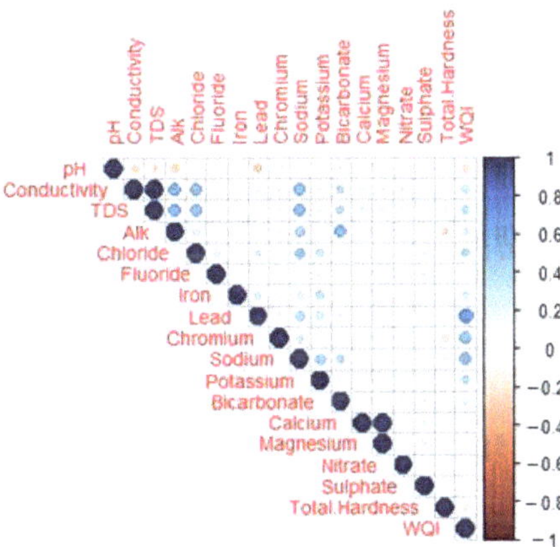

Figure 7. Correlation plot between various groundwater quality parameters.

Based on the WQI, the sample area's drinking water quality was divided into four categories. No sample was observed to be unsuitable for drinking based on the analysis. Very poor water quality was observed from the Raikheda pond area due to a very high chromium concentration. Poor water quality was observed in significant parts of the Deogaon, Dhansuli, Bangoli, Amlitalab, and Khauna villages. Most areas of all the villages had good water quality. Excellent water quality was observed in Saragaon, Nilja, Dhansuli, Bangoli, Khauna, Baronda, and Pauniarea. The observed water qualities may suggest that most of the study area's water quality is satisfactory and there is no immediate danger for the population. However, the values of certain parameters, such as the chromium concentration, total hardness, and total dissolved solids, were alarmingly high for many areas and could become worse. This may significantly influence the present scenario of the water quality in the study area under consideration. Therefore, concerned authorities should note the situation and plan proper steps for maintaining or improving the current situation of the drinking water quality in the study area.

Furthermore, the averages and ranges of the values of different parameters corresponding to water quality are presented in a boxplot format in Figure 8a–p. The concentration of some parameters such as alkalinity, chloride, conductivity, chromium, iron, bicarbonate, sodium, and TDSs are found to be directly proportional and has much more significant impact on the WQI of the study area. These are, therefore, the parameters that have to be first taken care of when aiming to improve the water quality for the specific study area. The influences presented in Figure 8a–p are the combined effect of the concentration of each parameter and the relative weight of each parameter. Therefore, even if a parameter's relative weight is much less, it could make a significant impact if it had a very high concentration. However, these plots are strictly applicable to the present study area and no inference should be derived from these plots for any other samples. The boxplots and correlation plots can be extremely useful for conveying a detailed picture regarding the water quality of the study area and the influence of different parameters on the water quality.

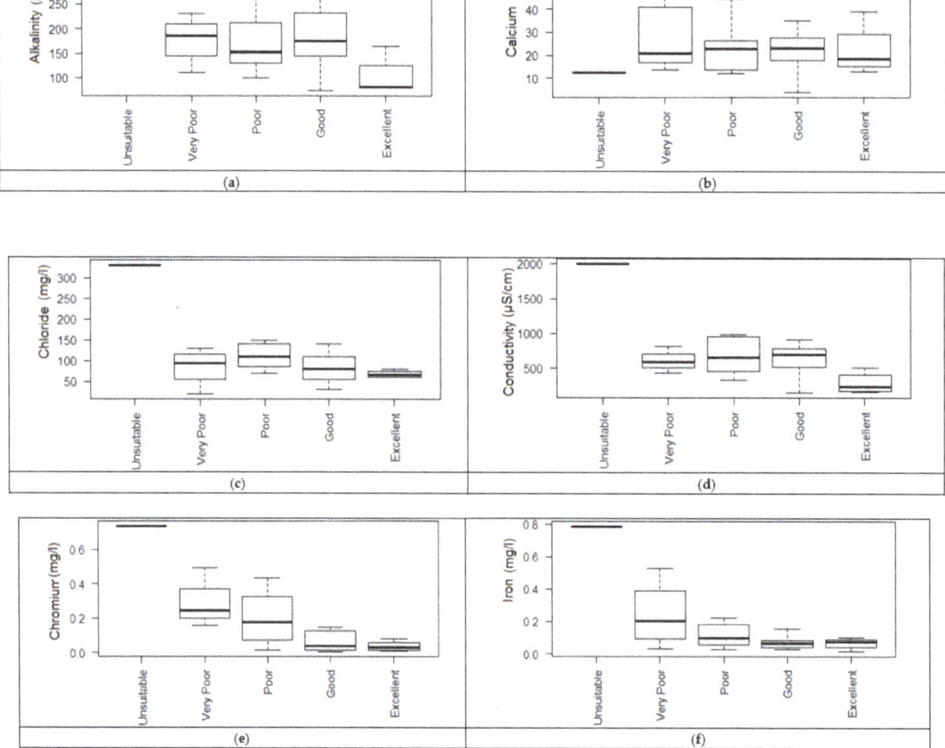

Figure 8. *Conts.*

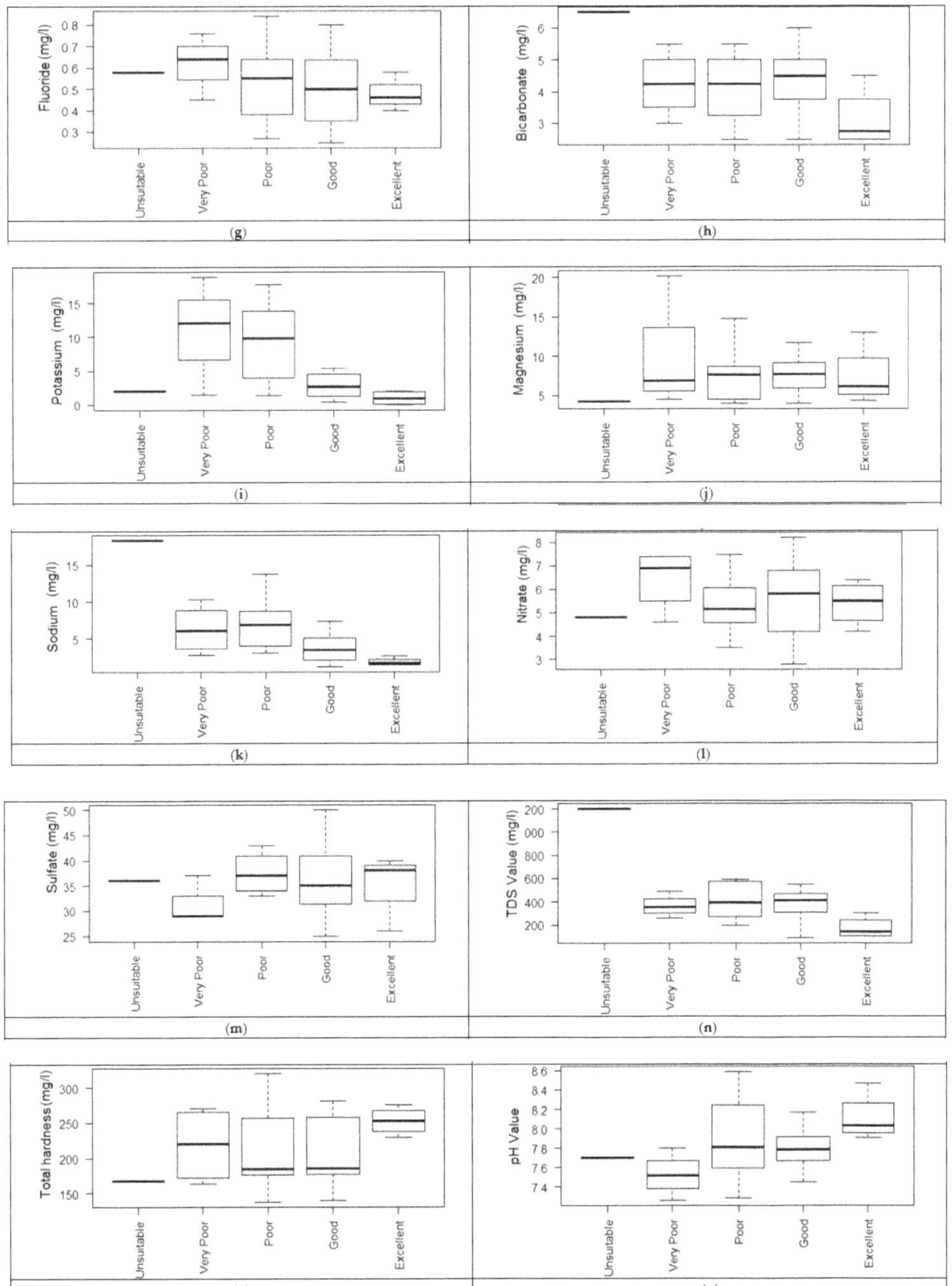

Figure 8. Ranges of various parameters corresponding to the water quality: (**a**) alkalinity, (**b**) calcium, (**c**) chloride, (**d**) conductivity, (**e**) chromium, (**f**) iron, (**g**) fluoride, (**h**) bicarbonate, (**i**) potassium, (**j**) magnesium, (**k**) sodium, (**l**) nitrate, (**m**) sulfate, (**n**) TDSs, (**o**) total hardness, and (**p**) pH.

4.2. Result from the PSO–SVM Study

The performance of the model is presented using the confusion matrix in Figure 9a. The confusion matrix is used to explain the model's classification and overall performance on the testing datasets whose original labels are known. The instances in a predicted class and actual class are represented in every row and each column respectively (or vice versa). In Figure 9a, the rows from the top to the bottom correspond to the excellent, good, poor, very poor, and unfit for drinking water qualities, respectively, as predicted using the SVM classifier.

Furthermore, the columns from left to right follow a similar arrangement of the target class (actual classifications based on the WQI values). Each column related to these classes had 50 variable vectors (water quality class from excellent to unfit for drinking), totaling 250 variable vectors. In the first row, 50 variable vectors are presented, indicating 50 excellent water class WQIs, where the system predicted them all as being in the excellent category. Similarly, in the second, third, fourth, and fifth rows, a sum of 61, 54, 69, and 16 variable vectors are presented, respectively. The result indicates that the algorithm predicted 61 samples as good quality, 54 as poor quality, 69 as very poor quality, and 16 as unfit for the drinking category. The prediction accuracies corresponding to each class are also presented in the last column from the left-hand side. The overall accuracy of the algorithm was found to be 77.60%. Furthermore, a difference between the classifications based on actual values of the WQI and the predicted classification based on the SVM classifier is presented in Figure 9b.

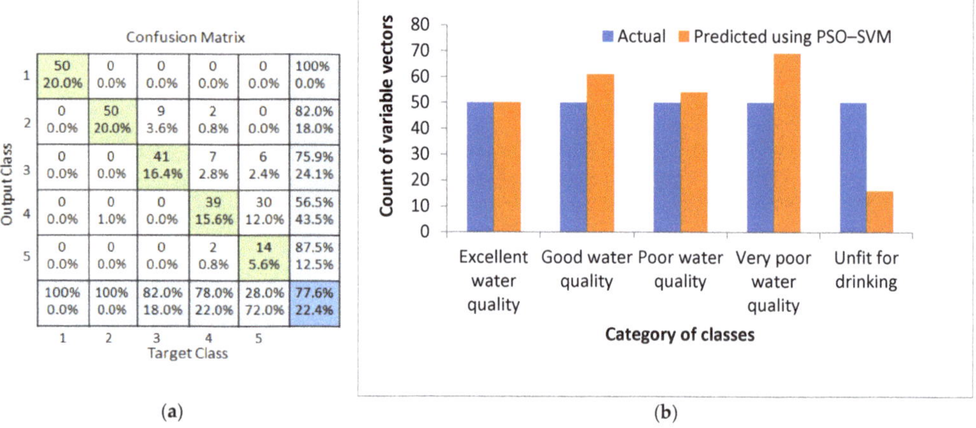

<table>
<tr><td>(a)</td><td>(b)</td></tr>
</table>

Figure 9. Comparison between the predicted class and target class using the SVM approach: (**a**) confusion matrix and (**b**) column plots.

4.3. Discussion of the PSO–NBC Approach

The PSO–NBC study was carried out by considering the same dataset as in the PSO–SVM approach. The test accuracy is discussed using the confusion matrix presented in Figure 10a. The rows and columns marked as 1 to 5 indicate the excellent (1), good (2), poor (3), very poor (4), and unfit for drinking (5) water qualities. The 51 variable vectors in the first row indicate that the algorithm identified 51 variable vectors as excellent water quality when there were 50 actual excellent water categories (1 more due to misclassification). Similarly, in the second (50 variable vectors of good water quality), third (50 variable vectors of poor water quality), fourth (50 variable vectors of very poor water quality), and fifth rows (50 variable vectors of unfit for drinking water quality), the algorithm placed 57 (good water quality), 46 (poor water quality), 51 (very poor water quality), and 45 (unfit water quality) variable vectors. The prediction accuracy of the algorithm corresponding to each class is presented in the sixth column. The total accuracy of the algorithm was observed to be 92.80%.

The comparisons of the model-predicted outcomes against the actual WQI values are graphically represented in Figure 10b.

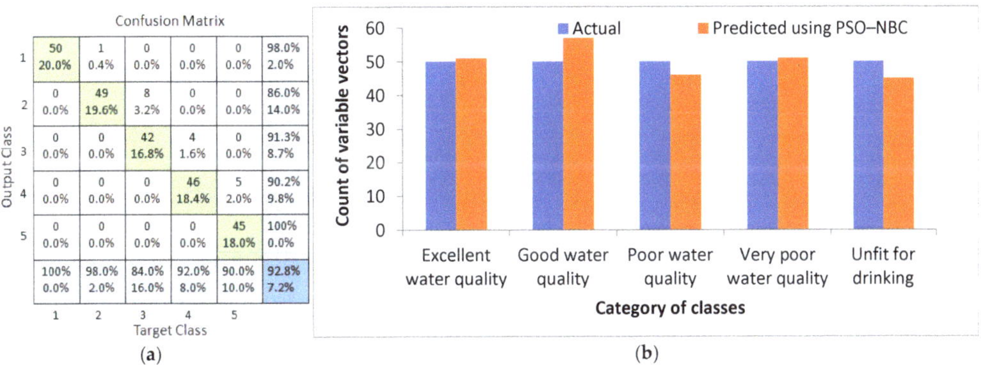

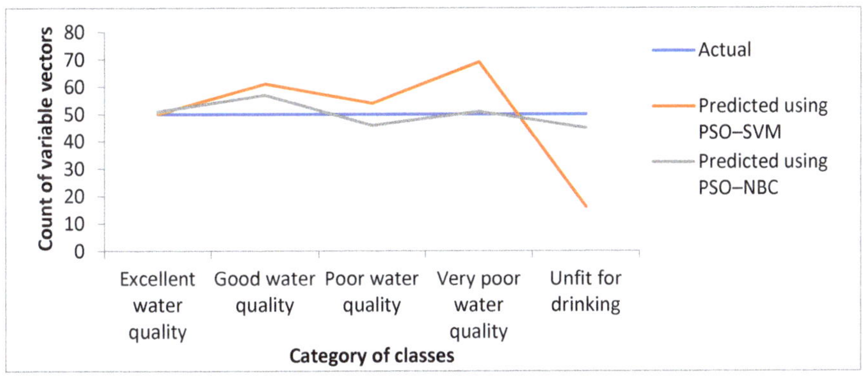

Figure 10. Comparison between the predicted class and target class using the NBC approach: (**a**) confusion matrix and (**b**) column plots.

4.4. Comparison between the PSO–SVM and PSO–NBC Approaches

The performances of the PSO–SVM and PSO–NBC approaches used in the present study are presented inFigure 11.

Figure 11. Comparison of the predicted outcomes using the PSO–SVM and PSO–NBC approaches.

The figure indicates that the PSO–SVM algorithm predicted some classes (excellent and poor water categories) with significant accuracy; however, significant deviations were observed in the model's performance for the other categories. On the other hand, the prediction accuracies of PSO-NBC were much higher for all the classes and did not distinctly deviate for any specific categories. Therefore, a naive Bayes classifier aided by particle swarm optimization can be efficiently used to construct a machine learning model to classify water for drinking purposes.

5. Conclusions

The process of WQI estimation is often associated with handling large quantities of identical data. This can create significant confusion during the calculation process and make decision making difficult. A machine-learning-based predictive model can assemble the necessary information and predict the groundwater quality with significant accuracies. This study aimed to utilize modern machine learning techniques for the prediction of water quality for drinking. The groundwater samples collected from parts of the Pindrawan tank command area were used for testing and validation of the developed model. The collected samples were tested for different parameters of water quality and the subsequent values of WQI were computed. Conclusions derived from the present work are as follows:

1. The calculated WQI values suggested that 32.43% and 43.24% of the water samples of the study area represented excellent and good water qualities, respectively. Similarly, it can also be observed that 21.62% and 2.71% of the water in the study area were of poor and very poor drinking water qualities. Very poor water quality was observed from the Raikheda pond area due to very high chromium concentration. Poor water quality was observed in significant parts of the Deogaon, Dhansuli, Bangoli, Amlitalab, and Khauna villages.

2. The major cation and anion data revealed that all anions were within the limits, except for potassium, where 13% of the samples exceeded the limit. However, the heavy metals pollution in the area due to mining activities could be a cause for concern soon. A total of 48.6% of the samples from the area exceeded the permissible limits of chromium, which can cause conditions such as hearing loss, blood disorders, hypertension, and death at high levels.

3. The study further suggests that ensemble machine learning algorithms can be used for the estimation and prediction of a WQI with significant accuracies. In the present study, a particle swarm optimization approach coupled with a naive Bayes classifier provided a 92.8% accurate prediction of the WQI indices. Therefore, with the help of a user interface, this algorithm can be efficiently utilized for the estimation of WQIs, which can save significant effort and time.

The general outcomes from the present research indicate the benefits of using ensemble machine learning techniques, where outcomes from several different algorithms can be combined and used to achieve predictions with enhanced accuracies. Finally, with the help of a user interface, the algorithm developed in the present study can be used for water quality estimation in different regions across the globe.

The classification in the present study was carried out by taking the synthetic dataset that was generated using particle swarm optimization. However, the developed approach can be further improved if more real data is available. Therefore, the authors suggest using a larger field dataset to obtain better accuracy, though this is often a difficult undertaking provided the painstaking process of sample collection and laboratory analysis for all the water quality parameters. The developed algorithm can be further improved by studying its performance and fine-tuning it with different input parameters.

Author Contributions: Conceptualization, P.A., A.S. and S.P.; data curation, P.A.; formal analysis, P.A. and S.K.; investigation, P.A., S.K. and A.B.; methodology, P.A., A.S., S.K., A.A., A.B. and S.P.; project administration, A.S., V.G.K.V. and S.P.; resources, P.A., A.A. and J.S.; software, P.A., A.B., C.S.R.A. and R.D.; supervision, A.S., J.S. and S.P.; validation, P.A., A.B., C.S.R.A. and R.D.; visualization, A.S., A.A., V.V.R.D. and S.P.; writing—original draft, P.A., A.S., A.B., C.S.R.A. and J.S.; writing—review and editing, A.S., S.K., A.A., V.G.K.V., V.V.R.D., J.S. and S.P. All authors have read and agreed to the published version of the manuscript.

Funding: This research received no external funding.

Institutional Review Board Statement: Not applicable.

Informed Consent Statement: Not applicable.

Data Availability Statement: The data may be obtained from the authors upon request.

Acknowledgments: The authors would like to sincerely thank the Indian Institute of Technology (Indian School of Mines) authorities, Dhanbad, for extending their support and allowing the use of facilities from various engineering departments, i.e., Environmental Science and Engineering, Civil Engineering, Mining Engineering, and Computer Science Engineering Departments. The authors would also like to acknowledge the support received and facilities used from the Water Resources Department, Government of Chhattisgarh, and Indira Gandhi Krishi Vishwavidyalaya, Raipur, in carrying out this research work. A.A. acknowledges the infrastructural support provided by the Indian Institute of Technology Roorkee.

Conflicts of Interest: The authors declare no conflict of interest.

Appendix A

Table A1. Locations used for the groundwater samples.

Sample No.	Lat	Long	Sample No.	Lat	Long	Sample No.	Lat	Long
1	81.8282	21.3761	13	81.8367	21.3994	25	81.8155	21.4273
2	81.8023	21.3764	14	81.8373	21.3977	26	81.8124	21.4226
3	81.8077	21.371	15	81.834	21.4001	27	81.8152	21.4252
4	81.7961	21.3815	16	81.828	21.3760	28	81.8584	21.4041
5	81.8028	21.3801	17	81.8258	21.3736	29	81.8377	21.4311
6	81.7961	21.3815	18	81.8282	21.3761	30	81.8566	21.4033
7	81.8391	21.4107	19	81.7824	21.3942	31	81.8001	21.4089
8	81.8353	21.4134	20	81.7807	21.3896	32	81.8426	21.4000
9	81.8371	21.4103	21	81.7837	21.3985	33	81.8405	21.3729
10	81.8383	21.4010	22	81.7837	21.4066	34	81.8384	21.4329
11	81.842	21.3943	23	81.8001	21.4089	35	81.8384	21.4325
12	81.8433	21.4002	24	81.8056	21.4119	36	81.819	21.4177
						37	81.8145	21.4183

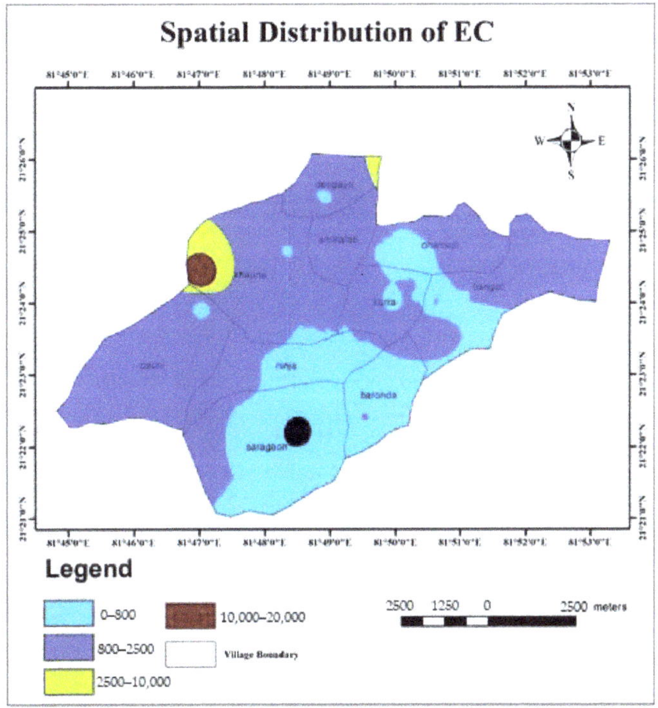

Figure A1. Spatial distribution of EC.

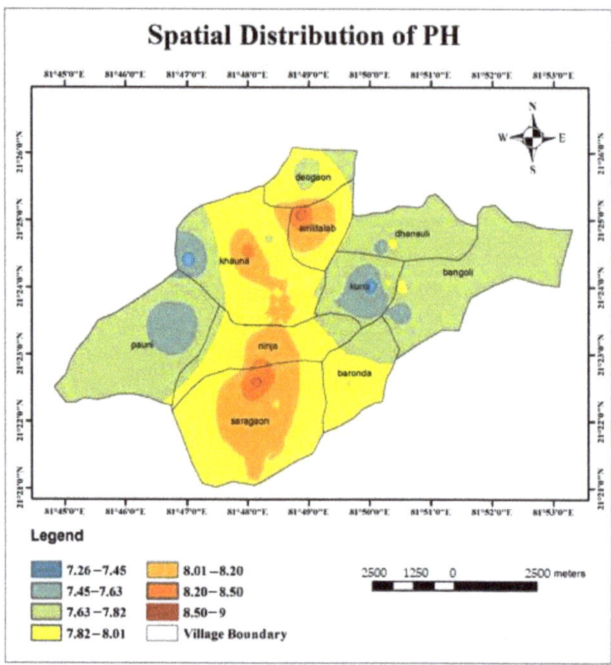

Figure A2. Spatial distribution of PH.

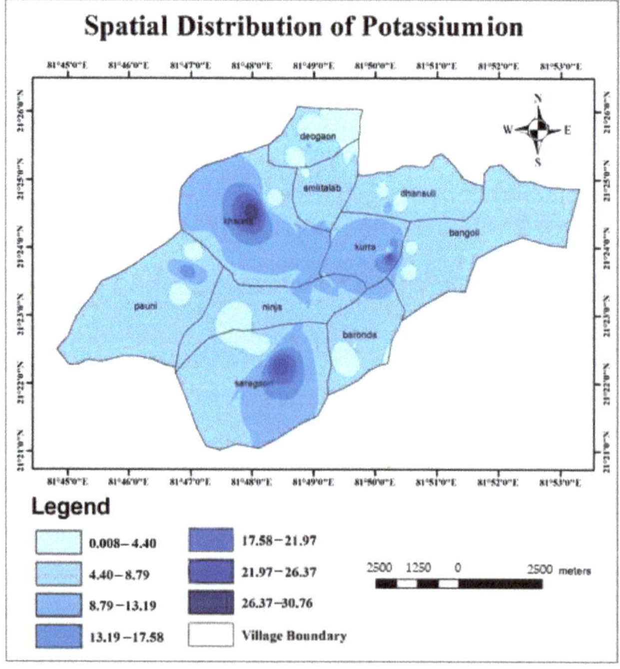

Figure A3. Spatial distribution of potassium.

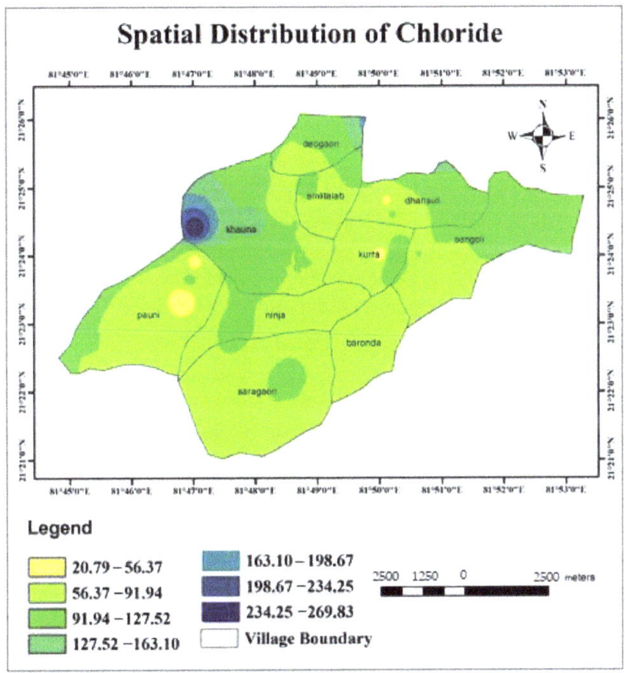

Figure A4. Spatial distribution of chloride.

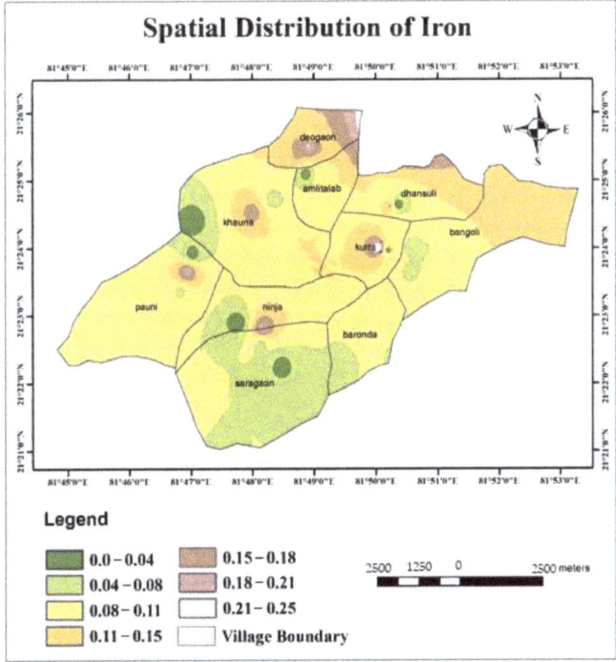

Figure A5. Spatial distribution of iron.

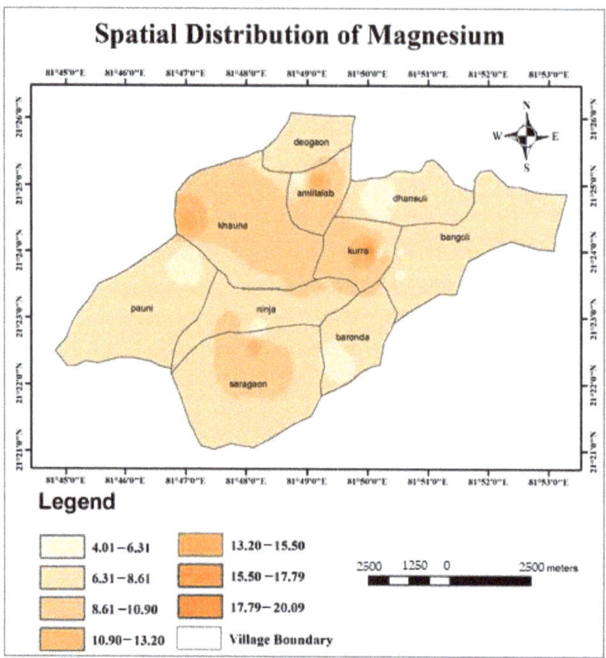

Figure A6. Spatial distribution of magnesium.

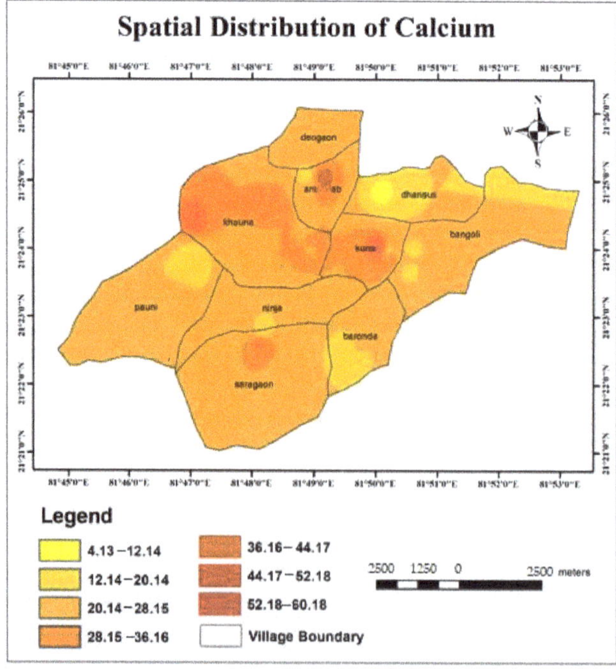

Figure A7. Spatial distribution of calcium.

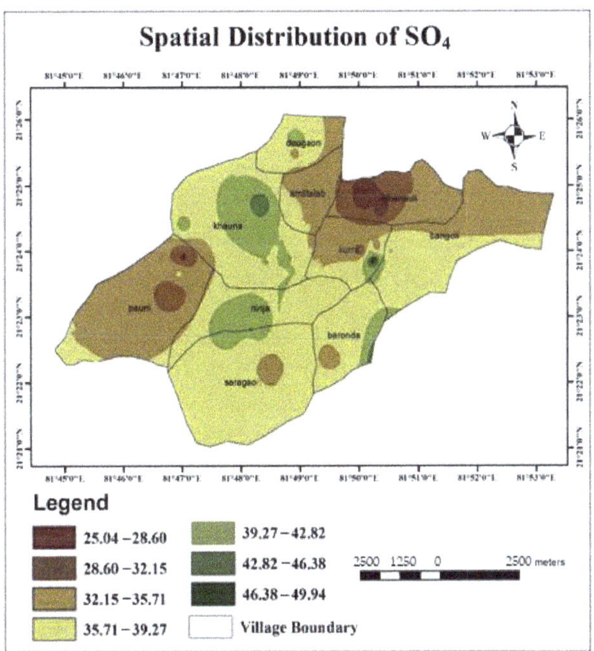

Figure A8. Spatial distribution of SO$_4$.

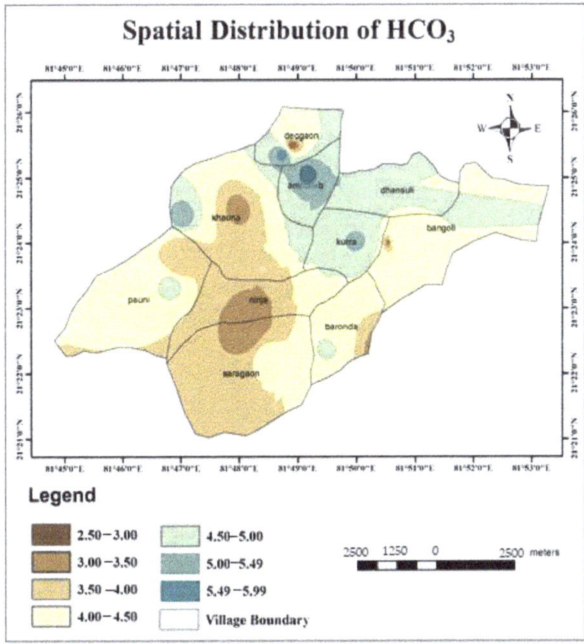

Figure A9. Spatial distribution of HCO$_3$.

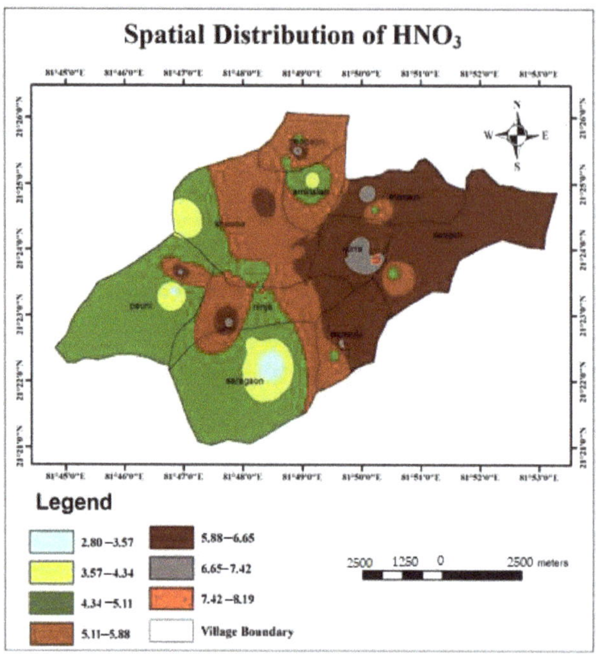

Figure A10. Spatial distribution of HNO$_3$.

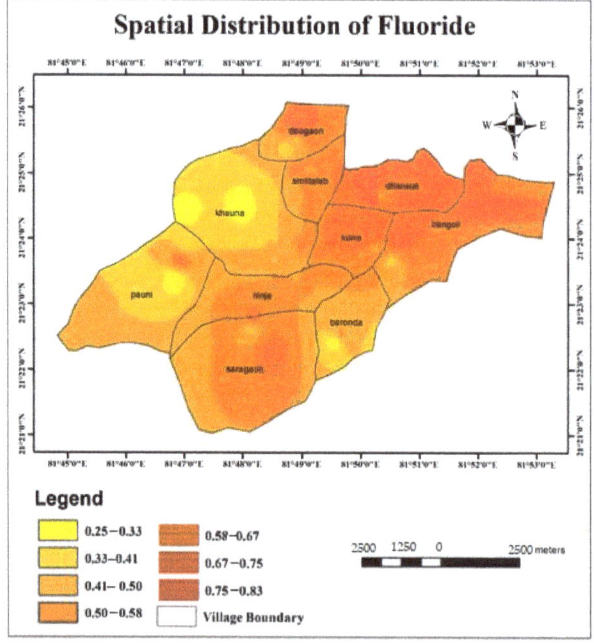

Figure A11. Spatial distribution of fluoride.

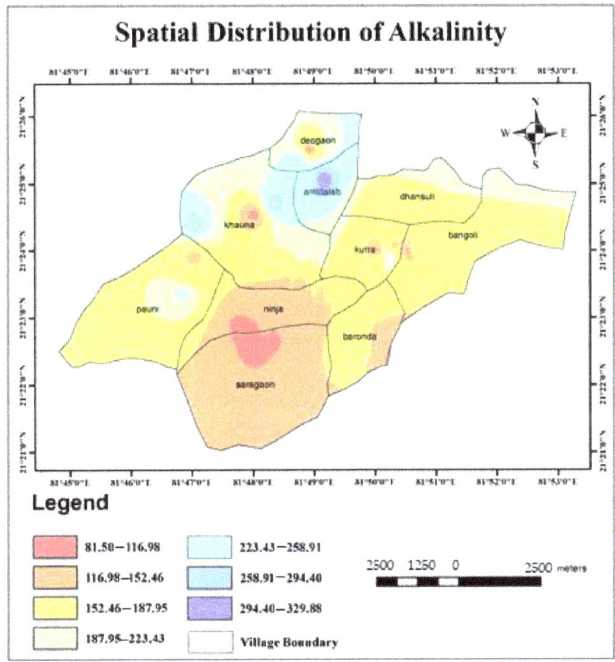

Figure A12. Spatial distribution of alkalinity.

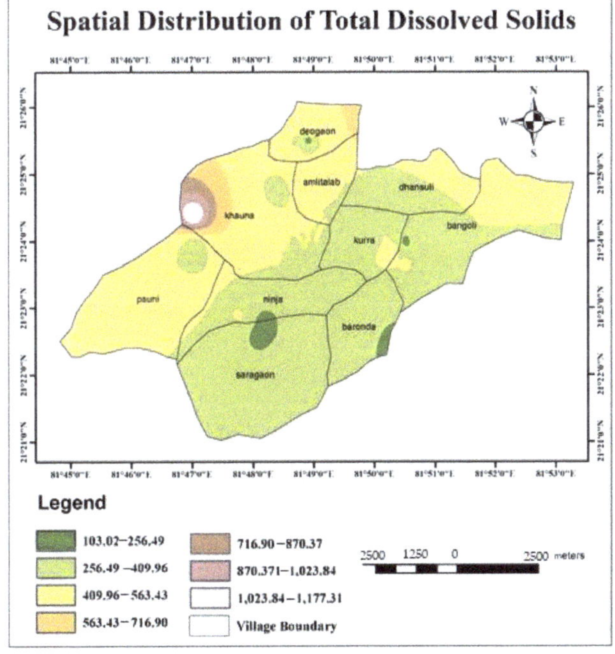

Figure A13. Spatial distribution of TDSs.

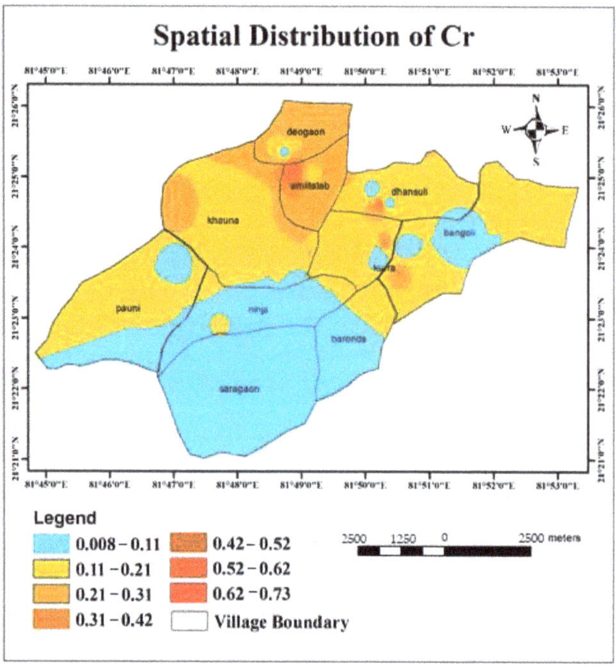

Figure A14. Spatial distribution of Cr.

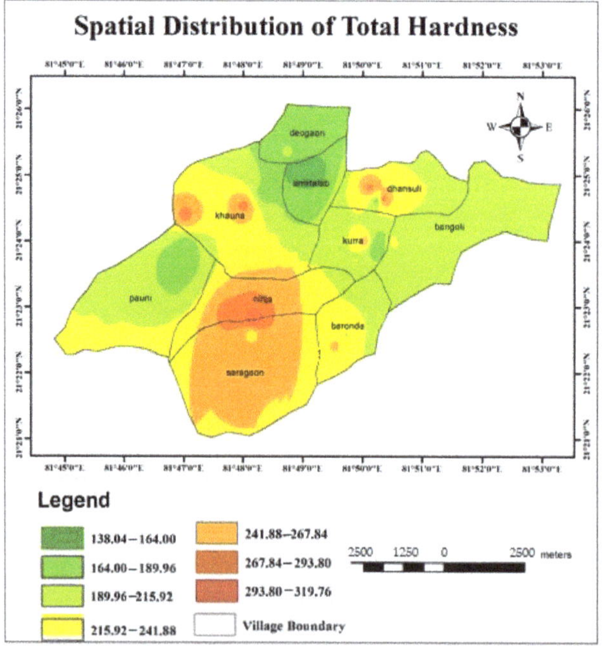

Figure A15. Spatial distribution of TH.

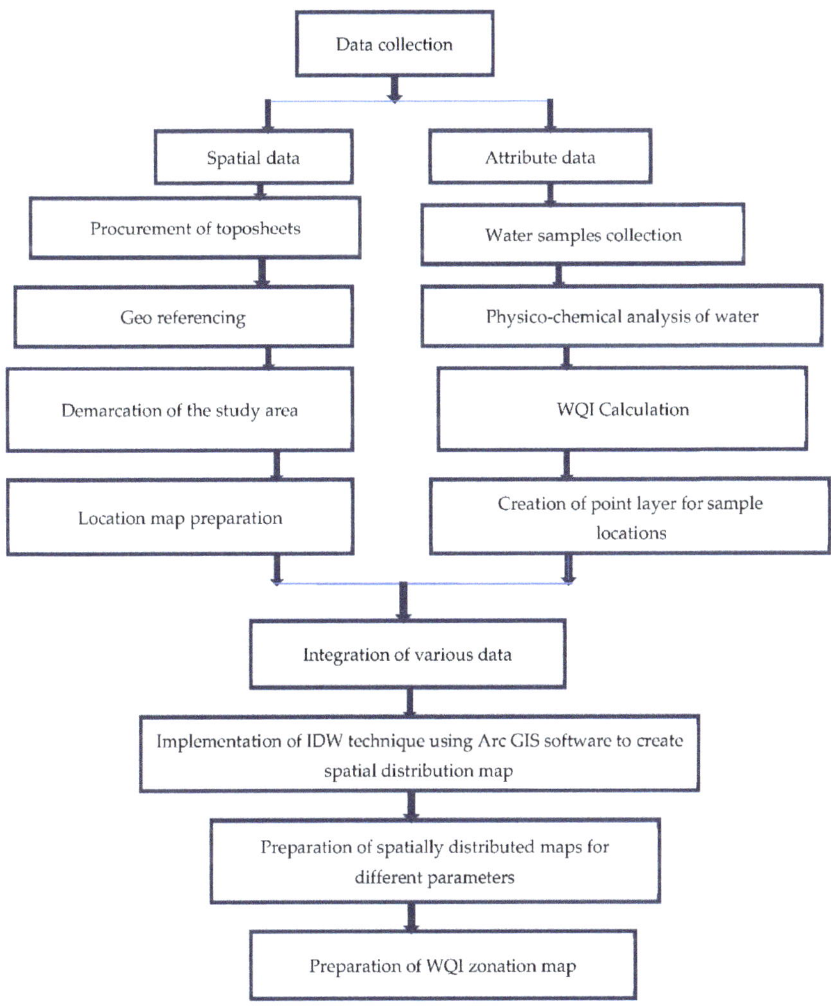

Figure A16. Flowchart of the procedure followed in the study.

References

1. Islam, S.; Rasul, T.; Bin Alam, J.; Haque, M.A. Evaluation of Water Quality of the Titas River Using NSF Water Quality Index. *J. Sci. Res.* **2010**, *3*, 151. [CrossRef]
2. Al-Zahrani, M.A.; Abo-Monasar, A. Urban Residential Water Demand Prediction Based on Artificial Neural Networks and Time Series Models. *Water Resour. Manag.* **2015**, *29*, 3651–3662. [CrossRef]
3. Elkiran, G.; Ergil, M. The assessment of a water budget of North Cyprus. *Build. Environ.* **2006**, *41*, 1671–1677. [CrossRef]
4. Shalby, A.; Elshemy, M.; Zeidan, B.A. Assessment of Climate Change Impacts on Water Quality Parameters of Lake Burullus, Egypt. *Environ. Sci. Poll. Res.* **2019**, *27*, 1–22. [CrossRef] [PubMed]
5. Kavitha, R.; Elangovan, K. Ground water quality characteristics at Erode district, Tamilnadu India. *Int. J. Environ. Sci.* **2010**, *1*, 163–175.
6. Sharma, D.; Kansal, A. Water quality analysis of River Yamuna using water quality index in the national capital territory, India (2000–2009). *Appl. Water Sci.* **2011**, *1*, 147–157. [CrossRef]
7. Smith, D.G. A better water quality indexing system for rivers and streams. *Water Res.* **1990**, *24*, 1237–1244. [CrossRef]
8. Kannel, P.R.; Lee, S.; Lee, Y.-S.; Kanel, S.R.; Khan, S.P. Application of Water Quality Indices and Dissolved Oxygen as Indicators for River Water Classification and Urban Impact Assessment. *Environ. Monit. Assess.* **2007**, *132*, 93–110. [CrossRef]

9.   Singh, R.P.; Nath, S.; Prasad, S.C.; Nema, A.K. Selection of Suitable Aggregation Function for Estimation of Aggregate Pollution Index for River Ganges in India. *J. Environ. Eng.* **2008**, *134*, 689–701. [CrossRef]

10.   Yadav, N.S.; Kumar, A.; Sharma, M. Ecological health assessment of Chambal River using water quality parameters. *J. Integr. Sci. Technol.* **2014**, *2*, 52–56.

11.   Agrawal, P.; Sinha, A.; Pasupuleti, S.; Nune, R.; Saha, S. Geospatial Analysis Coupled with Logarithmic Method for Water Quality Assessment in Part of Pindrawan Tank Command Area in Raipur District of Chhattisgarh. In *Climate Impacts on Water Resources in India*; Springer: Berlin, Germany, 2021; pp. 57–78.

12.   Hameed, M.; Sharqi, S.S.; Yaseen, Z.M.; Afan, H.A.; Hussain, A.; Elshafie, A. Application of artificial intelligence (AI) techniques in water quality index prediction: A case study in tropical region, Malaysia. *Neural Comput. Appl.* **2017**, *28*, 893–905. [CrossRef]

13.   Rufino, F.; Busico, G.; Cuoco, E.; Darrah, T.H.; Tedesco, D. Evaluating the Suitability of Urban Groundwater Resources for Drinking Water and Irrigation Purposes: An Integrated Approach in the Agro-Aversano Area of Southern Italy. *Environ. Monit. Assess.* **2019**, *191*, 1–17. [CrossRef] [PubMed]

14.   Vadiati, M.; Asghari-Moghaddam, A.; Nakhaei, M.; Adamowski, J.; Akbarzadeh, A. A fuzzy-logic based decision-making approach for identification of groundwater quality based on groundwater quality indices. *J. Environ. Manag.* **2016**, *184*, 255–270. [CrossRef] [PubMed]

15.   Bournaris, T.; Papathanasiou, J.; Manos, B.; Kazakis, N.; Voudouris, K. Support of irrigation water use and eco-friendly decision process in agricultural production planning. *Oper. Res.* **2015**, *15*, 289–306. [CrossRef]

16.   Sargaonkar, A.; Deshpande, V. Development of an Overall Index of Pollution for Surface Water Based on a General Classification Scheme in Indian Context. *Environ. Monit. Assess.* **2003**, *89*, 43–67. [CrossRef]

17.   Gazzaz, N.M.; Yusoff, M.K.; Aris, A.Z.; Juahir, H.; Ramli, M.F. Artificial neural network modeling of the water quality index for Kinta River (Malaysia) using water quality variables as predictors. *Mar. Pollut. Bull.* **2012**, *64*, 2409–2420. [CrossRef]

18.   Leong, W.C.; Bahadori, A.; Zhang, J.; Ahmad, Z. Prediction of water quality index (WQI) using support vector machine (SVM) and least square-support vector machine (LS-SVM). *Int. J. River Basin Manag.* **2019**, 1–8. [CrossRef]

19.   Iticescu, C.; Georgescu, L.P.; Murariu, G.; Topa, C.; Timofti, M.; Pintilie, V.; Arseni, M.; Timofti, M. Lower Danube Water Quality Quantified through WQI and Multivariate Analysis. *Water* **2019**, *11*, 1305. [CrossRef]

20.   Yaseen, Z.M.; Ramal, M.M.; Diop, L.; Jaafar, O.; Demir, V.; Kisi, O. Hybrid adaptive neuro-fuzzu models for water quality index estimation. *Water Resour. Manag.* **2018**, *32*, 2227–2245. [CrossRef]

21.   Diamantopoulou, M.J.; Papamichail, D.M.; Antonopoulos, V.Z. The use of a Neural Network technique for the prediction of water quality parameters. *Oper. Res.* **2005**, *5*, 115–125. [CrossRef]

22.   Khalil, B.; Ouarda, T.; St-Hilaire, A. Estimation of water quality characteristics at ungauged sites using artificial neural networks and canonical correlation analysis. *J. Hydrol.* **2011**, *405*, 277–287. [CrossRef]

23.   Gupta, R.; Singh, A.N.; Singhal, A. Application of ANN for Water Quality Index. *Int. J. Mach. Learn. Comput.* **2019**, *9*, 688–693. [CrossRef]

24.   Isiyaka, H.A.; Mustapha, A.; Juahir, H.; Phil-Eze, P. Water quality modelling using artificial neural network and multivariate statistical techniques. *Model. Earth Syst. Environ.* **2019**, *5*, 583–593. [CrossRef]

25.   Nourani, V.; Elkiran, G.; Abdullahi, J. Multi-station artificial intelligence based ensemble modeling of reference evapotranspiration using pan evaporation measurements. *J. Hydrol.* **2019**, *577*, 123958. [CrossRef]

26.   Gaya, M.S.; Abba, S.I.; Abdu, A.M.; Tukur, A.I.; Saleh, M.A.; Esmaili, P.; Wahab, N.A. Estimation of Water Quality Index Using Artificial Intelligence Approaches and Multi-Linear Regression. *Int. J. Artif. Intell. ISSN* **2020**, *2252*, 8938. [CrossRef]

27.   Najah, A.; El-Shafie, A.H.; Karim, O.A. Performance of ANFIS versus MLP-NN dissolved oxygen prediction models in water quality monitoring. *Environ. Sci. Pollut. Res.* **2014**, *21*, 1658–1670. [CrossRef] [PubMed]

28.   Ahmed, A.N.; Othman, F.B.; Afan, H.A.; Ibrahim, R.K.; Fai, C.M.; Hossain, S.; Ehteram, M.; Elshafie, A. Machine learning methods for better water quality prediction. *J. Hydrol.* **2019**, *578*, 124084. [CrossRef]

29.   Karim, S.A.A.; Kamsani, N.F. Water Quality Index Using Fuzzy Regression. In *Water Quality Index Prediction Using Multiple Linear Fuzzy Regression Model*; Springer: New York, NY, USA, 2020; pp. 37–53.

30.   Nayak, J.G.; Patil, L.; Patki, V.K. Development of water quality index for Godavari River (India) based on fuzzy inference system. *Groundw. Sustain. Dev.* **2020**, *10*, 100350. [CrossRef]

31.   Yasin, M.I.; Karim, S.A.A. A New Fuzzy Weighted Multivariate Regression to Predict Water Quality Index at Perak Rivers. In *Optimization Based Model Using Fuzzy and Other Statistical Techniques towards Environmental Sustainability*; Springer: New York, NY, USA, 2020; pp. 1–27.

32.   Abobakr Yahya, A.S.; Ahmed, A.N.; Binti Othman, F.; Ibrahim, R.K.; Afan, H.A.; El-Shafie, A.; Fai, C.M.; Hossain, M.S.; Ehteram, M.; Elshafie, A. Water Quality Prediction Model Based Support Vector Machine Model for Ungauged River Catchment under Dual Scenarios. *Water* **2019**, *11*, 1231. [CrossRef]

33.   Abba, S.I.; Pham, Q.B.; Saini, G.; Linh, N.T.T.; Ahmed, A.N.; Mohajane, M.; Khaledian, M.; Abdulkadir, R.A.; Bach, Q.-V. Implementation of data intelligence models coupled with ensemble machine learning for prediction of water quality index. *Environ. Sci. Pollut. Res.* **2020**, *27*, 41524–41539. [CrossRef]

34.   Elkiran, G.; Nourani, V.; Abba, S. Multi-step ahead modelling of river water quality parameters using ensemble artificial intelligence-based approach. *J. Hydrol.* **2019**, *577*, 123962. [CrossRef]

35. BIS (Bureau of Indian Standard). *Indian Standard Drinking Water–Specification*; Second Revision; Bureau of Indian Standards (BIS): New Delhi, India, 2012.

36. Chaurasia, A.K.; Pandey, H.K.; Tiwari, S.K.; Prakash, R.; Pandey, P.; Ram, A. Groundwater Quality assessment using Water Quality Index (WQI) in parts of Varanasi District, Uttar Pradesh, India. *J. Geol. Soc. India* **2018**, *92*, 76–82. [CrossRef]

37. WHO. *Guidelines for Drinking Water, Recommendations*; World Health Organization (WHO): Geneva, Switzerland, 2012.

38. Yisa, J.; Jimoh, T. Analytical studies on water quality index of river Landzu. *Am. J. Appl. Sci.* **2010**, *7*, 453. [CrossRef]

39. Tyagi, S.; Singh, P.; Sharma, B.; Singh, R. Assessment of Water Quality for Drinking Purpose in District Pauri of Uttarakhand, India. *Appl. Ecol. Environ. Sci.* **2014**, *2*, 94–99. [CrossRef]

40. Akter, T.; Jhohura, F.T.; Akter, F.; Chowdhury, T.R.; Mistry, S.K.; Dey, D.; Barua, M.K.; Islam, A.; Rahman, M. Water Quality Index for measuring drinking water quality in rural Bangladesh: A cross-sectional study. *J. Health Popul. Nutr.* **2016**, *35*, 1–12. [CrossRef]

41. Ramakrishnaiah, C.R.; Sadashivaiah, C.; Ranganna, G. Assessment of Water Quality Index for the Groundwater in Tumkur Taluk, Karnataka State, India. *E-J. Chem.* **2009**, *6*, 523–530. [CrossRef]

42. Eberhart, R.; Kennedy, J. *A New Optimizer Using Particle Swarm Theory*; IEEE: Piscataway, NJ, USA, 1995; pp. 39–43.

43. Gilani, S.-O.; Sattarvand, J.; Hajihassani, M.; Abdullah, S.S. A stochastic particle swarm based model for long term production planning of open pit mines considering the geological uncertainty. *Resour. Policy* **2020**, *68*, 101738. [CrossRef]

44. Yasin, Q.; Sohail, G.M.; Ding, Y.; Ismail, A.; Du, Q. Estimation of Petrophysical Parameters from Seismic Inversion by Combining Particle Swarm Optimization and Multilayer Linear Calculator. *Nat. Resour. Res.* **2020**, *29*, 3291–3317. [CrossRef]

45. Mehrabi, M.; Pradhan, B.; Moayedi, H.; Alamri, A. Optimizing an Adaptive Neuro-Fuzzy Inference System for Spatial Prediction of Landslide Susceptibility Using Four State-of-the-art Metaheuristic Techniques. *Sensors* **2020**, *20*, 1723. [CrossRef] [PubMed]

46. Sun, D.; Wen, H.; Wang, D.; Xu, J. A random forest model of landslide susceptibility mapping based on hyperparameter optimization using Bayes algorithm. *Geomorphol.* **2020**, *362*, 107201. [CrossRef]

47. Gafar, A.A.; Khayat, M.E.; Ahmad, S.A.; Yasid, N.A.; Shukor, M.Y. Response Surface Methodology for the Optimization of Keratinase Production in Culture Medium Containing Feathers by *Bacillus* sp. UPM-AAG1. *Catalysts* **2020**, *10*, 848. [CrossRef]

48. Bui, Q.-T.; Nguyen, Q.-H.; Nguyen, X.L.; Pham, V.D.; Nguyen, H.D.; Pham, V.-M. Verification of novel integrations of swarm intelligence algorithms into deep learning neural network for flood susceptibility mapping. *J. Hydrol.* **2020**, *581*, 124379. [CrossRef]

49. Pourghasemi, H.R.; Razavi-Termeh, S.V.; Kariminejad, N.; Hong, H.; Chen, W. An assessment of metaheuristic approaches for flood assessment. *J. Hydrol.* **2020**, *582*, 124536. [CrossRef]

50. Roshanravan, B.; Aghajani, H.; Yousefi, M.; Kreuzer, O. Particle Swarm Optimization Algorithm for Neuro-Fuzzy Prospectivity Analysis Using Continuously Weighted Spatial Exploration Data. *Nat. Resour. Res.* **2018**, *28*, 309–325. [CrossRef]

51. Engelbrecht, A.P. *Computational Intelligence: An Introduction*, 2nd ed.; Jonh Wiley & Sons, Ltd.: Chichester, UK, 2007; Volume 1.

52. Clerc, M.; Kennedy, J. The particle swarm—Explosion, stability, and convergence in a multidimensional complex space. *IEEE Trans. Evol. Comput.* **2002**, *6*, 58–73. [CrossRef]

53. Ciarelli, P.M.; Krohling, R.A.; Oliveira, E. Particle swarm optimization applied to parameters learning of probabilistic neural networks for classification of economic activities. In *Particle Swarm Optimization*; InTech: Rijeka, Croatia, 2009; pp. 313–327.

54. Scholkopf, B.; Sung, K.-K.; Burges, C.J.C.; Girosi, F.; Niyogi, P.; Poggio, T.; Vapnik, V. Comparing support vector machines with Gaussian kernels to radial basis function classifiers. *IEEE Trans. Signal Process.* **1997**, *45*, 2758–2765. [CrossRef]

55. Dibike, Y.B.; Velickov, S.; Solomatine, D.; Abbott, M.B. Model Induction with Support Vector Machines: Introduction and Applications. *J. Comput. Civ. Eng.* **2001**, *15*, 208–216. [CrossRef]

56. Li, C.-H.; Lin, C.-T.; Kuo, B.-C.; Ho, H.-H. An Automatic Method for Selecting the Parameter of the Normalized Kernel Function to Support Vector Machines. In Proceedings of the 2010 International Conference on Technologies and Applications of Artificial Intelligence, Hsinchu, Taiwan, 18–20 November 2010; pp. 226–232.

57. Shukla, R.; Chakraborty, A.; Sachdeva, K.; Joshi, P.K. Agriculture in the western Himalayas—An asset turning into a liability. *Dev. Pract.* **2017**, *28*, 318–324. [CrossRef]

58. Shukla, R.; Sachdeva, K.; Joshi, P.K. Demystifying vulnerability assessment of agriculture communities in the Himalayas: A systematic review. *Nat. Hazards* **2017**, *91*, 409–429. [CrossRef]

Article

Socio-Hydrological Approach to Explore Groundwater–Human Wellbeing Nexus: Case Study from Sundarbans, India

Soham Halder [1], Pankaj Kumar [2,*], Kousik Das [1], Rajarshi Dasgupta [2] and Abhijit Mukherjee [1,3,4]

[1] School of Environmental Science and Engineering, Indian Institute of Technology Kharagpur, Kharagpur 721302, India; sohamhalder07@iitkgp.ac.in (S.H.); kousik.das89@iitkgp.ac.in (K.D.); abhijit@gg.iitkgp.ac.in (A.M.)
[2] Natural Resources and Ecosystem Services, Institute for Global Environmental Strategies, Hayama, Kanagawa 240-0115, Japan; dasgupta@iges.or.jp
[3] Department of Geology and Geophysics, Indian Institute of Technology Kharagpur, Kharagpur 721302, India
[4] Applied Policy Advisory to Hydrogeosciences Group, Indian Institute of Technology Kharagpur, Kharagpur 721302, India
*   Correspondence: kumar@iges.or.jp

Citation: Halder, S.; Kumar, P.; Das, K.; Dasgupta, R.; Mukherjee, A. Socio-Hydrological Approach to Explore Groundwater–Human Wellbeing Nexus: Case Study from Sundarbans, India. *Water* **2021**, *13*, 1635. https://doi.org/10.3390/w13121635

Academic Editor: Alexander Yakirevich

Received: 7 May 2021
Accepted: 8 June 2021
Published: 10 June 2021

Publisher's Note: MDPI stays neutral with regard to jurisdictional claims in published maps and institutional affiliations.

Abstract: Coastal regions are the residence of an enormously growing population. In spite of rich biodiversity, coastal ecosystems are extremely vulnerable due to hydroclimatic factors with probable impact on socio-economy. Since the last few decades, researchers and policymakers were attracted towards the existing water demand–resource relationship to predict its future trends and prioritize better water resource management options. Water Evaluation And Planning (WEAP) serves the wholesome purpose of modeling diverse aspects of decision analysis using water algorithm equations for proper planning of water resource management. In this study, future groundwater demand (domestic, agricultural, and livestock sector) in the fragile Sundarbans ecosystem was estimated considering different human population growth rates (high, low, and current) for 2011–2050. The results showed that the sustainability of coastal aquifer-dependent rural livelihood is expected to face great danger in the near future. The total groundwater demand is expected to rise by approximately 17% at the current growth rate in the study area to fulfill the domestic and agricultural requirement, while this value goes up to around 35% for a higher growth rate and around 4% for a lower growth rate. The impact of increasing groundwater demand was analyzed further to identify any socio-economic shifts in this region.

Keywords: groundwater demand; Sundarbans; agriculture; WEAP; vulnerability; sensitivity loop; water–human wellbeing nexus

1. Introduction

The history of humankind can be scripted with regard to the human and water interactions and interrelationship [1]. Water is an indispensable natural resource. However, sustainable management of water resources requires a proper understanding of interactions between human and hydrological parameters [2]. Almost 70% of global water (both surface and groundwater) is utilized in the agricultural sector, making it the largest consumer of water resources worldwide [3]. According to the water bulletin published by Circle of Blue [4], nearly 80% of Indian livelihoods directly or indirectly rely on groundwater. Furthermore, rapid population expansion entails intensification of the food manufacturing sector, demanding a constant water supply. Rockstrom et al. (2001) defined a country as water-stressed where per capita per year (pcpy) water availability is less than 1700 m^3 [5]. As stated in the Ministry of Jal Shakti report [6], the annual water availability per capita was 1545 m^3 in 2011, which may further reduce to 1486 m^3 in 2021, signifying India as a water-stressed country. According to the study conducted by Vorosmarty et al. [7], nearly 2.4 billion people, which is 1/3rd of the world population, are residing in water-stressed

121

countries, whereas this value might increase to 2/3rd by the end of 2025. Moreover, the impact of climate change is intensifying over South Asian countries with adverse effects on the agricultural sector, especially in coastal regions. Water resources, agriculture, human health status, and the overall ecosystem are in ever-increasing stress due to inconsistent rainfall, rapid temperature increases, and the rise in sea level along with frequent intense extreme climatic events such as cyclonic storms in coastal regions of India [8]. Global climate change and increased demand for economic expansion with population growth at full tilt have led to the disproportionate exploitation of water resources causing declining water availability per capita specifically in developing countries such as India [9].

In the recent past, deserving emphasis was given to the term "vulnerability" by the scientific community. Vulnerability is described as the sensitivity of the system to any stressor variable, which is also dependent on the state, rate, and duration of exposure to the variable as well as capability to adapt to the stressful situation [10]. The lower part of the Gangetic delta comprising Sundarbans, the largest mangrove forest in the world, is a well-recognized hotspot of climate change [11]. The entire Sundarbans region is extremely fragile due to the low-lying coastal plain, and in addition to that, excessive reliance on rain-fed agro-economy affects the remunerative security of farmers as a result of inconsistent rainfall driven by climatic change, which has further worsened rural livelihood [12,13]. The sustainability of livelihood in Sundarbans and associated coastal zones is facing great insecurity due to over-dependence on the agricultural sector [14]. Additionally, increasing soil salinity due to proximity to the Bay of Bengal and frequent cyclonic storms (such as Aila and super cyclone Amphan in the recent past) along with the increasing prevalence of harmful pests and disease because of elevated temperature and humid conditions further deteriorates agricultural output [15]. Saline river water and brackish subsurface groundwater have led the residents of the Sundarbans region to depend on fresh groundwater abstracted from the confined aquifer at 160 to 300 m below ground level. The drinking water requirement of nearly 4.5 million people residing in this region is fulfilled by deep groundwater [16], whereas merely 32% of total households receive water from the piped groundwater facility [17], intensifying the hardship of local people to obtain one of the most important natural resources—water.

Socio-hydrology or hydro-sociology is a broad domain that combines both socio-economic aspects and environmental features of hydrology, concentrating on basic scientific principles of interrelationship, feedback mechanisms consisting of two separate loops of socio-economy and community sensitivity loops, and evolving human behavior. The sensitivity loop, a decisive component, depends solely on the community behavior and water management decisions which are driven by a community's social and environmental values and local action translating into direct or indirect impacts on any marginal change in water variables. Naturally, the sensitivity loop determines how behavioral response will impact on future available water quality as well as quantity [2]. Various tools are available to assess water resource management along with coupling with socio-economic parameters. Some of the widely used models are mentioned below. The Spatial Agro Hydro Salinity Model or SAHYSMOD (developed by International Institute for Land Reclamation and Improvement, Wageningen, Netherlands) is used to incorporate hydrological and physical consequences along with social and economic factors to understand the issues and approach to the sustainable development of any basin [18]. The ModSim, also named as Modular Simulator (developed by Colorado State University, Colorado, USA), is widely used as a decision support system (DSS) for both short and long-term planning in any river basin to develop advanced policy-making for better water allocation strategy [19]. Among many, Water Evaluation And Planning (WEAP) (developed by Stockholm Environment Institute's U.S. Center, Massachusetts) modeling has emerged as one of the best platforms in many catchments to understand and implicate water management strategies under varying climatic and socio-economic scenarios [20,21]. A major advantage of using WEAP lies in the simulation of the water situation in recent times, quantitative evaluation of water, and based on that, proper management scenarios for water demand and supply problems.

There are some previously set objects along with procedures provided by WEAP to solve any management-related problem of any stream, reservoir, watershed, or canal by applying a scenario-based approach [22]. Previously, the WEAP model has been successfully used to supervise the transboundary water resource-related issues associated with political conflicts in the Jordon River basin [23]. Ospina et al. (2011) [24] analyzed the adaptive strategies against increasing climate change to understand an effective supply–demand scenario in the Sinu-Caribe River basin, located in Columbia. In India, one major case study (Polavaram project) was conducted by Bharati et al. (2008) [25] to estimate the water availability–demand conflict from Godavari River to Krishna River. Nevertheless, there are no such studies on the hydrological impact on the socio-economy in the highly vulnerable Sundarbans region, where major risk factors are increasing groundwater salinity and depleting groundwater storage [26].

This study aimed to identify the basic drivers of the "sensitivity loop", the most important feedback loop in socio-hydrology [2], to understand the human behavior and management policies under a hydrological context in the Sundarbans region. This is one of the first studies to estimate the groundwater demand and its impact on socio-economic parameters with the application of the WEAP model in the Sundarbans region. Under this model, each scenario describes compatible plots of how a system might evolve in the future (within a particular period) under the specific socio-economic condition. Scenario analysis acts as an effective tool for developing proper policy amidst uncertainties [27,28].

More precisely, the research objectives are:

i. Analyzing the current socio-economic status (shift in occupation) over the decade under a hydrological context for the study area;
ii. Predicting the sector-wise (household, agricultural and livestock) groundwater demand for different growth scenarios in 2050;
iii. Identifying the way forward to achieve sustainable water resource management and human wellbeing.

The integrated hydro-socioeconomic approach is applied to provide scientific evidence to guide policy-makers in formulating better water management practices. Throughout the entire manuscript, the water should be regarded as groundwater and the agricultural crop is specified as paddy or rice.

2. Study Area

2.1. Description of the Study Area

Combined depositional activities of the world's largest river network, namely the Ganges–Brahmaputra–Meghna (GBM), shaped the Sundarbans delta, covering an area of nearly one million hectares, shared among India (38%) and Bangladesh (62%) [29]. Geographically, the Indian Sundarbans region lies within 21°30′ to 22°40′ N latitude and 88°05′ to 89°55′ E longitude and is bounded by the Dampier and Hodges line from the northern side, the Hoogli River (a distributary of the river Ganges and a part of the Bhagirathi–Hoogli river channel) from the west, and Bangladesh from the east, while the Bay of Bengal demarcates the southern boundary. The Sundarbans deltaic complex is characterized by extensive sand flats, beaches, coastal dunes, and estuaries, while creeks, mudflats, and mangrove swamps are other important morphotypes [30,31]. The warm-humid climate is prevalent throughout the hydrological year with an average annual precipitation of around 1600 mm in the study area [26]. Lithologically, the study area is categorized into three zones: (i) upper shallow aquifer with a depth range up to 60 mbgl; (ii) the main aquifer (up to a depth of 140–150 mbgl), which is characterized by semi-confined to locally unconfined comprising medium to coarse grained sand with alternate layers of gravel and varying in thickness up to 75 m in the southern part; and (iii) the aquifer beyond 150 m is regarded as a deeper aquifer which is separated from the main aquifer by a clay layer which consists of medium to coarse sands with inter-beds of clay and silt [32–34]. The Indian Sundarbans Biosphere Reserve comprises around 9630 sq. km., among which 5367 sq. km. of forest area is cleared for habitation extending over 19 community development or C.D. blocks

(13 blocks from South 24 Parganas and 6 from North 24 Parganas) [31,35]. The present study was conducted specifically in thirteen blocks of South 24 Parganas, which are further divided into three zones, i.e., Zone I (Patharpratima, Kakdwip, Namkhana, and Sagar), Zone II (Canning I, Canning II, Basanti, Gosaba), and Zone III (Jaynagar I, Jaynagar II, Mathurapur I, Mathurapur II, Kultali) (Figure 1). The southern region (Zone I in this study) is famous for its tourist attraction, which is mainly divided into three types, viz., religious tourism, beach tourism, and wildlife tourism. Sagar Island is well known as a Hindu pilgrimage site, attracting millions annually. Beach tourism is mostly prevalent in the southern beaches of the Namkhana community development block. Although they are elusive, tigers are the major attraction of wildlife tourism [36]. According to Mitra et al. (2009) [37], water quality in the Sundarbans region has been deteriorating over the last few decades due to the multifarious impact of climate change and anthropogenic activities. Furthermore, saline intrusion is evident from ionic ratios and salinity content of groundwater [26].

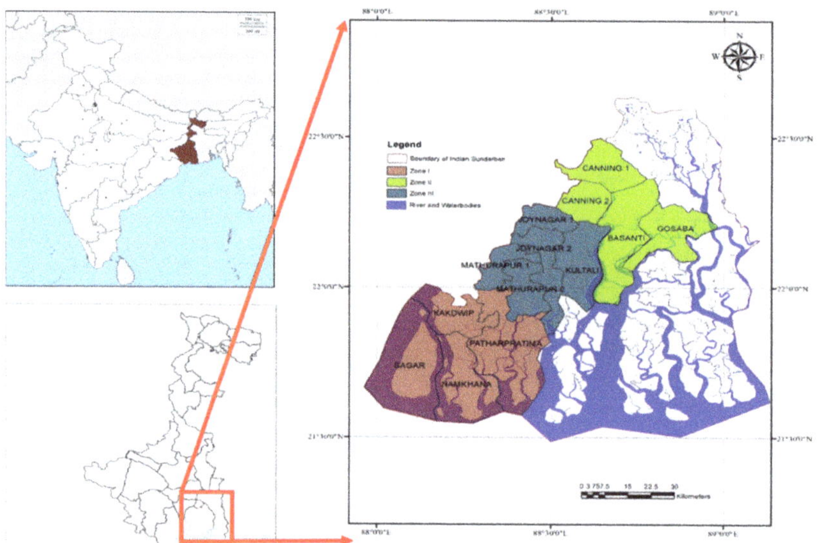

Figure 1. Location map of the study area, the Indian Sundarbans delta. The brown color signifies Zone I, the blue color designates Zone II and the green color represents Zone III.

2.2. Livelihood of Indian Sundarbans Region

Agriculture is the primary occupation of the inhabitants of Indian Sundarbans. Sánchez-Triana et al. (2014) [38] reported that among the total working population, 23.6% belongs to cultivators while 36.1% are agricultural laborers, summating around 60% of the total working population solely dependent on the agricultural sector. However, the indirect dependence on agriculture may well reach over 90%, which also supports the backbone of an agrarian economy. A large proportion of the inhabited area, measuring about 2648 sq. km, is dedicated to agriculture [39]. Due to inefficient irrigation amenities in non-monsoonal months, the mono-crop rainfed paddy (Aman rice) is cultivated during the rainy seasons (July–September). Farming sites are dominated in marginal areas. The second-largest occupation is fisheries and aquaculture [38]. Extensive traditional fishing is reported in Frazergunj, Sagar, Bakkhali, and Kalisthan [40]. Apart from these, rural livelihood is largely dependent on mangrove forest-derived products such as honey and timber. Eco-tourism is also promoted in Indian Sundarbans, which serves both remunerative and conservative purposes [41].

3. Required Dataset

Block-level census data (human) were obtained from the Office of the Registrar General and Census Commissioner, India, Ministry of Home Affairs, Government of India, Census Digital Library [42]. The livestock population was retrieved from the statistical handbook of South 24 Parganas under the Department of Planning and Statistics, Govt. of West Bengal [43]. The daily per capita groundwater (both from the deep aquifer for drinking purposes and shallow aquifers for household works) requirement was set at 40 L (14.5 cubic meters annually per capita) for humans, while it was kept at 10 L (5 cubic meters annually per capita) for livestock [44]. The groundwater level data were retrieved from the Central Groundwater Board Annual Report [45]. The agricultural statistics were extracted from the official website of the Directorate of Economics and Statistics under the supervision of the Department of Agriculture, Cooperation and Farmers Welfare [46]. The inland and marine fisheries data were retrieved from the Handbook of Fisheries Statistics (many editions) [47]. Another socio-economic parameter (literacy rate) was retrieved from census data and the Department of Planning and Statistics (Govt. of West Bengal) [42]. The entire dataset in the reference scenario is provided in Table 1.

Table 1. List of parameters considered for this study.

Parameter	Value	Reference
Base year for modeling	2011	
End year for modeling	2050	
Human Population (cap)	3,309,526	Census India, 2011 [42]
Agricutural land irrigated by groundwater (Ha)	26,000	District statistical abstract [43]
Livestock Population (cap)	4,082,384	District statistical abstract [43]
Human population growth rate	1.5%	Census India, 2011 [42]
Livestock population growth rate	1%	District statistical abstract [43]
Agricultural worker (%)	76% (1991), 56% (2001), 54% (2011)	Census India, 2011 [42]
Non-Agricultural worker (%)	24% (1991), 44% (2001), 46% (2011)	Census India, 2011 [42]
Socioeconomic parameter (Literacy rate)	Described in Section 6.4	Census India, 2011 [42]
Hydroclimatic parameters (rainfall, groundwater level)	Described in Sections 6.2 and 6.3	Central Groundwater Board Annual Report [45]

4. Methodology and Model Set Up

The WEAP model was applied in this study due to its robustness and utility based on data availability as it is useful in performing both aggregated and disaggregated forms of water management analysis in multiple sites [28]. To estimate the groundwater demand and unmet demand which indicates the water shortage, the study area (three zones comprising 13 blocks) was divided into three demand sectors: domestic, agriculture, and livestock. The primary intention was to incorporate the management problems regarding the availability, consumption, and conservation of water resources for comprehensive policy development. The WEAP model was used because it implements a coordinated approach to simulate the present and future scenarios with the consequent optimization of water resources, ultimately aligning them into policy construction [28,48,49]. The water resource system is represented by both hydrological and water quality modules within the WEAP model. Further, these modules can be represented by various elements such as administrative boundaries, river networks, groundwater, reservoir withdrawals, ecosystem requirements, wastewater treatment facilities, etc., depending on our research objectives [28]. The methodology flowchart is provided in Figure 2.

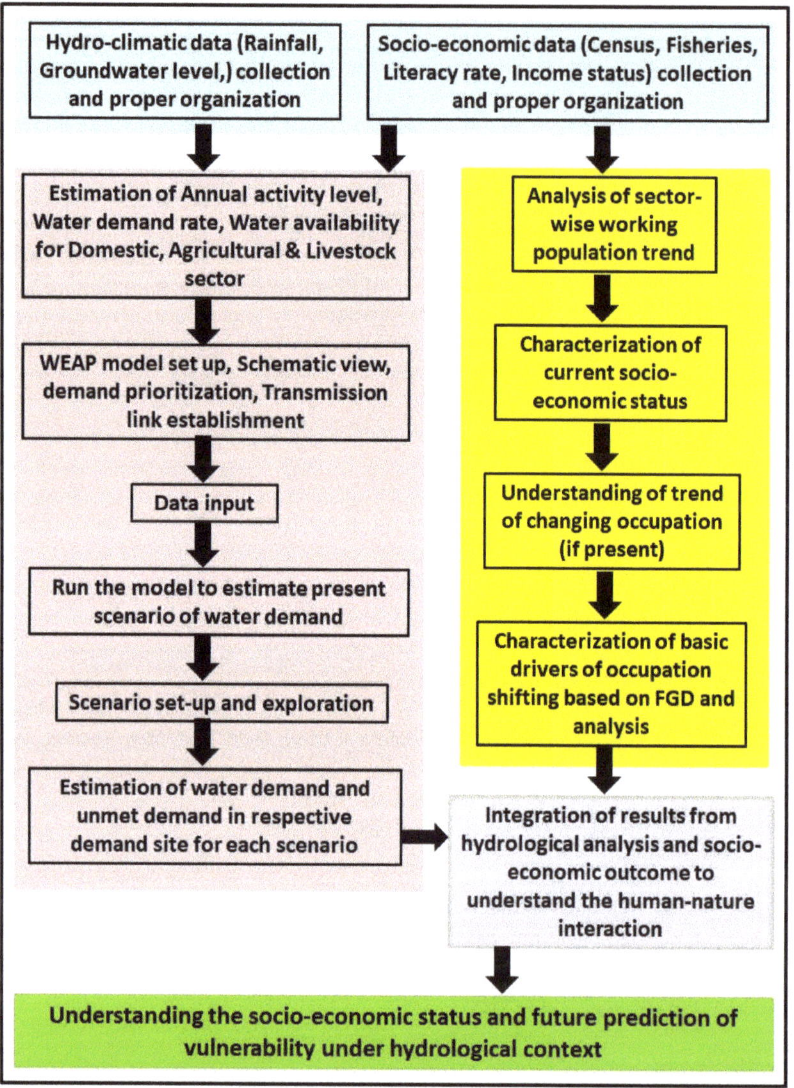

Figure 2. Flowchart of methodology (Blue color: Secondary data, Pink color: Hydrological data analysis in WEAP, Yellow color: Socio-economic parameter analysis, Grey color: Incorporation of results, Green color: Conclusion).

The year 2011 was considered as the base year. The simulation on the socio-economic scenario (e.g., human population growth, water consumption rate) continued until 2050, which was specified as the end year of modeling. Water quantity is calculated for each node and link on a regular (monthly or annually) interval in the selected system. The overall groundwater demand for domestic purposes is a function of human population size and the rate of utilization. Two fundamental parameters, annual activity level and annual water use rate, were defined for the demand sectors. The key assumptions made for these two parameters for the human population, agricultural status, and livestock population are given in Table 2.

Table 2. Table for initial parameters to set up model.

	Table for Initial Parameters to Set Up the Model								
		Domestic			Agriculture			Livestock	
	Zone I	Zone II	Zone III	Zone I	Zone II	Zone III	Zone I	Zone II	Zone III
Annual Activity	1,008,653 (no.)	1,140,562 (no)	1,160,311 (no.)	8000 (Ha)	5000 (Ha)	13,000 (Ha)	1,301,355 (no.)	1,622,139 (no.)	1,158,890 (no.)
Annual water use	14.5 (m^3)	14.5 (m^3)	14.5 (m^3)	10,000 (m^3)	10,000 (m^3)	10,000 (m^3)	5 (m^3)	5 (m^3)	5 (m^3)
Consumption	25%	25%	25%	80%	80%	80%	25%	25%	25%

According to the specified demand priorities, WEAP allocates the available water to satisfy the different demand sites [27,28]. In this study, the domestic and agricultural demand sites are given a priority value of one as these are the major groundwater consuming sites, whereas livestock water demand is described with a priority value of two for lesser consumption. The description of the hydrological model setup is provided in Figure 3. The annual activity level for domestic, agriculture, and livestock defines the total human population, land irrigated by groundwater, and livestock population, respectively, in 2011 for three zones. The domestic groundwater utilization in the study area was determined by five Focused Group Discussions (FGD) involving cultivators conducted during February, 2020 in Zone I of the study area [50]. The consumption rate or the amount of groundwater that is not returned to the system is estimated as 25%, 80%, and 25% for the respective annual activity levels [44,51,52]. The human and livestock population were presumed to increase over the years (2011–2050). The annual rural population growth rate was estimated from the previous census data, which was 1.44%, 1.93%, and 1.73%, respectively, for Zone I, II, and III. For the simulations, a minimum growth rate of 1% and maximum rate of 2.5% were determined (according to various growth simulation model outputs reported in macrotrends [53] and several print media [54]). This was further confirmed by the rural population growth rate according to World Bank data [55]. The livestock population increase rate (1%) was retrieved from Vikashpedia [56].

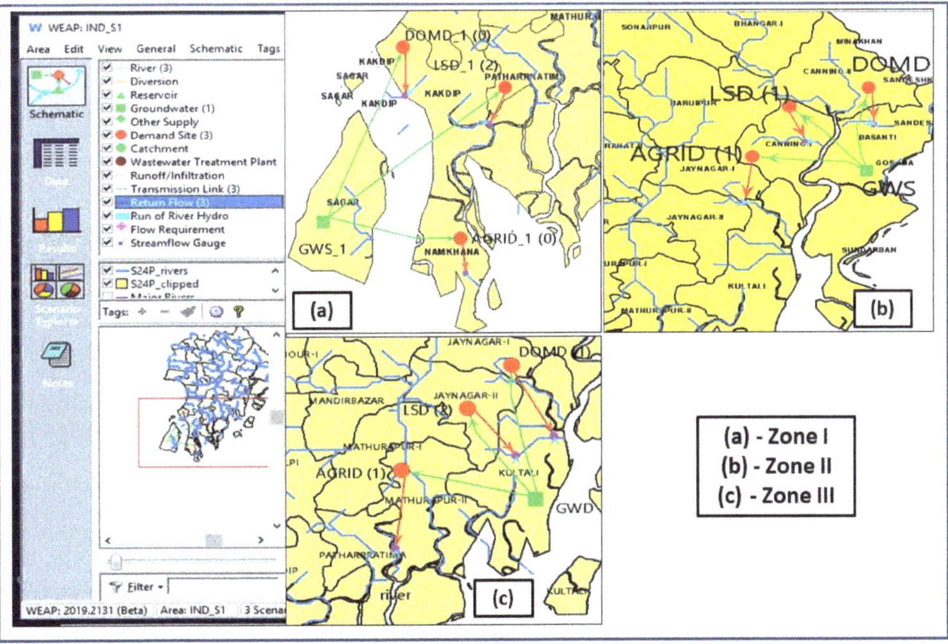

Figure 3. Hydrological simulation model in the study area (Sundarbans).

The groundwater use rate for agricultural purposes was assumed to decrease due to certain improvements in technologies and adaptation measures. The rainwater harvesting potential for the Sundarbans region was already estimated by Bhadra et al. (2018) [35]. Along with these data, the seasonal effective agricultural land area was estimated (District statistical abstract, many editions); further groundwater requirement for these areas was calculated [57]. Additionally, the potentially stored rainwater amount was subtracted from the total water demand. It was estimated that technological improvements for agriculture as well as awareness among communities in the Sundarbans region might cause a decrement in groundwater demand by 0.2 cubic meters with an interval of 5 years. The agricultural land area was considered constant throughout the study period as only the impact of varying human growth rates on water demand was studied. The overall activity was multiplied by the rate of water use to calculate the water consumption. The activity level was adopted for socio-economic demand analysis. The annual water use rate defines the mean water consumption rate per unit of activity. The total groundwater demand was calculated by multiplying the activity level and the rate of water consumption. It is worth mentioning that the annual water use rate does not indicate the total amount used, rather this represents the average annual water consumption per unit of activity.

$$D = A \times C \tag{1}$$

where: D = Groundwater demand at each node;
A = Activity level;
C = Water consumption rate at each node.

The reference scenario is described as the ongoing status of the system without any futuristic strategy and policy management. This is also helpful in distinguishing the demand sites and where more focus is to be given [28,58]. Prior to the simulations, the demand priorities were defined to specify the importance of groundwater requirement in-demand sites and assure that demands are met properly [28,48]. Therefore, the three scenarios selected for this study are:

(a) Higher human growth rate (2.5%), livestock growth rate 1%, and agricultural groundwater demand decrease by 0.2 cubic meters with an interval of 5 years.
(b) Current growth rate of respective zones, livestock growth rate 1%, and agricultural groundwater demand decrease by 0.2 cubic meters with an interval of 5 years.
(c) Lower human growth rate (1%), livestock growth rate 1%, and agricultural groundwater demand decrease by 0.2 cubic meters with an interval of 5 years.

5. Results

5.1. Decadal Trend of Census Data

According to the Indian Census, the working population is divided into two major groups, viz, main workers, who work a substantial portion (more than 180 days) of any year, and marginal workers, who work less than 180 days in a year. Furthermore, the main worker population is divided into four classes, namely, agricultural cultivators, agricultural laborers, household industry workers, and other workers (Indian Census, 2011). For simplicity, cultivators and agricultural laborers are considered as one group "Agriculture-dependent population". Block-wise census data for two decades (1991–2011) show a hike in marginal worker percentage compared to main workers. A steady decline in the total main worker population (26.09% in 1991 and 22.9% in 2011) was observed in the study area. The male main worker population in the study area has decreased from 47.63% to 34.85% during 1991–2011, while the opposite trend was observed for the female main worker population with an increasing trend (2.76% in 1991 and 5.44% in 2011 of total population). The male marginal worker population in the study area has increased from 1.56% to 18.72% during 1991–2011. A similar trend was observed for the female marginal worker population, which has shown an increasing pattern (6.09% in 1991 and 12.85% in 2011) (Figure 4a). These data are also significant as men are drifting away from main to marginal status due to mass outward migration of male workers to urban areas as well as different states or even overseas. Furthermore, the non-agriculture occupation is becoming an important sector to absorb the increasing labor force.

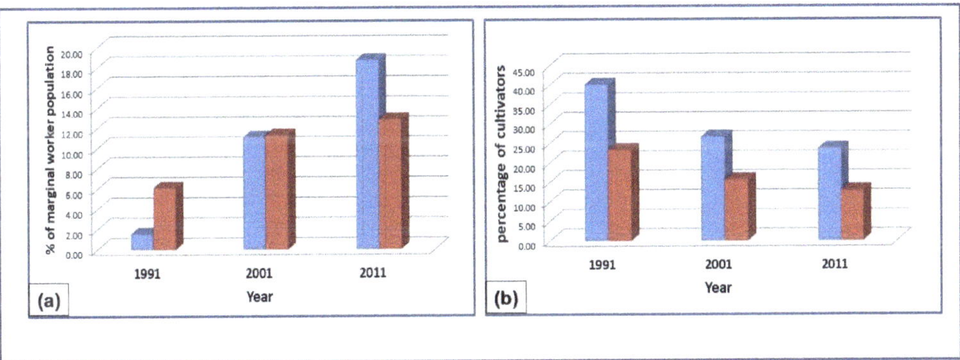

Figure 4. (**a**) Gender-wise percentage of marginal worker population and (**b**) the percentage of cultivator population in the study area (blue bar—male, brown bar—female).

The occupation preference of the total workforce is still inclined towards agriculture in the study area as 54% of workers are strongly dependent on it according to 2011 census reports. Another noteworthy observation is that the percentage of self-cultivators is declining sharply (39.61% in 1991 and 22.22% in 2011) along with a steady decrease in agriculture laborers (35.44% in 1991 and 32.34% in 2011) (Figure 4b). Since 1991, increasing participation (around 24% in 1991 to 46% in 2011) in the non-agricultural sector has been observed. Nevertheless, this high growth rate in non-agriculture has decelerated, and during the recent decade (2001–2011), the shift was recorded to be 2%, while the shift in the past decade (1991–2001) was 20%. As a result, the considerable change in the structure of the workforce towards non-agriculture observed in the 1990s was reduced in the 2000s. This is in contrast to the growth of GDP in India in its non-agriculture sector that had registered its highest ever during 2001–2011. This indicates a glitch in the structural change of the workforce.

Moreover, the study area is famous for its inland and marine fishery products [40]. Extensive inland fishing is reported in all the blocks of the study area (Figure 5a,b). The fisheries data from 1992 to 2018 have shown that there is an observable decrease in marine fish production [47]. Although various new techniques are available for marine fish capture nowadays, increased cyclones, death rates, and other associated geopolitical complications have forced the marine fishermen to shift into inland fisheries or any other occupation. Subsequently, the total marine fish production has decreased steadily from 20% to 10% over the above-mentioned period (Figure 5c).

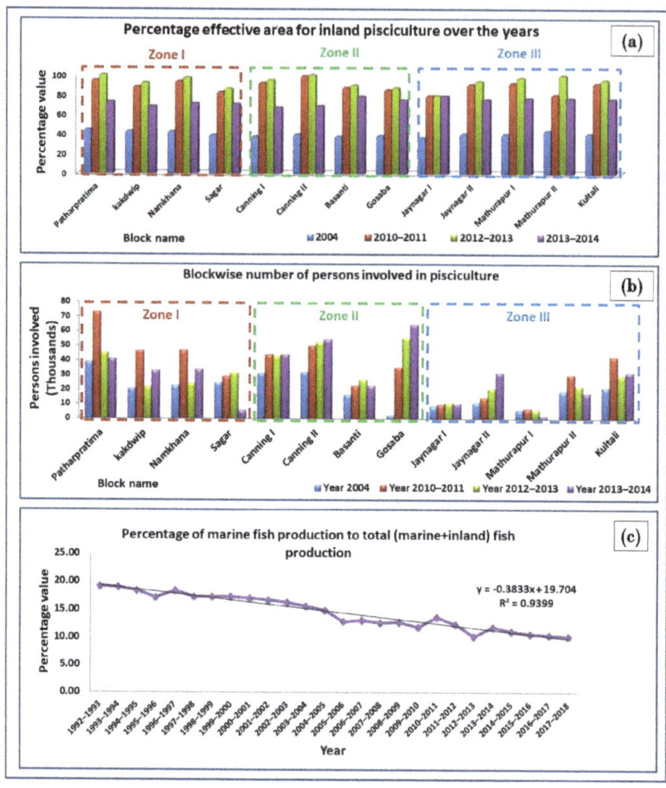

Figure 5. Statistics of fisheries. (**a**) Represents the effective area for inland fisheries over the years in the study area. (**b**) Represents the number of persons involved in inland fisheries. (**c**) Representation of contribution of marine fisheries to total fish production in the study area over the years (1992–2018).

5.2. Water Demand Derived from WEAP

5.2.1. Domestic Groundwater Demand

The total groundwater demand under different scenarios is shown in Figure 6. The domestic water demand for humans is expected to increase with population growth as groundwater demand for domestic purposes is a function of human population size and the rate of utilization. The water requirement under three scenarios for three zones is shown in Figure 7. According to the reference scenario, the domestic groundwater demand (annually) at the start of modeling (2011) was estimated as 14.62, 16.54 and 16.82 million cubic meters (mcm) for class I, II and III, respectively. According to the current growth rate in respective areas, the water demand for the domestic purpose for those respective zones is calculated as 26.13, 35.80, and 68.24 cubic meters in 2050. When the scenarios are considered, at a higher population growth rate (2.5%), the groundwater demand in 2050 showed an expected hike up to 31.65, 43.32, and 124.72 cubic meters in respective zones. At a lower growth rate (1%), the water demand decreased to 21.55, 24.38, and 34.08 mcm in respective areas in 2050. The higher domestic groundwater demand in Zone III is due to more population than the other two zones. The sector-wise groundwater demand is provided in Table 3.

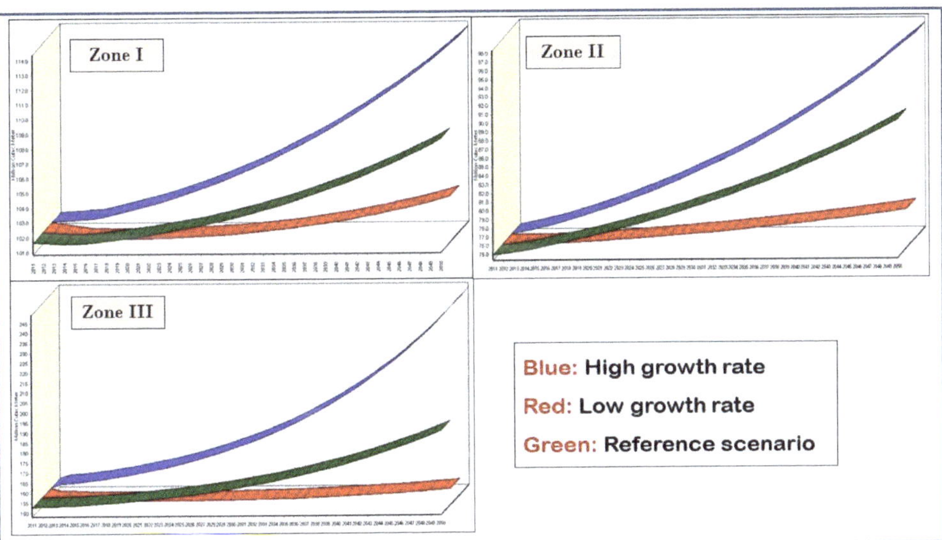

Figure 6. Total groundwater demand under different scenarios in the study area.

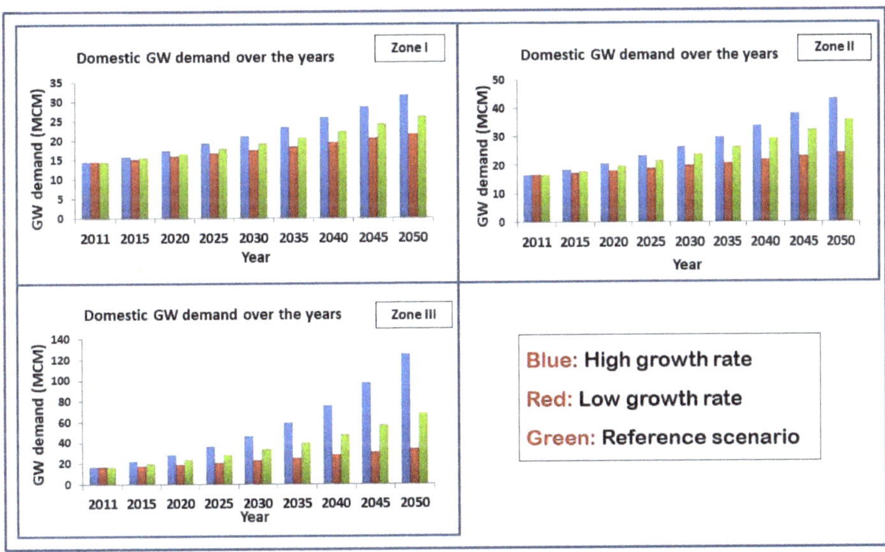

Figure 7. Domestic groundwater demand under different scenarios.

Table 3. Table of sector-wise groundwater demand (in mcm) in 2050 (HGR—High growth rate, LGR—Low growth rate).

	Zone 1			Zone 2			Zone 3		
	Reference	HGR	LGR	Reference	HGR	LGR	Reference	HGR	LGR
Domestic	26.13	31.65	21.55	35.8	43.32	24.38	68.24	124.72	34.08
agriculture	67.2	67.2	67.2	42	42	42	109.2	109.2	109.2
Livestock	15.25	15.25	10.38	12.44	12.44	12.44	13.1	13.1	13.1
Total	108.58	114.1	99.13	90.24	97.76	78.82	190.54	247.02	156.38

5.2.2. Agricultural Groundwater Demand

Sundarbans, a prominent agro-economic region, contributes over 60% of the district's (South 24 Parganas) annual rice production (calculated from district statistical abstract). Due to the non-availability of data for areas irrigated by groundwater only, the total canal irrigated area was subtracted from the total agricultural area for respective zones. The study on future trends in groundwater abstraction for irrigation aims to understand the relationship between evolving agricultural demand and the consequent groundwater withdrawal rate. The groundwater demand for respective zones in the base year (2011) was 80, 50 and 120.6 mcm. It was considered that the water consumption per hectare would decrease yearly due to technological improvements (e.g., rainwater harvesting and artificial groundwater recharge) and irrigational efficiency. Based on this scenario, the rate of groundwater use was estimated to decrease by 0.2 cubic meters every 5 years. With this assumption, the groundwater demand is likely to decline up to 67.2, 42, 109.2 mcm in 2050 for the respective zones (Figure 8).

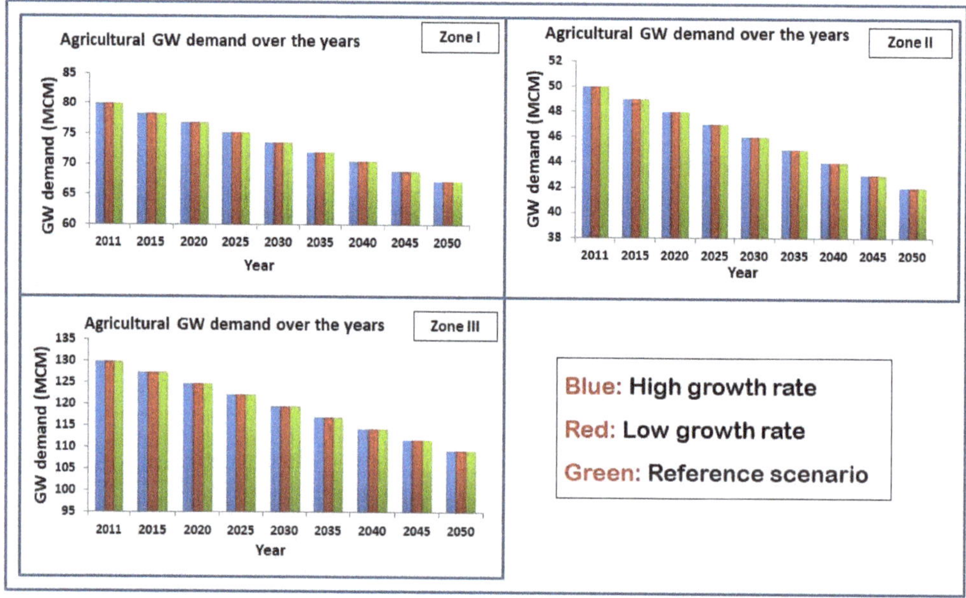

Figure 8. Agricultural groundwater demand under different scenarios.

5.2.3. Livestock Groundwater Demand

Cattle, buffalo, goat and poultry birds are the major livestock animals in the study area. Cattle and buffalos are reared because of milk and meat while the poultry farms are reared for meat only. Cattles and buffalos are the major water consuming livestock followed by poultry in the study area. The water consumption rate for poultry farms was retrieved from The Poultry Site [59] and Water Consumption in Broiler [60]. The groundwater requirements for dairy animals are only attributed to drinking purposes, whereas pond water is used for washing. The livestock water demand is estimated to double from 7.04 mcm (in 2011) to 15.25 mcm in Zone I in 2050, while a steady increase in demand is observed in Zone II (8.44 mcm in 2011 to 12.44 mcm in 2050). In Zone III, the livestock water demand is also estimated to double from 6.03 mcm in 2011 to 13.1 mcm in 2050 (Figure 9). This anomaly is due to the presence of more cattle and buffalo in Zone I than the other two regions, while the steady increase in water demand in Zone II is due to more poultry farms contributing to lesser water demand.

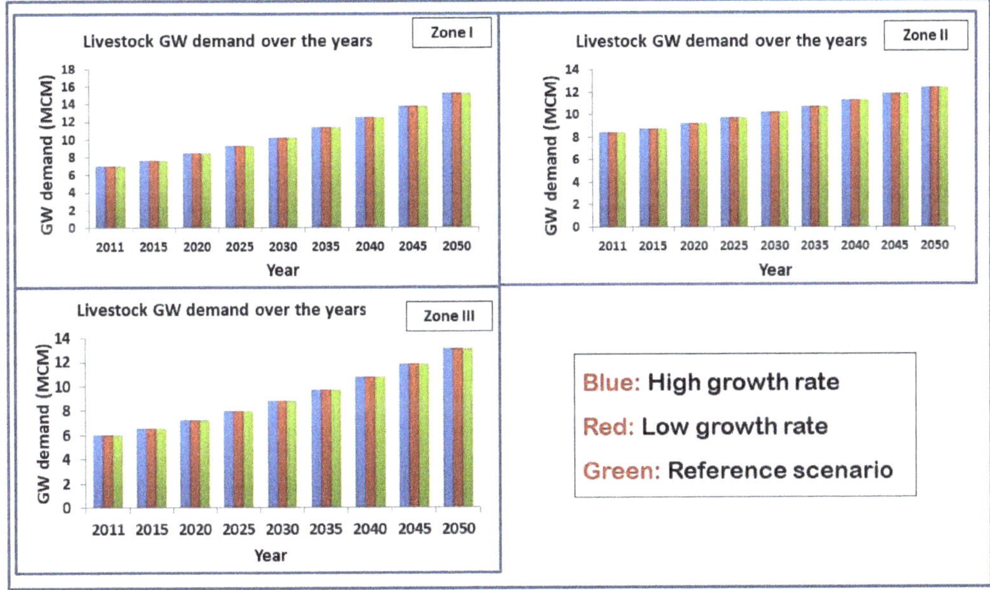

Figure 9. Livestock groundwater demand under different scenarios.

5.3. Unmet Groundwater Demand Derived from WEAP

The unmet demand represents the gap between the allocated water and the demand in each node, or simply the shortage of water. It was calculated by subtracting the WEAP allocated water amount from the demand values in each node. The estimated value of groundwater flow in deep aquifers is 18.25 mcm annually in the study area [35]. The groundwater abstracted from shallow tube wells was estimated to be around 24 mcm (calculated from the water abstraction rate and a number of shallow tube wells installed as mentioned in district statistical abstract and CGWB annual reports). The WEAP model was capable of simulation and the results indicated strong seasonal changes in required groundwater utilization over the years. The sectoral division of unmet demand is compiled in Table 4 and represented in Figure 10. Based on unmet demand values, it is clearly indicated that water stress-related issues (described in the next section) are going to rise at any growth rate scenario in Zone III due to higher agricultural and domestic demand.

Table 4. Table of sector-wise groundwater unmet demand (in mcm) in 2050.

	Zone 1			Zone 2			Zone 3		
	Reference	HGR	LGR	Reference	HGR	LGR	Reference	HGR	LGR
Domestic	12.14	21.45	10.55	15.8	21.45	12.8	18.5	30.7	16.46
agriculture	34.2	37.6	32.44	24.45	26.65	21.38	69.52	73.16	66.33
Livestock	7.65	8.45	6.84	5.25	5.82	4.97	4.65	4.65	4.65
Total	53.99	67.5	49.83	45.5	53.92	39.15	92.67	108.51	87.44

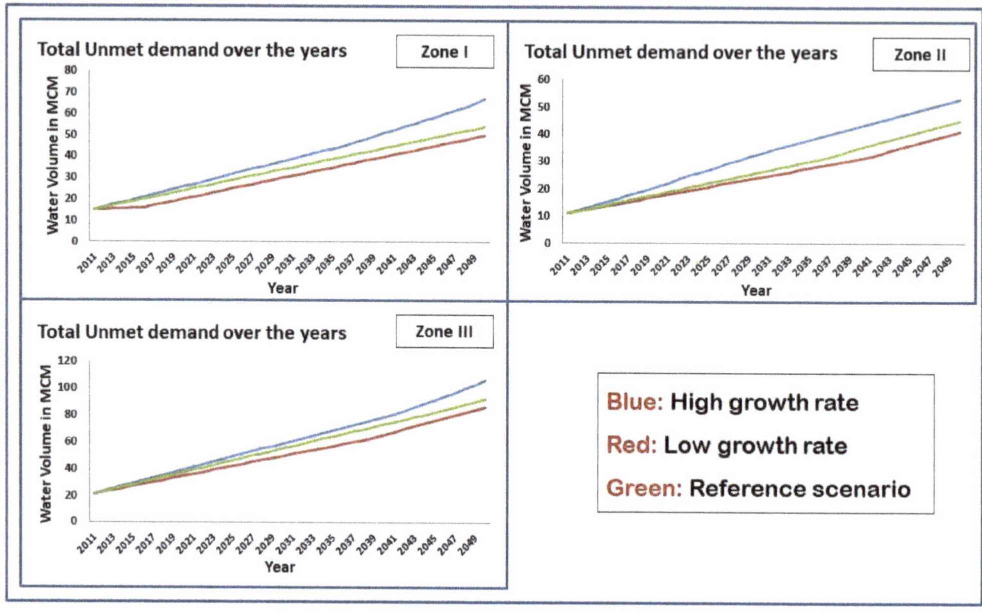

Figure 10. Zone-wise unmet demand under different scenarios.

6. Discussion

6.1. Implications of Increasing Groundwater Demand

The future sectoral water demand and unmet demand in three separate zones gave an idea about the groundwater storage vulnerability. Based on unmet water demand, the zones can be segregated into three major criteria: more vulnerable (Zone III), medium vulnerable (Zone I), and less vulnerable (Zone II). Zone III is comprised of five community development blocks, with a decadal human population growth rate of 1.71% along with a larger land area irrigated by groundwater, which is translated into higher groundwater demand. The decadal human growth rate (1.44%) is much lesser in Zone I, but being the major tourist attraction, the water demand is expected to rise at a greater rate with a subsequent depletion in groundwater level. The groundwater-dependent agricultural land area is much lesser in Zone II compared to the other two zones. This is significant as, despite a higher decadal population growth rate (1.93%), the overall demand stayed at the lower level. Along with population size, an increasing trend of population density (573 in 1991 to 819 person/km^2 in 2011) [42] is observed in the study area, impacting on the groundwater abstraction rate significantly. From the monthly demand, it was confirmed that the groundwater demand rises in the pre-monsoon months (March–June, when rainfall is not sufficient) during the sowing of Aman paddy, and again it rises in November during the sowing of Boro paddy, indicating a strong negative correlation with monthly rainwater availability (Figure 11). In other words, groundwater demand increases when rainfall is lower and demand decreases when rainfall is adequate. During FGD, it was confirmed that in any extreme climatic event (cyclone), hydrological changes can trigger serious repercussions on economic output and associated mental stress among communities. This leads to severe consequences such as restlessness, anxiety, and frequent domestic violence and suicides too [50].

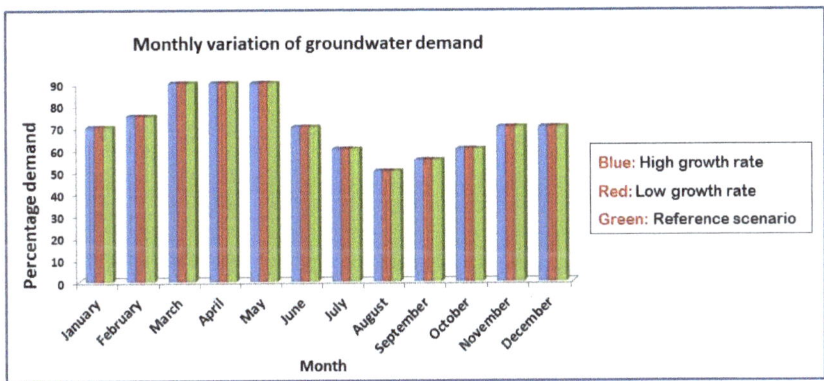

Figure 11. The monthly variation of groundwater demand in the study area.

6.2. Relationship between Agricultural Activities and Climate

Climatic parameters (rainfall) play a prominent role in agricultural activities in the study area. The monthly variation of rainfall amount in the study area is depicted in Figure 12a. This rainfall pattern is significant as agricultural groundwater demand increases during non-monsoonal months. Any variability in rainfall might cause a significant impact on crop yield. Decreasing rainfall [26] prompts the farmers to extract more groundwater. This was confirmed by FGD in the study area. Furthermore, the reach of groundwater-based irrigation is not available to all farmers, thus, it greatly acts as a limiting factor of economic growth. Increasing population density [42] and decreasing rainfall and groundwater recharge rate [26] might be important reasons behind the declining groundwater level in Figure 12b. This leads to the extraction of water from deeper wells, which increases the electricity cost as well as the total cost of production. According to Mandal et al. (2015) [15], the Sundarbans coastal region has been experiencing a delay in the onset of monsoon and late recession, while sometimes-heavy rainfall was observed during the harvest period of the monsoonal crop. The depletion of the groundwater level causes a higher rate of seawater intrusion, resulting in the qualitative deterioration and quantitative depletion of groundwater resources [26,50]. Additionally, the changes in hydrological parameters have a significant aftermath due to the increased rate of pest attacks with a consequent decrease in crop productivity [61].

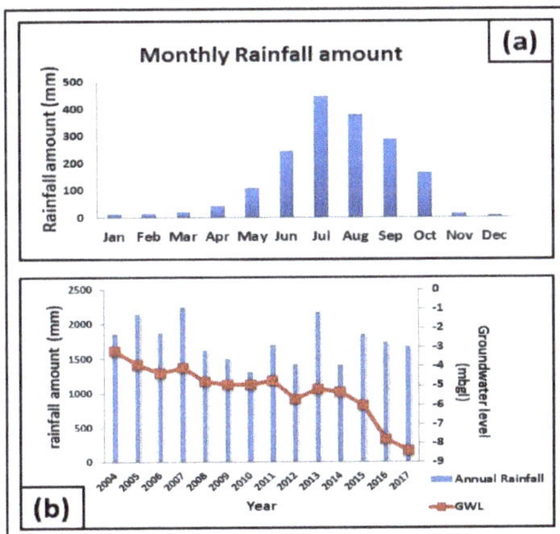

Figure 12. (**a**) Monthly rainfall amount in the study area, (**b**) Relationship between annual rainfall amount and changing groundwater level over the years (2004–2017) in the study area.

6.3. Status of the Agricultural Economy

For the last couple of decades, the contribution of the agricultural sector to GDP has been decreasing significantly in India (from over 50% during the 1950s to 15.4% in the year 2015–2016), while an increasing contribution from the industrial (manufacturing) and service sectors was observed (constant prices considered) [62]. According to the reports of PRS Legislative Research [62], the major influencing factors of agricultural productivity are over-dependence on monsoonal rainfall, landholdings, deteriorating soil nutrients due to improper use of fertilizers, deprivation from formal credit, and the disproportionate use of technology, etc. Compared to other rice-producing states, the lowest growth rate (−1.3%) was observed in West Bengal, as mentioned in the NSSO 2015 report [63]. Almost 95% of farming households who cultivate <4 hectares were incapable of fulfilling their basic consumption need [64]. These statistics are highly significant as nearly 84% of total farmland is cultivated by small and marginal farmers in West Bengal. Additionally, West Bengal belongs to those states where the average monthly income of farmers lies below 5000 rupees [64].

When plotting the groundwater level data (GWL) and the cost of cultivation in the study area, a significant inverse relationship was found. As the GWL started declining, the cost of cultivation started increasing (Figure 13a). This may be due to electricity consumption during agricultural water abstraction. The relationship between groundwater and electricity is straightforward, as electricity is consumed for groundwater abstraction from aquifers. Unlike other states, the farmers of West Bengal do not receive free electricity for agriculture or tariff subsidies [65], and as a result, the cost of production increases along with the depleting groundwater level. This trend is also significant as the cost of cultivation has shown an inverse relationship with the rainfall amount in 2013–2018 time periods (Figure 13b). As such, these data imply that when rainfall is lower, there is a higher groundwater demand in agriculture, which leads to excessive electricity costs and an associated increase in the overall cost of paddy production (Figure 13c).

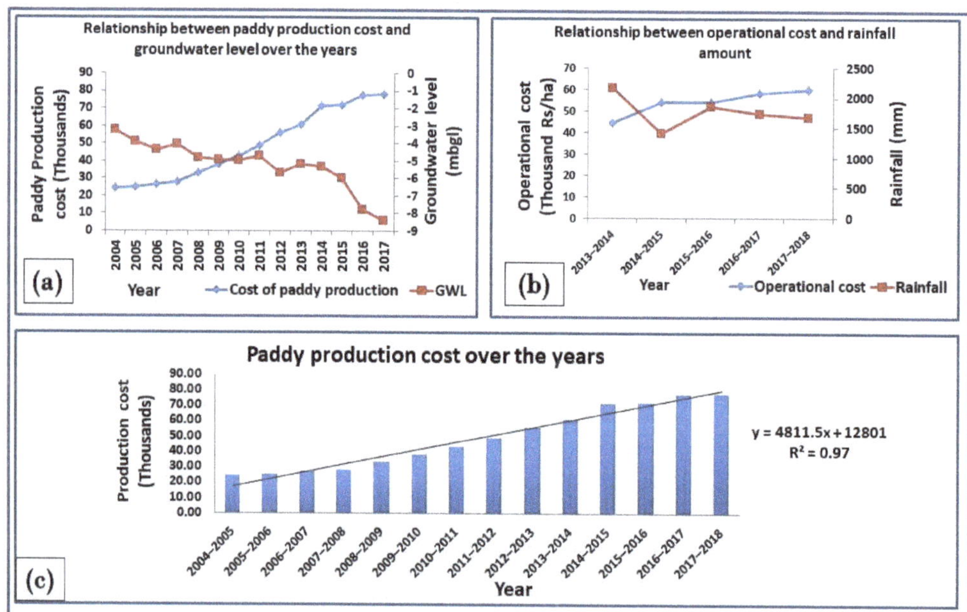

Figure 13. (**a**) Relationship between groundwater level (GWL) and cost of paddy production (2004–2017), (**b**) Relationship between annual rainfall and operational cost of paddy production (2013–2018), (**c**) Overall cost of paddy production over the years.

6.4. Impact of Literacy Rate on Economic Wellbeing

Literacy is one of the key factors of the economic prosperity of any region (country/state/district). Therefore, increasing the literacy rate is suggestive of better employment opportunities [66]. According to the Indian Census, literacy is defined as the capability of reading, writing, and interpreting in any language [42]. It is worth mentioning that, along with the described literacy definition, it is important to have certain technical skills to survive in the modern labor market. This will eventually decrease occupational passivity in the workforce with an improving rate of employability. Therefore, a certain level of educational qualification and technical knowledge helps in the attainment of better salary jobs in India [67]. The literacy rate in the study area showed an upsurging trend; 39%, 54%, and 75% in 1991, 2001, and 2011, respectively. An increasing literacy rate is associated with a shift in occupation from agriculture (lower wage rate) to other industrial sectors (higher wage rate). The daily wage rate for unskilled agricultural laborers is 260 rupees (without food), whereas any unskilled industrial (bakery, automobile manufacturing and repairing, biscuit manufacturing, etc.) worker gets around 338 rupees (without food) daily, as mentioned in a recent report published by Labour Commissionerate, Labour Department, Government of West Bengal [68]. As the economic benefit from agriculture is very low in West Bengal (with respect to other states), the occupation shift will be more prominent in the coming years. The relationship between the decadal literacy rate and trend in occupation shifting is provided in Figure 14.

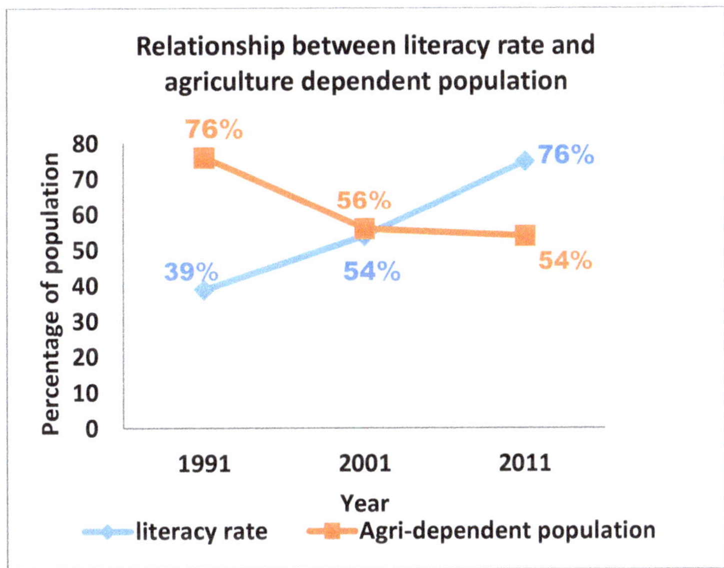

Figure 14. Relationship between literacy rate (percentage value) and agriculture-dependent population.

6.5. Overall Socio-Economic Status under the Hydrological Context

The increasing salinity content of surface water has compelled the people in Sundarbans to depend solely on groundwater as a drinking water source [26]. Apart from surface water salinization, several constraints in agricultural groundwater utilization have led to economic hardship and the associated migration of local people [50]. According to Bhanja et al. (2017) [69], unsustainable abstraction of groundwater and anthropogenic impacts are the consequences of the inaccurate implementation of management policies. Unproductive agricultural lands are transformed into brackish water aquaculture, possibly leading to increased salinization of shallow aquifers [26,70]. Therefore, improper knowledge about the hydrology of this area and a lack of convenient awareness among communities have resulted in the over-abstraction of deep groundwater, ultimately leading to the disruption of groundwater dynamics [26].

From the above analysis, it was observed that there is conspicuous shifting in professional dependency. Occupational shifting is a broad term that is controlled by multifarious factors starting from the resource availability, ecosystem vulnerability, and socio-economic status of the area [71]. Sundarbans is exposed to various highly impactful climatic events such as cyclones [72,73] with an additional higher rate of coastal aquifer salinization [26,34], leading to the serious vulnerability of the ecosystem. One significant outcome of the analysis is changing occupational trends as the portion of cultivators is declining with an increase in agricultural laborers in the workforce. This suggests that self-cultivators are quitting farming activities and becoming laborers in the agriculture field. Additionally, a trend in occupation shifting was also observed in the fisheries-dependent population, where inland fisheries are blooming at a considerable rate with a slight decrease in marine fish production. The prevalence of inland aquaculture practice was also reported in Dubey et al. [70]. Even though marine fish production can be affected by various other factors (quality of seawater, climatic change, etc.), the rate of overall production is significant as less yield may lead to a shift in occupation. Although more than half of the total working people are dependent on agriculture (Figure 15), a higher rate of conversion is expected in the coming years if the situation does not improve significantly. The principal drivers of hydrological parameters and associated socio-economic status in the study area are described in Figure 16.

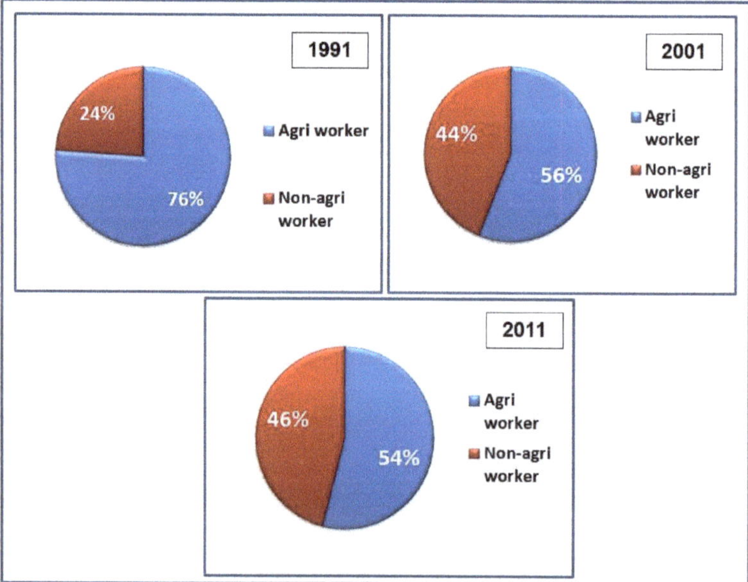

Figure 15. Overall trend of occupation shifting in the study area.

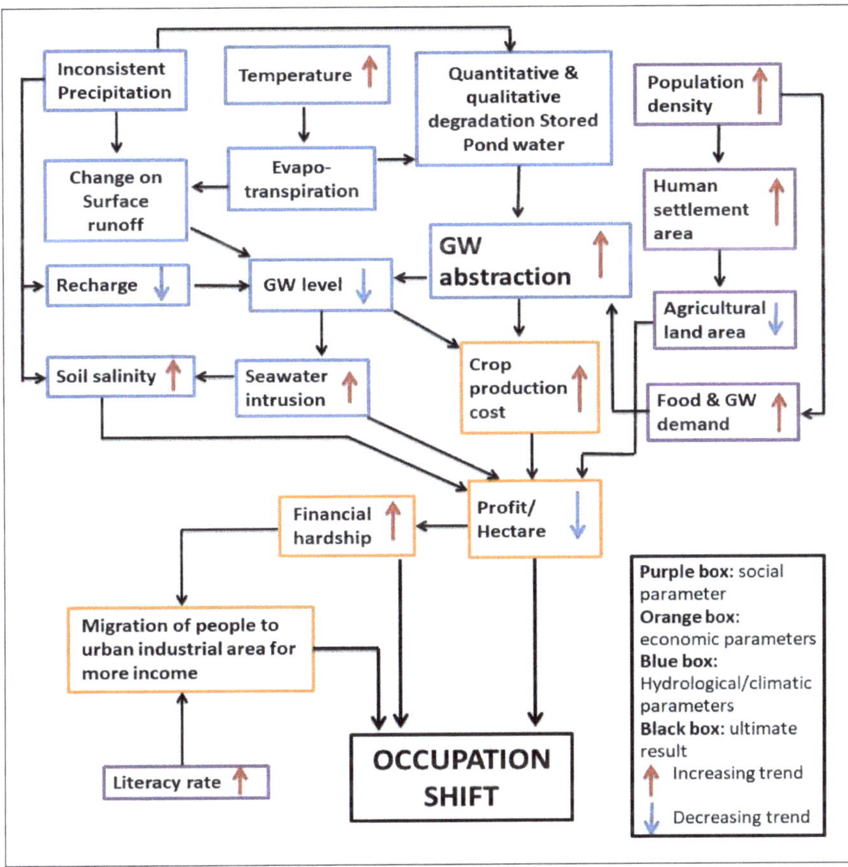

Figure 16. The drivers of socio-economic trend/status under hydroclimatic context in the study area.

6.6. Limitations and Way Forward

Previously, no data were published on the contribution of groundwater and agricultural growth in the Sundarbans region. Due to this, the net area irrigated by groundwater (only) was calculated by subtracting the total canal/pond irrigated area from the total agricultural area. Moreover, the non-availability of data prompted us to use various estimated parameters from printed media and non-governmental association data. The increasing pattern of groundwater is reflected through human population growth rate only, and a further increment due to urbanization is not considered in this study due to the lack of data. Additionally, the complex hydrostratigraphy of the study area was not considered during modeling as the zone-wise groundwater demand and unmet demand were only studied with respect to varying population growth rate. Although database limitations are there, substantial analysis has shown the importance of groundwater and associated climatic factors and social parameters on the overall economic status of the study area. There is always a gap between outcomes from scientific experiments and policy-makers' decisions. It is noteworthy to mention that innumerable feedback mechanisms exist between humans and the hydrological system. Combining all these factors is beyond the scope of this study. The presented model and findings solely portrayed a snapshot of future trends in water use and human response for defined scenarios only. This study identified separate zones based on water demand and unmet demand to guide policy-makers for the implementation of location-specific policies to accomplish an impactful water management strategy to maintain the socio-economic development of humans. This is one of the first studies to acknowledge the complex human–water interactions in the Sundarbans region and further studies are required to predict the overall social dynamics under the hydrological context.

7. Conclusions

This study addresses crucial hydrological queries regarding human–nature interaction along with socio-economic issues in a highly complex coastal ecosystem. Despite the history of Sundarbans narrating the transformation for mangrove forests into agricultural land, gradually, it is becoming uninhabitable owing to changing hydrological dynamics, including the rapid salinization and contamination of freshwater aquifers. With these irreversible impacts, crop productivity and aquaculture yield are further expected to decrease in the near future. Sundarbans is possibly the classic case where groundwater availability is a great limiting factor to socio-economic growth. The relationship between hydroclimatic factors and socio-economic vulnerability is prominent, and such a relationship is further magnified by low per capita income, uneven allocation of natural resources, an inefficient health care system, improper education, and several associated factors, ultimately leading to an inadequate adaptive response to stressors. As a result, the major occupation sectors (e.g., agriculture, fishery) are affected considerably, posing challenges to populations living there. The impact of rapid changes in hydroclimatic factors on natural resources can endanger the utilization of natural resources. Both qualitative and quantitative securities of natural resources are important for sustainable socio-economic wellbeing. The probable factors for occupation shifting are: firstly, the ever-increasing cost of irrigation with reducing profitability in paddy production. Secondly, being a tourism attraction, there is a rapid bloom in real estate with urbanization that ultimately leads to a decrease in land area for cultivation. Thirdly, climate change (inadequate and inconsistent rainfall, cyclones, storms and increasing surface water temperature) acts as proper slow-burn with ever-lasting impact to promote this shifting. These factors either individually or in combination act as the push factors. Therefore, the early estimation of vulnerability is crucial to mitigate the aftermath of stressors. Further analysis is required to estimate the groundwater exploitation with increasing demand and understand the impact of climate change. Additional studies are also required to design effective adaptation strategies, both for agriculture and other employment sectors.

Author Contributions: Conceptualization: S.H., P.K., A.M.; methodology: S.H., P.K.; formal analysis: S.H., P.K.; field investigation: S.H., P.K., K.D., R.D.; writing—original draft preparation: S.H., P.K.; writing—review and editing: S.H., P.K., K.D., R.D., A.M. All authors have read and agreed to the published version of the manuscript.

Funding: This work is supported by Asia Pacific Network for Global Change Research (APN) under Collaborative Regional Research Programme (CRRP) project with project reference number CRRP2019-01MY-Kumar.

Institutional Review Board Statement: Not applicable.

Informed Consent Statement: Not applicable.

Data Availability Statement: Not applicable.

Acknowledgments: The authors also appreciate the support from the people of Namkhana, Kakdwip and Patharpratima community development blocks during FGD.

Conflicts of Interest: The authors declare no conflict of interest.

References

1. Biswas, A.K. *History of Hydrology*; North-Holland Publishing Company: Amsterdam, The Netherlands, 1970.
2. Elshafei, Y.; Sivapalan, M.; Tonts, M.; Hipsey, M.R. A prototype framework for models of socio-hydrology: Identification of key feedback loops and parameterisation approach. *Hydrol. Earth Syst. Sci.* **2014**, *18*, 2141–2166. [CrossRef]
3. WWAP. *The United Nations World Water Development Report 2017: Wastewater, the Untapped Resource*; WWAP (United Nations World Water Assessment Programme); UNESCO: Paris, France, 2017.

4. Cicle of Blue Website. Available online: https://www.circleofblue.org/2018/world/groundwater-scarcity-pollution-set-india-on-perilous-course (accessed on 28 May 2021).

5. Rockström, J.; Gordon, L. Assessment of green water flows to sustain major biomes of the world: Implications for future ecohydrological landscape management. *Phys. Chem. Earth Part B Hydrol. Ocean. Atmos.* **2001**, *26*, 843–851. [CrossRef]

6. Ministry of Jal Shakti. Available online: https://pib.gov.in/PressReleasePage.aspx (accessed on 28 April 2021).

7. Vörösmarty, C.J.; McIntyre, P.B.; Gessner, M.O.; Dudgeon, D.; Prusevich, A.; Green, P.; Glidden, S.; Bunn, S.E.; Sullivan, C.A.; Liermann, C.R.; et al. Global threats to human water security and river biodiversity. *Nature* **2010**, *467*, 555–561. [CrossRef] [PubMed]

8. Wheeler, T.; Von Braun, J. Climate change impacts on global food security. *Science* **2013**, *341*, 508–513. [CrossRef]

9. Bakken, T.H.; Modahl, I.S.; Raadal, H.L.; Bustos, A.A.; Arnøy, S. Allocation of water consumption in multipurpose reservoirs. *Water Policy* **2016**, *18*, 932–947. [CrossRef]

10. Bryan, E.; Deressa, T.T.; Gbetibouo, G.A.; Ringler, C. Adaptation to climate change in Ethiopia and South Africa: Options and constraints. *Environ. Sci. Policy* **2009**, *12*, 413–426. [CrossRef]

11. Hazra, S.; Ghosh, T.; DasGupta, R.; Sen, G. Sea level and associated changes in the Sundarbanss. *Sci. Cult.* **2002**, *68*, 309–321.

12. Chowdhury, A.N.; Mondal, R.; Brahma, A.; Biswas, M.K. Culture and stigma: Ethnographic case studies of tiger-widows of Sundarbans Delta, India. *World Cult. Psychiatry Res. Rev.* **2014**, *9*, 99–122.

13. Mandal, S.; Choudhury, B.U.; Satpati, L.N. Monsoon variability, crop water requirement, and crop planning for kharif rice in Sagar Island, India. *Int. J. Biometeorol.* **2015**, *59*, 1891–1903. [CrossRef]

14. Solomon, S. IPCC (2007): Climate change the physical science basis. In *AGU Fall Meeting Abstracts*; Intergovernmental Panel on Climate Change (IPCC): Geneva, Switzerland, 2007; Volume 2007, p. U43D-01.

15. Jakariya, M.; Alam, M.S.; Rahman, M.A.; Ahmed, S.; Elahi, M.L.; Khan, A.M.S.; Saad, S.; Tamim, H.M.; Ishtiak, T.; Sayem, S.M.; et al. Assessing climate-induced agricultural vulnerable coastal communities of Bangladesh using machine learning techniques. *Sci. Total Environ.* **2020**, *742*, 140255. [CrossRef]

16. Bhadra, T.; Hazra, S.; Ray, S.S.; Barman, B.C. Assessing the groundwater quality of the coastal aquifers of a vulnerable delta: A case study of the Sundarbans Biosphere Reserve, India. *Groundw. Sustain. Dev.* **2020**, *11*, 100438. [CrossRef]

17. World Bank Publications. *Turn Down the Heat: Confronting the New Climate Normal*; World Bank Publications: Herndon, VA, USA, 2014.

18. Inam, A.; Adamowski, J.; Prasher, S.; Halbe, J.; Malard, J.; Albano, R. Coupling of a distributed stakeholder-built system dynamics socio-economic model with SAHYSMOD for sustainable soil salinity management. Part 2: Model coupling and application. *J. Hydrol.* **2017**, *551*, 278–299. [CrossRef]

19. Vaghefi, S.A.; Mousavi, S.J.; Abbaspour, K.C.; Srinivasan, R.; Arnold, J.R. Integration of hydrologic and water allocation models in basin-scale water resources management considering crop pattern and climate change: Karkheh River Basin in Iran. *Reg. Environ. Chang.* **2015**, *15*, 475–484. [CrossRef]

20. Yates, D.; Purkey, D.; Sieber, J.; Huber-Lee, A.; Galbraith, H.; West, J.; Herrod-Julius, S.; Young, C.; Joyce, B.; Rayej, M. Climate driven water resources model of the Sacramento Basin, California. *J. Water Resour. Plan. Manag.* **2009**, *135*, 303–313. [CrossRef]

21. Hum, N.N.M.F.; Talib, S.A. Modeling water supply and demand for effective water management allocation in Selangor. *J. Teknol.* **2016**, *78*. [CrossRef]

22. Al-Amin, S.; Berglund, E.Z.; Mahinthakumar, G.; Larson, K.L. Assessing the effects of water restrictions on socio-hydrologic resilience for shared groundwater systems. *J. Hydrol.* **2018**, *566*, 872–885. [CrossRef]

23. Hoff, H.; Bonzi, C.; Joyce, B.; Tielbörger, K. A water resources planning tool for the Jordan River Basin. *Water* **2011**, *3*, 718–736. [CrossRef]

24. Ospina-Noreña, J.E.; Gay-García, C.; Conde, A.C.; Sanchez-Torres Esqueda, G. Water availability as a limiting factor and optimization of hydropower generation as an adaptation strategy to climate change in the Sinú-Caribe river basin. *Atmósfera* **2011**, *24*, 203–220.

25. Bharati, L.; Anand, B.K.; Smakhtin, V. *Analysis of the Inter-Basin Water Transfer Scheme in India: A Case Study of the Godavari-Krishna link*; (No. 614-2016-40883); International Water Management Institute: New Delhi, India, 2008.

26. Das, K.; Mishra, A.K.; Singh, A.; Agrahari, S.; Chakrabarti, R.; Mukherjee, A. Solute exchanges between multi-depth groundwater and surface water of climatically vulnerable Gangetic delta front aquifers of Sundarbanss. *J. Environ. Manag.* **2021**, *284*, 112026. [CrossRef]

27. Vogel, R.M.; Sieber, J.; Archfield, S.A.; Smith, M.P.; Apse, C.D.; Huber-Lee, A. Relations among storage, yield, and instream flow. *Water Resour. Res.* **2007**, *43*. [CrossRef]

28. Yates, D.; Sieber, J.; Purkey, D.; Huber-Lee, A. WEAP21—A demand-, priority-, and preference-driven water planning model: Part 1: Model characteristics. *Water Int.* **2005**, *30*, 487–500. [CrossRef]

29. Sahana, M.; Rehman, S.; Sajjad, H.; Hong, H. Exploring effectiveness of frequency ratio and support vector machine models in storm surge flood susceptibility assessment: A study of Sundarbans Biosphere Reserve, India. *Catena* **2020**, *189*, 104450. [CrossRef]

30. Ghosh, A.; Mukhopadhyay, S. Quantitative study on shoreline changes and Erosion Hazard assessment: Case study in Muriganga–Saptamukhi interfluve, Sundarbans, India. *Modeling Earth Syst. Environ.* **2016**, *2*, 75. [CrossRef]

31. Chaudhuri, A.B.; Choudhury, A. *Mangroves of the Sundarbans. Volume 1: India;* International Union for Conservation of Nature and Natural Resources (IUCN): Gland, Switzerland, 1994.
32. Mukherjee, A.; Fryar, A.E.; Howell, P.D. Regional hydrostratigraphy and groundwater flow modeling in the arsenic-affected areas of the western Bengal basin, West Bengal. India. *Hydrogeol. J.* **2007**, *15*, 1397. [CrossRef]
33. Mukherjee, A.; Fryar, A.E.; Thomas, W.A. Geologic, geomorphic and hydrologic framework and evolution of the Bengal basin, India and Bangladesh. *J. Asian Earth Sci.* **2009**, *34*, 227–244. [CrossRef]
34. Das, K.; Mukherjee, A. Depth-dependent groundwater response to coastal hydrodynamics in the tropical, Ganges river mega-delta front (the Sundarbans): Impact of hydraulic connectivity on drinking water vulnerability. *J. Hydrol.* **2019**, *575*, 499–512. [CrossRef]
35. Bhadra, T.; Das, S.; Hazra, S.; Barman, B.C. Assessing the demand, availability and accessibility of potable water in Indian Sundarbans biosphere reserve area. *Int. J. Recent Sci. Res.* **2018**, *9*, 25437–25443.
36. Danda, A.A.; Rahman, M. Transformation of the Sundarbans Eco-region: Lessons from Past Approaches and Suggested Development Options. In *The Sundarbanss: A Disaster-Prone Eco-Region;* Springer: Cham, Switzerland, 2019; pp. 61–89.
37. Mitra, A.; Gangopadhyay, A.; Dube, A.; Schmidt, A.C.; Banerjee, K. Observed changes in water mass properties in the Indian Sundarbans (northwestern Bay of Bengal) during 1980–2007. *Curr. Sci.* **2009**, *97*, 1445–1452.
38. Sánchez-Triana, E.; Paul, T.; Ortolano, L.; Ruitenebeek, J. *Building Resilience for the Sustainable Development of the Sundarbanss;* Strategy Report No. 88061-1N; South Asia Region Sustainable Development Department Environment Water Resources Management Unit, The World Bank: Washington, DC, USA, 2014.
39. DasGupta, R.; Hashimoto, S.; Okuro, T.; Basu, M. Scenario-based land change modelling in the Indian Sundarbans delta: An exploratory analysis of plausible alternative regional futures. *Sustain. Sci.* **2019**, *14*, 221–240. [CrossRef]
40. Datta, D.; Chattopadhyay, R.N.; Deb, S. Prospective livelihood opportunities from the mangroves of the Sunderbans, India. *Res. J. Environ. Sci.* **2011**, *5*, 536.
41. Sen, H.S.; Ghorai, D. The Sundarbanss: A flight into the wilderness. In *The Sundarbanss: A Disaster-Prone Eco-Region;* Springer: Cham, Switzerland, 2019; pp. 3–28.
42. Ministry of Home Affairs, Government of India, Census Digital Library. Available online: https://censusindia.gov.in/DigitalLibrary/MapsCategory.aspx (accessed on 17 May 2020).
43. Department of Planning Statistics [Govt. of West Bengal]. Available online: http://www.wbpspm.gov.in/publications/District%20Statistical%20Handbook (accessed on 26 May 2020).
44. Phansalkar, S.J. Livestock-water interaction: Status and issues. In Proceedings of the 5th IWMI-Tata Annual Partners Meet, Gujarat, India, 8–10 March 2006.
45. Central Groundwater Board Annual Report. Available online: http://cgwb.gov.in/reportspublished.html (accessed on 19 May 2020).
46. Directorate of Economics and Statistics under the Supervision of Department of Agriculture, Cooperation and Farmers Welfare. Available online: https://eands.dacnet.nic.in/latest_20011.htm (accessed on 7 July 2020).
47. Handbook of Fisheries Statistics. Available online: http://dof.gov.in/statistics (accessed on 14 June 2020).
48. Kou, L.; Li, X.; Lin, J.; Kang, J. Simulation of urban water resources in Xiamen based on a WEAP model. *Water* **2018**, *10*, 732. [CrossRef]
49. Javadinejad, S.; Ostad-Ali-Askari, K.; Eslamian, S. Application of multi-index ecision analysis to management scenarios considering climate change prediction in the Zayandeh Rud River Basin. *Water Conserv. Sci. Eng.* **2019**, *4*, 53–70. [CrossRef]
50. Kumar, P.; Avtar, R.; Dasgupta, R.; Johnson, B.A.; Mukherjee, A.; Ahsan, M.N.; Nguyen, D.C.; Nguyen, H.Q.; Shaw, R.; Mishra, B.K. Socio-hydrology: A key approach for adaptation to water scarcity and achieving human wellbeing in large riverine islands. *Prog. Disaster Sci.* **2020**, *8*, 100134. [CrossRef]
51. Popkin, B.M.; D'Anci, K.E.; Rosenberg, I.H. Water, hydration, and health. *Nutr. Rev.* **2010**, *68*, 439–458. [CrossRef]
52. Dhawan, V. Water and agriculture in India. In *Background Paper for the South Asia Expert Panel during the Global Forum for Food and Agriculture (GFFA). German Asia-Pacific Business Association, German Agri-Business alliance, TERI;* 2017; Volume 28, Available online: https://www.oav.de/fileadmin/user_upload/5_Publikationen/5_Studien/170118_Study_Water_Agriculture_India.pdf (accessed on 26 March 2021).
53. Macrotends Data. Available online: https://www.macrotrends.net/countries/IND/india/rural-population (accessed on 19 March 2021).
54. India's Population to Cross 152 Crore in Next 16 Years, More People Will Be of Working Age. Available online: https://theprint.in/india/indias-population-to-cross-152-crore-in-next-16-years-more-people-will-be-of-working-age/481018/ (accessed on 19 March 2021).
55. World Bank Data. Available online: https://data.worldbank.org/indicator/SP.RUR.TOTL.ZS?locations=IN (accessed on 19 March 2021).
56. Vikashpedia. Available online: https://vikaspedia.in/agriculture/agri-directory/reports-and-policy-briefs/20th-livestock-census (accessed on 2 April 2021).
57. Masser, M.P.; Jensen, J.W.; Brunson, M.W. *Calculating Area and Volume of Ponds and Tanks;* Southern Regional Aquaculture Center: Stoneville, MI, USA, 1991.
58. Toure, A.; Diekkrüger, B.; Mariko, A.; Cissé, A.S. Assessment of groundwater resources in the context of climate change and population growth: Case of the Klela Basin in Southern Mali. *Climate* **2017**, *5*, 45. [CrossRef]

59. The Poultry Site, Broiler Water Consumption. Available online: https://www.thepoultrysite.com/articles/broiler-water-consumption (accessed on 3 April 2021).
60. Water Consumption in Broiler. Available online: https://www.fivetanimalhealth.com/management-articles/water-consumption-broilers (accessed on 4 April 2021).
61. Leal Filho, W.; Esilaba, A.O.; Rao, K.P.; Sridhar, G. *Adapting African Agriculture to Climate Change: Transforming Rural Livelihoods*; Springer: Berlin, Germany, 2015.
62. PRS Legislative Research. Available online: https://prsindia.org/policy/analytical-reports/state-agriculture-india (accessed on 15 April 2021).
63. Government of India. *Some Characteristics of Agricultural Households in India*; January–December 2013; National Sample Survey Office, Ministry of Statistics and Programme Implementation: Puram, India, 2015.
64. Bhalla, G.S. Agricultural growth and regional variations. In *India in a Globalising World: Some Aspects of Macro Economy, Agriculture and Poverty, Essays in Honour of Prof. C. H. Hanumantha Rao*; Centre for Economic and Social Studies, And Academic Foundation: New Delhi, India, 2006; pp. 309–346.
65. Mukherji, A.; Das, A. The political economy of metering agricultural tube wells in West Bengal, India. *Water Int.* **2014**, *39*, 671–685. [CrossRef]
66. Yeoh, E.; Chu, K.M. Literacy, Education and Economic Development in Contemporary China. 2012. Available online: https://papers.ssrn.com/sol3/papers.cfm?abstract_id=2207559 (accessed on 8 April 2021).
67. Bloom, D.E. *Population Dynamics in India and Implications for Economic Growth*; Working Paper no. 65; Harvard University: Cambridge, MA, USA, 2011; Available online: www.hsph.harvard.edu/pgda/WorkingPapers/2011/PGDA_WP_65.pdf (accessed on 8 April 2021).
68. Labour Commissionerate, Labour Department, Government of West Bengal. Available online: https://wblc.gov.in/synopsys/January/2021 (accessed on 19 April 2021).
69. Bhanja, S.N.; Mukherjee, A.; Rodell, M.; Wada, Y.; Chattopadhyay, S.; Velicogna, I.; Pangaluru, K.; Famiglietti, J.S. Groundwater rejuvenation in parts of India influenced by water-policy change implementation. *Sci. Rep.* **2017**, *7*, 1–7. [CrossRef]
70. Dubey, S.K.; Chand, B.K.; Trivedi, R.K.; Mandal, B.; Rout, S.K. Evaluation on the prevailing aquaculture practices in the Indian Sundarbans delta: An insight analysis. *J. Food Agric. Environ.* **2016**, *14*, 133–141.
71. Panori, A.; Psycharis, Y.; Ballas, D. Spatial segregation and migration in the city of Athens: Investigating the evolution of urban socio-spatial immigrant structures. *Popul. Space Place* **2019**, *25*, e2209. [CrossRef]
72. Ghosh, A.; Schmidt, S.; Fickert, T.; Nüsser, M. The Indian Sundarbans mangrove forests: History, utilization, conservation strategies and local perception. *Diversity* **2015**, *7*, 149–169. [CrossRef]
73. Dutta, D.; Das, P.K.; Paul, S.; Sharma, J.R.; Dadhwal, V.K. Assessment of ecological disturbance in the mangrove forest of Sundarbanss caused by cyclones using MODIS time-series data (2001–2011). *Nat. Hazards* **2015**, *79*, 775–790. [CrossRef]

Article

Did the COVID-19 Lockdown-Induced Hydrological Residence Time Intensify the Primary Productivity in Lakes? Observational Results Based on Satellite Remote Sensing

Ram Avtar [1,*], Pankaj Kumar [2], Hitesh Supe [1], Dou Jie [3], Netranada Sahu [4,5], Binaya Kumar Mishra [6] and Ali P. Yunus [7,8]

[1]   Faculty of Environmental Earth Science, Hokkaido University, Sapporo 060-0810, Japan; hiteshsupe@eis.hokudai.ac.jp
[2]   Natural Resources and Ecosystem Services, Institute for Global Environmental Strategies, Hayama, Kanagawa 240-0115, Japan; kumar@iges.or.jp
[3]   Department of Civil and Environmental Engineering, Nagaoka University of Technology, Kami-Tomioka, Nagaoka 1603-1, Japan; douj888@gmail.com
[4]   Department of Geography, Delhi School of Economics, University of Delhi, New Delhi 110007, India; nsahu@geography.du.ac.in
[5]   Disaster Prevention Research Institute, Kyoto University, Kyoto 611-0011, Japan
[6]   School of Engineering, Pokhara University, Lekhnath 22700, Nepal; bkmishra@pu.edu.np
[7]   State Key Laboratory for Geohazard Prevention and Geoenvironment Protection, Chengdu University of Technology, Chengdu, Sichuan 610059, China; pulpadan.yunusali@nies.go.jp
[8]   Center for Climate Change Adaptation, National Institute for Environmental Studies, Tsukuba, Ibaraki 305-8056, Japan
*   Correspondence: ram@ees.hokudai.ac.jp; Tel.: +81-011-706-2261

Received: 6 August 2020; Accepted: 10 September 2020; Published: 15 September 2020

Abstract: The novel coronavirus pandemic (COVID-19) has brought countries around the world to a standstill in the early part of 2020. Several nations and territories around the world insisted their population stay indoors for practicing social distance in order to avoid infecting the disease. Consequently, industrial activities, businesses, and all modes of traveling have halted. On the other hand, the pollution level decreased 'temporarily' in our living environment. As fewer pollutants are supplied in to the hydrosphere, and human recreational activities are stopped completely during the lockdown period, we hypothesize that the hydrological residence time (HRT) has increased in the semi-enclosed or closed lake bodies, which can in turn increase the primary productivity. To validate our hypothesis, and to understand the effect of lockdown on primary productivity in aquatic systems, we quantitatively estimated the chlorophyll-a (Chl-a) concentrations in different lake bodies using established Chl-a retrieval algorithm. The Chl-a monitored using Landsat-8 and Sentinel-2 sensor in the lake bodies of Wuhan, China, showed an elevated concentration of Chl-a. In contrast, no significant changes in Chl-a are observed for Vembanad Lake in India. Further analysis of different geo-environments is necessary to validate the hypothesis.

Keywords: hydrological residence time (HRT); lake; COVID; waterbodies

---

1. Introduction

The residence time is a fundamental descriptor in hydrology that provides information on the timescales of a molecule of water spend in a specific system. Hydrological residence time (HRT) is estimated as the amount of time the water spent in any section of the connected network [1]. The longer a parcel of water remains in a specific system (river, lake, ponds, etc.), the longer is its residence time, and vice versa. HRT has got important applications in a wide range of hydrological fields including water quality analysis, stratification, habitat ecology, age dating, water mixing and circulation, microbiological contaminants, etc., [2–4]. For example, Zwart et al. [4] showed that lakes with short HRT had higher dissolved organic carbon and greater net heterotrophy. Hein et al. [5] and others noticed that a prolonged residence time increases the primary productivity in aquatic systems. Similar observations have been noticed in several other works. For instance, León et al. [6] reported that chlorophyll-*a* was directly related to the HRT. Stumpner et al. [7] showed that zones of longer HRT (15–60 days) have higher Chl-*a* concentrations, and ones of shorter HRT (1–14 days) have lower Chl-a concentrations.

The SARS-CoV-2, or popularly the Coronavirus Disease 2019 (COVID-19) that affected the world population in early 2020, caused several nations and territories to a stand-still. Over 25 million infected persons and more than 860,000 deaths have been reported worldwide caused by the COVID-19 as of 1 September 2020 [8]. Since no vaccine or cure has developed to protect the body against the COVID-19, complete or partial lockdown has been induced in many countries and territories to stop the chain of infection. As vehicular movement halts, construction is put on hold, and industries stop production, the levels of pollution level has come down both in atmosphere and hydrosphere [9,10].

The industrial sewage input to the lakes through inlets, recreational activities in lakes such as boating, fishing, etc., stopped temporarily in aquatic systems during COVID-19-induced lockdown. We hypothesize that the hydrological residence time in closed or semi-enclosed lakes has increased, which may, in turn, increase the primary productivity. To validate our hypothesis and to understand the effect of lockdown on primary productivity in aquatic systems, this study explores to quantify the level of Chl-*a* before and during the lockdown period using remote sensing techniques. Chl-*a* derived from OC3 algorithm [11] is selected for comparing the eutrophication status before and during the lockdown.

While, the hydrological residence time is not the only factor that influence the primary productivity, other factors such as temperature, increased light exposure, oxygen level, initial nutrients levels, etc., also could cause an increase or decrease in the level of Chl-*a* concentration in the aquatic environments [12–14]. For instance, Castelao et al. [15] using geostationary satellite data showed the seasonal development of coastal upwelling in which the maximum was peaked in the summer and minimum during the winter. In oligotrophic waters, maximum phytoplankton production often occurs near the top of a nutricline [16]. In addition, in inland water bodies the intensity of monsoon rainfall and landscape heterogeneity tremendously influenced the Chl-*a* concentrations [17]. Further, in places of upwelling areas, the water surface temperatures are often cooler than nearby waters, resulting in an increased chlorophyll concentration. However, we assume in this study that the boundary conditions remain unchanged during the observation period for both the study areas and, hence, examined the effect of HRT on water quality.

2. Study Area

The stringent and biggest of all lockdown was imposed in two places, (i) Wuhan, China, the epicenter of COVID-19, and (ii) India, where 1.3 billion people have been staying home since 25 March 2020. Hence, we selected some lake bodies in Wuhan city (Figure 1a) and one lake body in India (Vembanad Lake, the longest freshwater lake in India) (Figure 1b). These lake bodies are also preferred because of the availability of cloud-free remote sensing images, and expected longer HRT caused by the stringent lockdown in these areas. Both the selected cases experienced severe pollution by wastewater disposal, industrial effluents, heavy metal concentration, and micro-plastics before the lockdown [18,19].

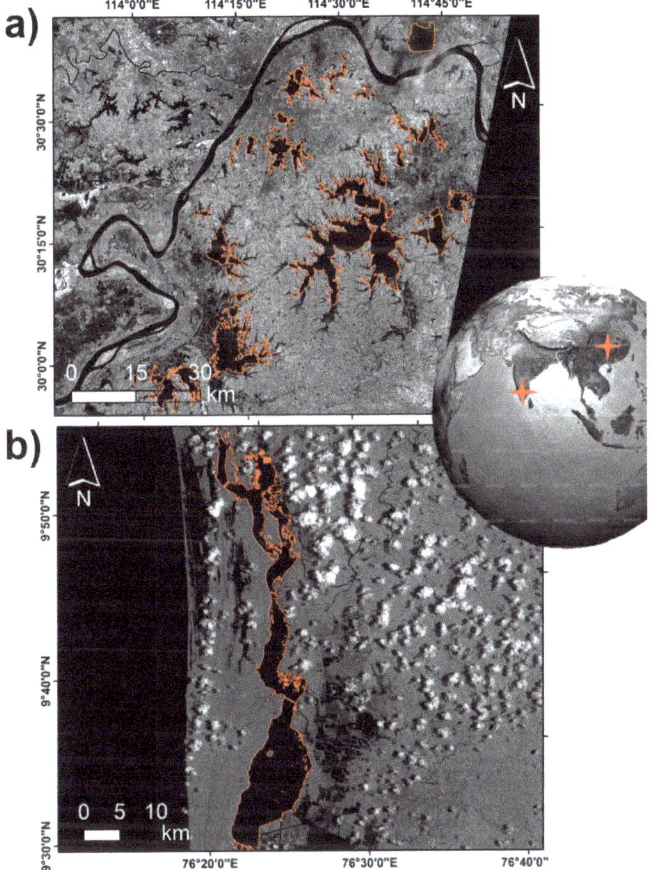

Figure 1. Location of the study area (**a**) lakes of Wuhan, China, and (**b**) Vembanad Lake, India.

3. Data and Methodology

3.1. Theoretical Framework

The amount of water spent in any section of the water body is an important consideration for many water quality problems [1]. In general, the regions near the inlet are having less residence time than the far places. Further, in a closed water body or semi-closed lakes, the movement of water is largely constrained, therefore having longer residence time. Dickman, [20] showed that increased water residence time more likely increases the algae bloom, especially for small reservoirs. Several other works showed that Chl-*a* increases with increasing residence time and decreases with increasing discharge [21,22]. The hydrological residence time in our study area was expected to increase during the lockdown period owing to following reasons: (i) sewage disposal to the lakes has completely stopped, causing reduced discharge via inlets, and (ii) all anthropogenic activities, including boating has stopped during the lockdown period, causing still waters. The lockdown induced by COVID-19 in this ecosystem with a long retention time, thus offering an opportunity to study the development of phytoplankton that are otherwise adapted to a turbid environment.

While HRT has been traditionally measured through dye-tracer experiments in the field or estimating the ratio of the volume of the domain of interest to an outgoing flux [23]. Recently, physically based hydrodynamic modeling is employed to estimate the residence [24]. In this study, we assumed that the HRT is the longest during the lockdown period.

3.2. Image Acquisition and Data Processing

Landsat 8 OLI images and Sentinel-2 images of the immediate pre-lockdown period (December 2019 and January 2020) and during the lockdown period (February–April 2020) were downloaded (Table 1) from the United States Geological Survey (USGS) website (earthexplorer.usgs.gov). All scenes had undergone terrain correction within prescribed tolerances. Table 1 shows the details of satellite images used in this study for Chl-*a* mapping. The Level 1 Landsat-8 and Sentinel-2 images were further treated using ACOLITE software for radiometric calibration (Top of Atmosphere Reflectance) and atmospheric correction (Surface Reflectance). ACOLITE, developed by Royal Belgian Institute of Natural Science, employs a "dark spectrum fitting" (DSF) approach [25,26] for atmospheric correction. For detailed procedure on atmospheric correction for Landsat-8 OLI in an ACOLITE environment, readers are referred to the following references [27,28]. For validation purpose, we used the satellite images of 2017–2019 March–April images. The validation dataset was directly used in Google Earth Engine (GEE) platform with the help of chlorophyll index algorithm [14]. The GEE codes are provided in the Supplementary File S1.

The variability in the meteorological conditions during the study period may affect the chlorophyll concentrations. While the regional air temperatures were increasing from December to April by about 10° Celsius for Wuhan, the average temperature difference was only about 2° Celsius for Vembanad region (source: https://www.timeanddate.com). The precipitation condition was normal for both case areas with occasional rainy days observed during the study period (Figure 2).

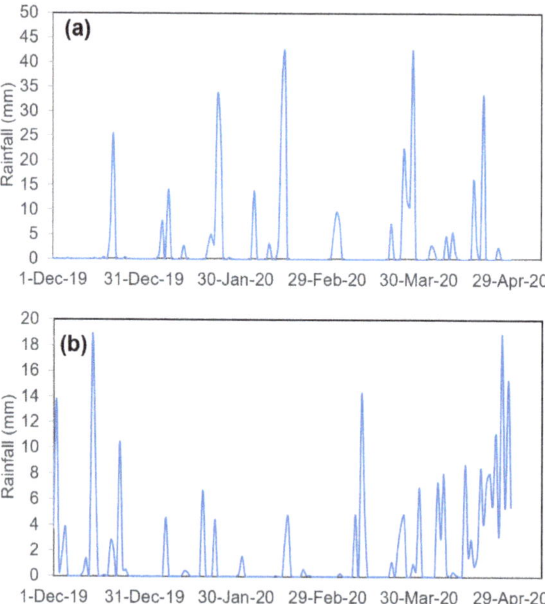

Figure 2. Daily precipitation (mm) time series chart for (**a**) Wuhan and (**b**) Vembanad (Cochin) during the study period (source: CHIRPS Daily: Climate Hazards Group InfraRed Precipitation with Station Data (version 2.0)).

Table 1. Details of the satellite images used for mapping chlorophyll-*a* (Chl-*a*) before and during the lockdown period in 2020.

S. No	Wuhan Lakes	Vembanad Lake
1	ID: LC08_L1TP_123039_20191207_20191217_01_T1 Acquisition Date: 2019-12-07, Path: 123 Row: 39	ID: LC08_L1TP_144053_20200228_20200313_01_T1 Acquisition Date: 2020-02-28, Path: 144, Row: 53
2	ID: L1C_T50RKU_A023910_20200120T030612 Acquisition Date: 2020/01/20, Tile Number: T50RKU	ID: LC08_L1TP_144053_20200315_20200325_01_T1 Acquisition Date: 2020-03-15
3	ID: L1C_T50RKU_A024053_20200130T030534 Acquisition Date: 2020/01/30	ID: LC08_L1TP_144053_20200331_20200410_01_T1 Acquisition Date: 2020-03-31
4	ID: LC08_L1TP_123039_20200209_20200211_01_T1 Acquisition Date: 2020-02-09	ID: L1C_T43PFL_A024755_20200319T051246 Acquisition Date: 2020/03/19, Tile Number: T43PFL
5	ID: L1C_T50RKU_A015788_20200315T030729 Acquisition Date: 2020/03/15	ID: L1C_T43PFL_A015918_20200324T052110 Acquisition Date: 2020/03/24
6	ID: L1C_T50RKU_A024768_20200320T030130 Acquisition Date: 2020/03/20	D: LC08_L1TP_144053_20200331_20200410_01_T1 Acquisition Date: 2020-03-31
7	ID: L1C_T50RKU_A025054_20200409T030244 Acquisition Date: 2020/04/09	ID: L1C_T43PFL_A016061_20200403T052351 Acquisition Date: 2020/04/03
8	ID: LC08_L1TP_123039_20200413_20200422_01_T1 Acquisition Date: 2020-04-13	ID: LC08_L1TP_144053_20200416_20200423_01_T1 Acquisition Date: 2020-04-16
9	ID: L1C_T50RKU_A025340_20200429T030455 Acquisition Date: 2020/04/29	
10	ID: LC08_L1TP_123039_20200429_20200509_01_T1 Acquisition Date: 2020-04-29	

3.3. Chlorophyll-a Retrieval

The reflectance ratio of blue and green wavelengths in the electromagnetic spectrum was recognized to correlate well with the distribution of chlorophyll in surface waters [29–31]. Several studies have supported the usage of blue-green bands on the assumption that any changes in these wavelengths are driven by changes in phytoplankton concentrations [32,33]. The performance of blue-green ratioed algorithms for retrieving Chl-*a* was tested independently in different environments [34–38]. We, therefore, employed the OC3 algorithm [11], which uses the water leaving reflectance (R_{rs}) in wavelength 443, 482, and 561 for Landsat 8 (Equations (1)–(4)), and 490 and 560 for Sentinel 2 sensor (Equations (1)–(3), and (5)). Mathematically, OC3 Chl-*a* algorithm is expressed as:

$$Chl_{OC3} = 10^y \tag{1}$$

$$y = a_0 + a_1 x + a_2 x^2 + a_3 x^3 + a_4 x^4 \tag{2}$$

$$x = \log_{10}(R) \tag{3}$$

$$R = \frac{\max(R_{rs}(443, 482))}{R_{rs}561} \tag{4}$$

$$R = \frac{R_{rs}490}{R_{rs}560} \tag{5}$$

The coefficients 0.2412, −2.0546, 1.1776, −0.5538, −0.4570 are, respectively, used for a_0 to a_4 (https://oceancolor.gsfc.nasa.gov/atbd/chlor_a/). The performance evaluation of Chl-*a* retrievals using Sentinel-2 data and OCx algorithms based on Acolite was found within the root mean squared logarithmic error (RMSLE) of 1.2–1.3 [39]. In another study, by employing OC3 algorithm for Indonesian seas, the RMSE of in situ vs. satellite Chl-*a* was found within the range of 0.04–0.05 [40], suggesting superior performance of satellite retrievals of Chl-*a* in aquatic systems.

4. Results and Discussion

4.1. Lakes in Wuhan

Figures 3 and 4a presents the Chl-*a* concentration maps of lakes in Wuhan for pre-lockdown (7 December 2019; 20 January 2020), during the lockdown (30 January 2020; 9 February 2020; 15 and 20 March 2020) and post-lockdown periods (9, 13 and 29 April 2020). It can be seen that the mean Chl-*a* concentrations were very low in the pre-lockdown period (2.77 and 2.95 µg/L) and immediately after the lockdown period (2.25 µg/L). By 9 February, the mean concentration gradually increased to 3.05 µg/L. Note that the lockdown was imposed on 23 January in Wuhan. The peak Chl-*a* was observed in the March months (6.57 and 5.88 µg/L), which also corresponds to the peak period of the quarantine period in Wuhan. The lockdown ended on April 8; the mean Chl-*a* for April shows a gradually decreasing trend (5.06, 4.49, and 4.74 µg/L).

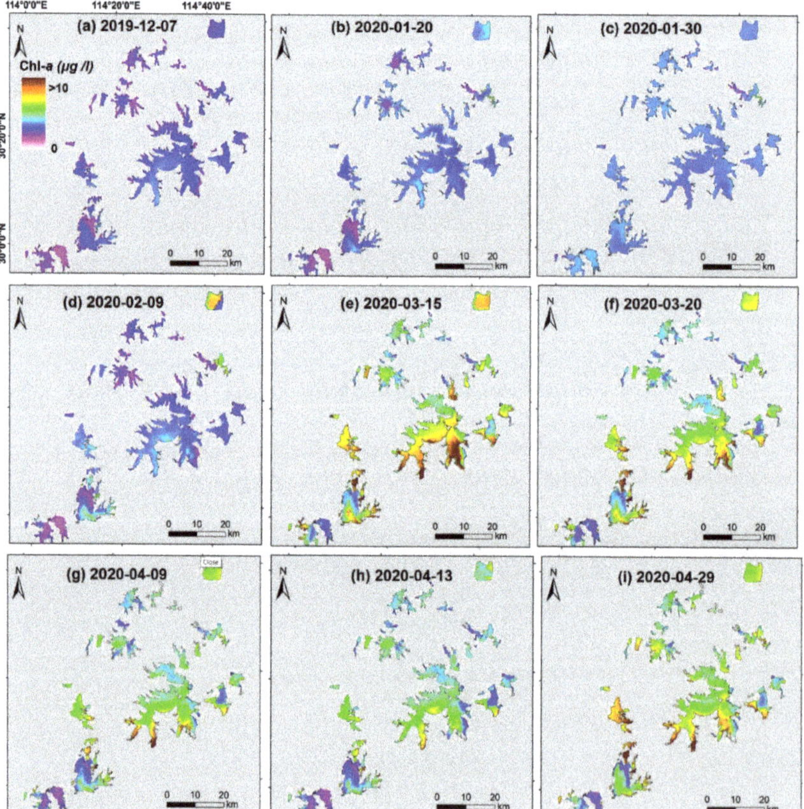

Figure 3. Chlorophyll-*a* concentration before, during, and post the lockdown period estimated using OC3 algorithm for Wuhan scenic lakes.

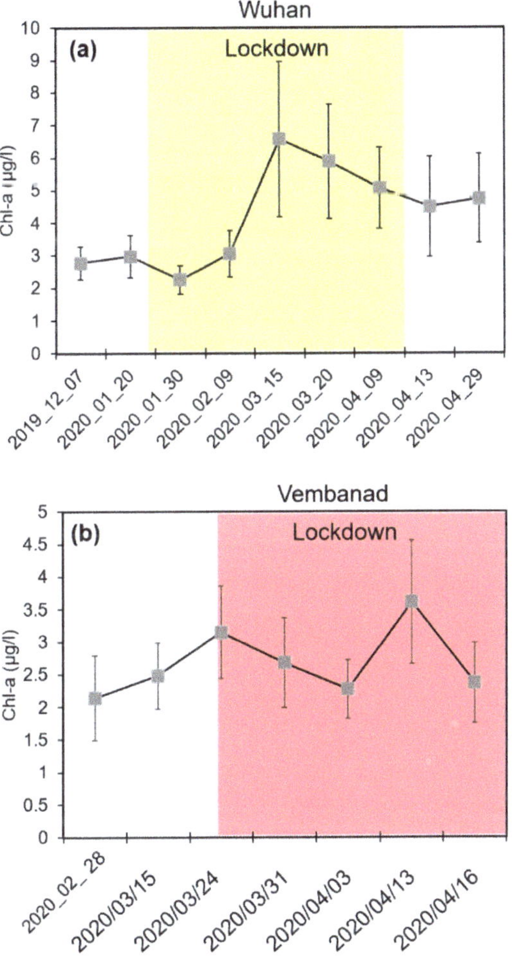

Figure 4. Mean chlorophyll-*a* during pre, during, and post lockdown in (**a**) Wuhan Lake and (**b**) Vembanad Lake (error bar shows ± standard deviations).

4.2. Vembanad Lake, India

Figures 4b and 5 present the results of Chl-*a* concentration before (28 February 2020; 15 March 2020; 24 March 2020) and during the lockdown period (31 March 2020; 4 April 2013; 13 April 2013; 16 April 2020). Note that the lockdown started on 25 February in India and was still ongoing during the study period. Contrary to the former analyzed area, Lake Vembanad does not show any significant increase in primary productivity during the lockdown period. Nevertheless, the concentration was not decreased during this period (Figure 4b).

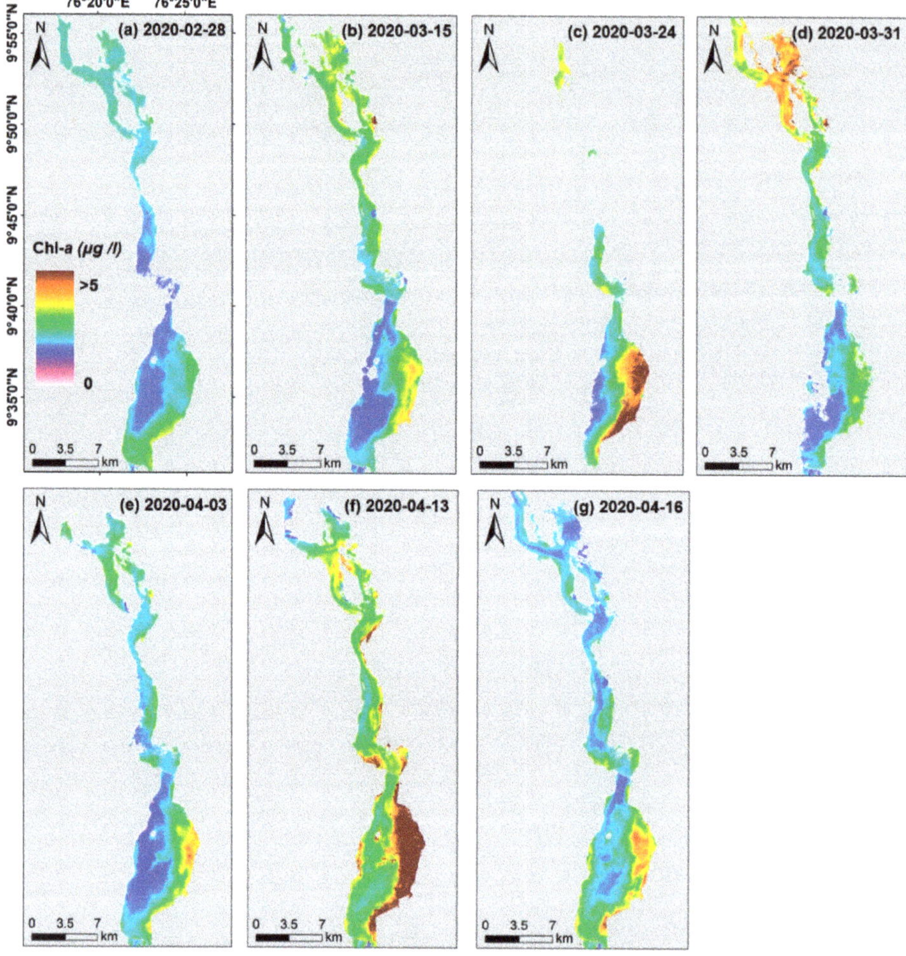

Figure 5. Chlorophyll-*a* concentrations before and during the lockdown period estimated using OC3 algorithm for Vembanad Lake waters.

4.3. Validation

Since there is an ~10° celcius increase in temperature from December to April in Wuhan climatology, it is expected that the Chl-*a* also increased during this time period. In order to validate the hypothesis that the increased Chl-*a* in the lakes of Wuhan during the lockdown period in 2020 is because of the increased HRT, we monitored the Chl-*a* for the previous years (2017, 2018, and 2019) during the same time period. The comparative maps of mean Chl-*a* for 2020 (March–April) and those of previous years (2017–2019) during the same time period are presented in Figure 6. It can be seen that the Chl-*a* during 2020 March–April is the maximum among the study years, especially in the lakes far away from the city center. This implies that hydrological residence time was maximum in places where the influence of human activity was minimum and, indeed, strengthen our results presented in Section 4.1.

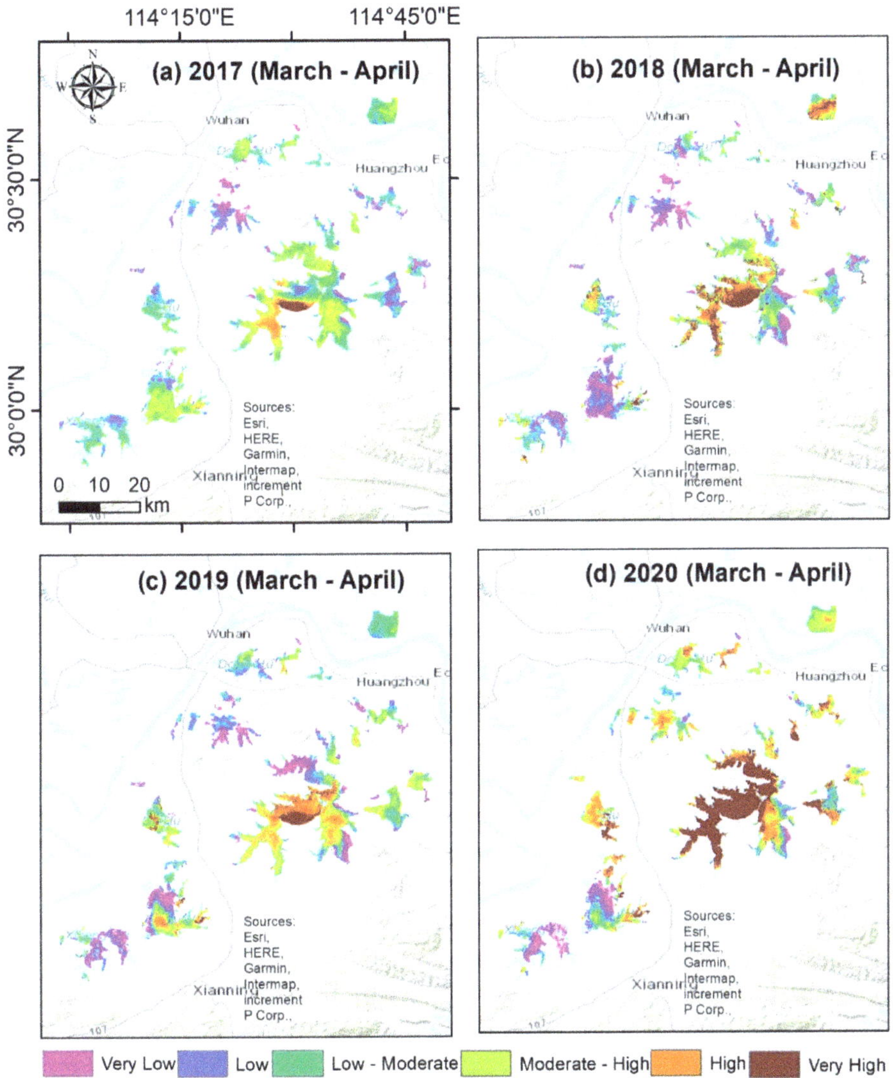

Figure 6. Mean chlorophyll-*a* concertation mapped for (**a**) 2017, (**b**) 2018, (**c**) 2019, and (**d**) 2020 during April–May months (for simplicity, the chlorophyll index algorithm [14] is used in Google Earth Engine (GEE) to derive the low to high classes).

5. Discussion and Concluding Remarks

As the pollutant discharges into lakes and human activities (boating and fishing) stopped or reduced during the COVID-19 lockdown period, we investigated whether the residence time in lakes also increased in associated with it? We tried to answer the problem mentioned above by analyzing the primary productivity in the lakes. Reynolds, [41], and others reported that increased residence time favors Chl-*a* and biomass accumulation in aquatic systems [7]. In our study, we noticed an elevated level of Chl-*a* for the lakes of Wuhan during the initial phase of the lockdown period, followed by a decreasing trend (Figure 4a). This elevated Chl-*a* indicates an increased HRT in the lakes of Wuhan.

The decreasing trend followed by the increase may be because a prolonged HRT can settle down the surface phytoplankton [42].

On the other hand, the Vembanad Lake in India does not show any significant changes in Chl-*a* concentration during the lockdown period. A possible explanation is that the Vembanad Lake is not a closed lake, unlike the lakes in Wuhan. Seven rivers are draining into it, plus it has an opening to the Arabian Sea in the north-west, which carries salt water up to 26 km inside the lake during the high-tide period. Thus, it may be because the discharge from the rivers (without pollutants) and tidal action cause the HRT to be insufficient for enhancing primary productivity in the Vembanad Lake. It is noteworthy to mention here that both cases, i.e., Wuhan lakes and the Vembanad Lake, have shown a significant decrease in suspended particulate matter during the lockdown period [10] (Supplementary Materials Figures S1 and S2).

One may, though, argue that the increased Chl-*a* in Wuhan lakes can also be associated with increased water temperature. However, time series Chl-*a* maps compared for different years (2017–2020) show that the year 2020 experienced the maximum value of chlorophyll in the lakes of Wuhan. This demonstrates that Chl-*a* concentration during the lockdown has increased in a closed lake system, i.e., for Wuhan, whereas the Chl-*a* remains unchanged in an open lake system such as the one demonstrated for Vembanad Lake in India. The hydrological residence time induced during the lockdown is by large the influencing factor on increased Chl-*a*, in that it can describe the prolonged residence time increases the primary productivity in closed systems.

The Chl-*a* retrieval using NASA's OC3 algorithm, however, was not validated in this study with spatiotemporally matched field-derived measurements of chlorophyll because of stringent lockdown measures in both cities. However, the capability of OC3 for Chl-*a* retrieval in inland lakes and ocean waters in previous studies show a near one-to-one relationship and can be accounted for the Chl-*a* variability up to upper 10 m of the water column [43]. In addition, the research framework also does not incorporate the variability of Chl-*a* caused by other natural phenomena's, and the amount of water contaminant flow into lakes from household wastes, which are usually difficult to model in satellite-based bio-physical parameter estimations. Nevertheless, the methodology and results presented in our study can help in understanding the influence of lockdown on water quality parameters, especially phytoplankton concentrations. Thus, although the research framework can offer important insights into short-term changes in the hydrosphere, additional analysis incorporating time-series data, and similar studies in closed lakes in other climatic environments is necessary to further validate our hypothesis.

Supplementary Materials: The following are available onlsupine at http://www.mdpi.com/2073-4441/12/9/2573/s1, Figure S1: Decreased suspended particulate matter in different lakes of Wuhan during the lockdown period, Figure S2: Decreased suspended particulate matter in Vembanad lake during the lockdown period, Text S1: Codes for chlorophyll index using Sentinel-2 images in GEE.

Author Contributions: Conceptualization: R.A., A.P.Y.; methodology: R.A., A.P.Y.; validation: R.A., A.P.Y.; writing—original draft: R.A., P.K., H.S., D.J., N.S., B.K.M., A.P.Y.; writing—review and editing: R.A., P.K., H.S., D.J., N.S., B.K.M., A.P.Y.; funding acquisition, R.A., A.P.Y. All authors have read and agree to the published version of the manuscript.

Funding: This research received no external funding.

Acknowledgments: Authors would like to thank Hokkaido University L-station and SOUSEI support for Young Researcher. Furthermore, the authors are thankful to the United States Geological Survey (USGS) and Copericus hub for providing satellite data and ACOLITE software provided by RBINS. We also acknowledge the support of Dr. Masago Yoshifumi and appreciate the contribution made by the anonymous reviewers.

Conflicts of Interest: The authors declare no conflict of interest.

References

1. Leray, S.; Engdahl, N.B.; Massoudieh, A.; Bresciani, E.; McCallum, J. Residence time distributions for hydrologic systems: Mechanistic foundations and steady-state analytical solutions. *J. Hydrol.* **2016**, *543*, 67–87. [CrossRef]

2. Baker, M.A.; Dahm, C.N.; Valett, H.M. Anoxia, anaerobic metabolism biogeochemistry of the stream water-ground water interface. In *Streams and Ground Waters*; Academic Press: San Diego, CA, USA, 2000; pp. 259–284.

3. Neumann, R.B.; LaBolle, E.M.; Harvey, C.F. The Effects of Dual-Domain Mass Transfer on the Tritium-Helium-3 Dating Method. *Environ. Sci. Technol.* **2008**, *42*, 4837–4843. [CrossRef]

4. Zwart, J.A.; Sebestyen, S.D.; Solomon, C.T.; Jones, S.E. The influence of hydrologic residence time on lake carbon cycling dynamics following extreme precipitation events. *Ecosystems* **2017**, *20*, 1000–1014. [CrossRef]

5. Hein, T.; Baranyi, C.; Heiler, G.; Holarek, C.; Riedler, P.; Schiemer, F. Hydrology as a major factor determining plankton development in two floodplain segments and the River Danube, Austria. *Large Rivers* **1999**, *11*, 439–452. [CrossRef]

6. León, J.G.; Beamud, S.G.; Temporetti, P.F.; Atencio, A.G.; Diaz, M.M.; Pedrozo, F.L. Stratification and residence time as factors controlling the seasonal variation and the vertical distribution of chlorophyll-*a* in a subtropical irrigation reservoir. *Int. Rev. Hydrobiol.* **2016**, *101*, 36–47. [CrossRef]

7. Stumpner, E.B.; Bergamaschi, B.A.; Kraus, T.E.; Parker, A.E.; Wilkerson, F.P.; Downing, B.D.; Dugdale, R.C.; Murrell, M.C.; Carpenter, K.D.; Orlando, J.L.; et al. Spatial variability of phytoplankton in a shallow tidal freshwater system reveals complex controls on abundance and community structure. *Sci. Total Environ.* **2020**, *700*, 134392. [CrossRef] [PubMed]

8. Worldometers. Coronavirus Updates. Available online: https://www.worldometers.info/ (accessed on 3 May 2020).

9. ESA COVID-19: Nitrogen Dioxide over China. Available online: https://www.esa.int/Applications/ Observing_the_Earth/Copernicus/Sentinel-5P/COVID-19_nitrogen_dioxide_over_China (accessed on 17 April 2020).

10. Yunus, A.P.; Masago, Y.; Hijioka, Y. COVID-19 and surface water quality: Improved lake water quality during the lockdown. *Sci. Total Environ.* **2020**, *731*, 139012. [CrossRef]

11. O'Reilly, J.E.; Maritorena, S.; Mitchell, B.G.; Siegel, D.A.; Carder, K.L.; Garver, S.A.; Kahru, M.; McClain, C. Ocean color chlorophyll algorithms for SeaWiFS. *J. Geophys. Res. Oceans* **1998**, *103*, 24937–24953. [CrossRef]

12. Nukapothula, S.; Chen, C.; Yunus, A.P.; Wu, J. Satellite-based observations of intense chlorophyll-*a* bloom in response of cold core eddy formation: A study in the Arabian Sea, Southwest Coast of India. *Reg. Stud. Mar. Sci.* **2018**, *24*, 303–310. [CrossRef]

13. Gao, K.; Xu, J.; Gao, G.; Li, Y.; Hutchins, D.A.; Huang, B.; Wang, L.; Zheng, Y.; Jin, P.; Cai, X.; et al. Rising CO_2 and increased light exposure synergistically reduce marine primary productivity. *Nat. Clim. Change* **2012**, *2*, 519–523. [CrossRef]

14. Hu, C.; Lee, Z.; Franz, B. Chlorophyll *a* algorithms for oligotrophic oceans: A novel approach based on three-band reflectance difference. *J. Geophys. Res. Oceans* **2012**, *117*. [CrossRef]

15. Scofield, A.E.; Watkins, J.M.; Osantowski, E.; Rudstam, L.G. Deep chlorophyll maxima across a trophic state gradient: A case study in the Laurentian Great Lakes. *Limnol. Oceanogr.* **2020**. [CrossRef]

16. Castelao, R.M.; Mavor, T.P.; Barth, J.A.; Breaker, L.C. Sea surface temperature fronts in the California Current System from geostationary satellite observations. *J. Geophys. Res. Oceans* **2006**, *111*. [CrossRef]

17. Atique, U.; An, K.-G. Landscape heterogeneity impacts water chemistry, nutrient regime, organic matter and chlorophyll dynamics in agricultural reservoirs. *Ecol. Indic.* **2020**, *110*, 105813. [CrossRef]

18. Sruthy, S.; Ramasamy, E. Microplastic pollution in Vembanad Lake, Kerala, India: The first report of microplastics in lake and estuarine sediments in India. *Environ. Pollut.* **2017**, *222*, 315–322. [CrossRef]

19. Zhu, S.; Yang, G.; Dai, J. Sediment Management in East Lake, China: A Combined Bio-Physical and Socioeconomic Approach for Managing Sediments in a Polluted Lake System. *Pol. J. Environ. Stud.* **2018**, *27*, 1891–1900. [CrossRef]

20. Dickman, M. Some Effects of Lake Renewal on Phytoplankton Productivity and Species Composition 1. *Limnol. Oceanogr.* **1969**, *14*, 660–666. [CrossRef]

21. Gomes, L.; Miranda, L. Hydrologic and climatic regimes limit phytoplankton biomass in reservoirs of the Upper Paraná River Basin, Brazil. *Hydrobiologia* **2001**, *457*, 205–214. [CrossRef]
22. Lee, S.; Lee, S.; Kim, S.H.; Park, H.; Park, S.; Yum, K. Examination of critical factors related to summer Chlorophyll *a* concentration in the Sueo Dam Reservoir, Republic of Korea. *Environ. Eng. Sci.* **2012**, *29*, 502–510. [CrossRef]
23. Wan, Y.; Qiu, C.; Doering, P.; Ashton, M.; Sun, D.; Coley, T. Modeling residence time with a three-dimensional hydrodynamic model: Linkage with Chlorophyll *a* in a subtropical estuary. *Ecol. Model.* **2013**, *268*, 93–102. [CrossRef]
24. Liu, W.-C.; Chen, W.-B.; Kuo, J.-T.; Wu, C. Numerical determination of residence time and age in a partially mixed estuary using three-dimensional hydrodynamic model. *Cont. Shelf Res.* **2008**, *28*, 1068–1088. [CrossRef]
25. Vanhellemont, Q.; Ruddick, K. Atmospheric correction of metre-scale optical satellite data for inland and coastal water applications. *Remote Sens. Environ.* **2018**, *216*, 586–597. [CrossRef]
26. Vanhellemont, Q. Adaptation of the dark spectrum fitting atmospheric correction for aquatic applications of the Landsat and Sentinel-2 archives. *Remote Sens. Environ.* **2019**, *225*, 175–192. [CrossRef]
27. Caballero, I.; Stumpf, R.P. Retrieval of nearshore bathymetry from Sentinel-2A and 2B satellites in South Florida coastal waters. *Estuar. Coast. Shelf Sci.* **2019**, *226*, 106277. [CrossRef]
28. Pahlevan, N.; Roger, J.-C.; Ahmad, Z. Revisiting short-wave-infrared (SWIR) bands for atmospheric correction in coastal waters. *Opt. Exp.* **2017**, *25*, 6015–6035. [CrossRef]
29. Clark, D.K.; Baker, E.T.; Strong, A.E. Upwelled spectral radiance distribution in relation to particulate matter in sea water. *Bound. Layer Meteorol.* **1980**, *18*, 287–298. [CrossRef]
30. Gordon, H.R.; Clark, D.K. Remote sensing optical properties of a stratified ocean: An improved interpretation. *App. Opt.* **1980**, *19*, 3428–3430. [CrossRef]
31. McKee, D.; Cunningham, A.; Dudek, A. Optical water type discrimination and tuning remote sensing band-ratio algorithms: Application to retrieval of chlorophyll and Kd (490) in the Irish and Celtic Seas. *Estuar. Coast. Shelf Sci.* **2007**, *73*, 827–834. [CrossRef]
32. Bowers, D.G.; Harker, G.E.L.; Stephan, B. Absorption spectra of inorganic particles in the Irish Sea and their relevance to remote sensing of chlorophyll. *Int. J. Remote Sens.* **1996**, *17*, 2449–2460. [CrossRef]
33. O'Reilly, J.E.; Maritorena, S.; Siegel, D.A.; O'Brien, M.C.; Toole, D.; Mitchell, B.G.; Kahru, M.; Chavez, F.P.; Strutton, P.; Cota, G.F.; et al. Ocean color chlorophyll *a* algorithms for SeaWiFS, OC2, and OC4: Version 4. In *SeaWiFS Postlaunch Calibration and Validation Analyses*; NASA Center for AeroSpace Information: Hanover, MD, USA, 2000; Volume 3, pp. 9–23.
34. Jena, B. The effect of phytoplankton pigment composition and packaging on the retrieval of chlorophyll-*a* concentration from satellite observations in the Southern Ocean. *Int. J. Remote Sens.* **2017**, *38*, 3763–3784. [CrossRef]
35. Nagamani, P.; Hussain, M.; Choudhury, S.; Panda, C.; Sanghamitra, P.; Kar, R.; Das, A.; Ramana, I.; Rao, K. Validation of chlorophyll-*a* algorithms in the coastal waters of Bay of Bengal initial validation results from OCM-2. *J. Ind. Soc. Remote Sens.* **2013**, *41*, 117–125. [CrossRef]
36. O'Reilly, J.E.; Werdell, P.J. Chlorophyll algorithms for ocean color sensors-OC4, OC5 & OC6. *Remote Sens. Environ.* **2019**, *229*, 32–47.
37. Witter, D.L.; Ortiz, J.D.; Palm, S.; Heath, R.T.; Budd, J.W. Assessing the application of SeaWiFS ocean color algorithms to Lake Erie. *J. Gt. Lakes Res.* **2009**, *35*, 361–370. [CrossRef]
38. Yunus, A.P.; Dou, J.; Sravanthi, N. Remote sensing of chlorophyll-*a* as a measure of red tide in Tokyo Bay using hotspot analysis. *Remote Sens. Appl. Soc. Environ.* **2015**, *2*, 11–25. [CrossRef]
39. Pahlevan, N.; Smith, B.; Schalles, J.; Binding, C.; Cao, Z.; Ma, R.; Alikas, K.; Kangro, K.; Gurlin, D.; Hà, N.; et al. Seamless retrievals of chlorophyll-*a* from Sentinel-2 (MSI) and Sentinel-3 (OLCI) in inland and coastal waters: A machine-learning approach. *Remote Sens. Environ.* **2020**, *240*, 111604. [CrossRef]
40. Winarso, G.; Marini, Y. MODIS standard (OC3) chlorophyll-*a* algorithm evaluation in Indonesian seas. *Int. J. Remote Sens. Earth Sci.* **2017**, *11*, 11–20. [CrossRef]
41. Reynolds, C.S. *The Ecology of Phytoplankton*; Cambridge University Press: Cambridge, UK, 2006.

42. May, C.L.; Koseff, J.R.; Lucas, L.V.; Cloern, J.E.; Schoellhamer, D.H. Effects of spatial and temporal variability of turbidity on phytoplankton blooms. *Mar. Ecol. Prog. Ser.* **2003**, *254*, 111–128. [CrossRef]
43. Bennion, D.H.; Warner, D.M.; Esselman, P.C.; Hobson, B.; Kieft, B. A comparison of chlorophyll *a* values obtained from an autonomous underwater vehicle to satellite-based measures for Lake Michigan. *J. Gt. Lakes Res.* **2019**, *45*, 726–734. [CrossRef]

Communication

Exposure to SARS-CoV-2 in Aerosolized Wastewater: Toilet Flushing, Wastewater Treatment, and Sprinkler Irrigation

Muhammad Usman [1],*, Muhammad Farooq [2], Muhammad Farooq [3] and Ioannis Anastopoulos [4],*

1. PEIE Research Chair for the Development of Industrial Estates and Free Zones, Center for Environmental Studies and Research, Sultan Qaboos University, Muscat Al-Khoud 123, Oman
2. Department of Clinical Sciences, College of Veterinary and Animal Sciences, Jhang, University of Veterinary and Animal Sciences, Lahore 531, Pakistan; muhammad.farooq@uvas.edu.pk
3. Department of Plant Sciences, Sultan Qaboos University, Muscat Al-Khoud 123, Oman; farooqcp@squ.edu.om
4. Department of Electronics Engineering, School of Engineering, Hellenic Mediterranean University, 73100 Chania, Crete, Greece
* Correspondence: muhammad.usman@squ.edu.om (M.U.); anastopoulos_ioannis@windowslive.com (I.A.)

Abstract: The existence of SARS-CoV-2, the etiologic agent of coronavirus disease 2019 (COVID-19), in wastewater raises the opportunity of tracking wastewater for epidemiological monitoring of this disease. However, the existence of this virus in wastewater has raised health concerns regarding the fecal–oral transmission of COVID-19. This short review is intended to highlight the potential implications of aerosolized wastewater in transmitting this virus. As aerosolized SARS-CoV-2 could offer a more direct respiratory pathway for human exposure, the transmission of this virus remains a significant possibility in the prominent wastewater-associated bioaerosols formed during toilet flushing, wastewater treatment, and sprinkler irrigation. Implementing wastewater disinfection, exercising precautions, and raising public awareness would be essential. Additional research is needed to evaluate the survival, fate, and dissemination of SARS-CoV-2 in wastewater and the environment and rapid characterization of aerosols and their risk assessment.

Keywords: SARS-CoV-2; COVID-19; bioaerosol; aerosolized wastewater; environmental transmission; agriculture

Citation: Usman, M.; Farooq, M.; Farooq, M.; Anastopoulos, I. Exposure to SARS-CoV-2 in Aerosolized Wastewater: Toilet Flushing, Wastewater Treatment, and Sprinkler Irrigation. *Water* **2021**, *13*, 436. https://doi.org/10.3390/w13040436

Academic Editor: Pankaj Kumar
Received: 25 December 2020
Accepted: 5 February 2021
Published: 8 February 2021

Publisher's Note: MDPI stays neutral with regard to jurisdictional claims in published maps and institutional affiliations.

1. Introduction

Recent research has demonstrated that people with coronavirus disease 2019 (COVID-19), even those who do not develop symptoms, discharge its etiologic virus, severe acute respiratory syndrome coronavirus 2 (SARS-CoV-2), through their excrement [1]. The presence of SARS-CoV-2 RNA in sewage raises the possibility of analyzing wastewater for the epidemiological monitoring of COVID-19 [2]. Therefore, researchers in many countries are tracking SARS-CoV-2 in wastewater as a complementary approach to monitor the spread of COVID-19 [1].

The widespread existence of SARS-CoV-2 throughout wastewater systems could have important implications in the environmental transmission of COVID-19 [2–4]. Potential health concerns due to direct waterborne exposure to this virus in wastewater are well documented [5–7]. However, the aerosolized pathway must also be considered in this context as it could offer a more direct respiratory pathway for human exposure to SARS-CoV-2. Aerosolized viruses are often produced locally in buildings and on a larger scale during wastewater treatment or irrigation [8,9]. This short review is intended to highlight the potential implications of aerosolized SARS-CoV-2 generated from the top three wastewater-associated sources of aerosol: toilet flushing, wastewater treatment, and sprinkler irrigation. Consistent with WHO, and the literature, here the term aerosol is referred for the small breathable particles of <10 μm (PM_{10}) that can remain airborne with the capability of short- and long-range transport [10].

Inhalation of respiratory droplets/aerosols and/or interactions with contaminated surfaces are the main transmission routes of SARS-CoV-2, which is a highly contagious virus [11]. According to WHO (2020), airborne transmission of this virus is possible during aerosol-generating medical events. However, as reported in its scientific brief of 9 July, 2020, WHO is also evaluating the possibility of SARS-CoV-2 spread through aerosols in the absence of aerosol-producing processes [11]. Experimental studies involving aerosols of infectious samples found that SARS-CoV-2 can remain viable for up to several hours [12,13]. Moreover, aerosols are likely to contribute to longer-range transport and potential infection from the pathogens [10]. Therefore, it is critical to assess the potential of the aerosolized pathway where the probability of direct respiratory exposure to SARS-CoV-2 is substantially higher. In this context, SARS-CoV-2 in aerosolized wastewater becomes an important scenario in its exposure pathways that should not be ruled out.

2. Exposure to SARS-CoV-2 in Aerosolized Wastewater

Although there is currently no proof of wastewater-related exposure to SARS-CoV-2, the contribution of aerosolized wastewater as a transmission route was recognized during the SARS epidemic of 2003 [14]. Respiration of virus-laden aerosols, created through the defective plumbing and sewage system, was recognized as a potential transmission route within a housing complex in Hong Kong where 187 people were infected [14]. Therefore, the transmission of this virus in aerosolized wastewater remains a significant possibility in the following scenarios: toilet flushing, wastewater treatment, and sprinkler irrigation (Figure 1). These scenarios are briefly described in the following sections.

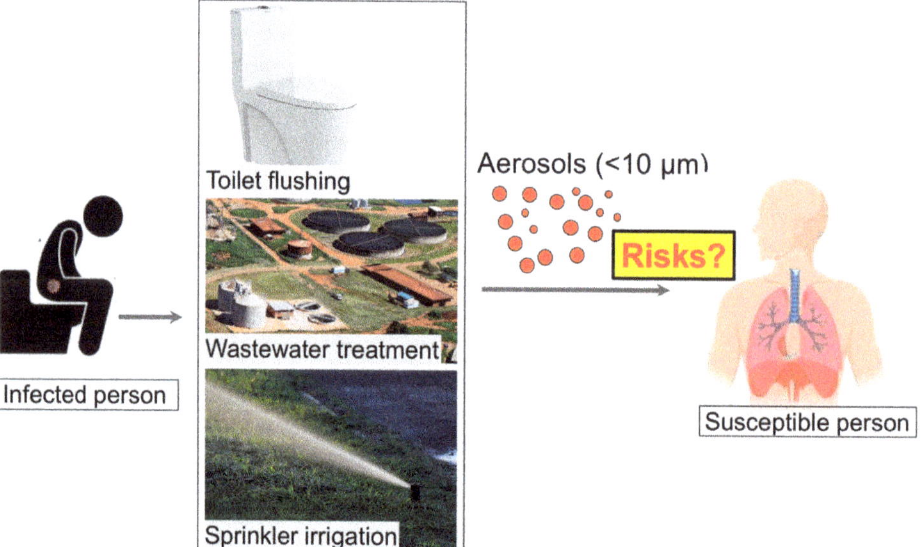

Figure 1. Overview of potential dissemination of SARS-CoV-2 (severe acute respiratory syndrome coronavirus 2) via aerosolized wastewater.

2.1. Bioaerosol Generation by Toilet Flushing

The toilet flushing creates a great deal of turbulence that generates bioaerosols containing pathogenic microorganisms [8]. For example, Wilson et al. [15] reported that toilet flushing increased the concentration of pathogens such as *Clostridioides difficile* in hospital air. Moreover, 95% of these droplets were small enough (<2 µm diameter and >99% <5 µm) to present an airborne infection concern [16,17]. A full-scale pilot experiment on a two-story wastewater plumbing system using *Pseudomonas putida* (as a model pathogen being flushed into the system) revealed that pathogens can be aerosolized and transmitted between rooms [18]. Droplet fallout contaminated the surfaces within the system and rooms. Recently, Li et al. [19] used computational fluid dynamics to model fluid flows to estimate how far aerosol particles may transport due to toilet flushing. The simulations show a substantial upward transport of virus particles, 40–60% of which rise above the toilet seat and may reach to a height of 106.5 cm from the ground. Moreover, these particles remain suspended in the air more than a minute after the flush [19]. An analysis in two hospitals from Wuhan, China revealed that the concentration of SARS-CoV-2 RNA was high in patient toilets, while it was very low in aerosols in ventilated patient rooms and isolation wards [20]. In a hospital for COVID-19 patients, Ding et al. [21] marked toilets as the high-risk area where most of the identified SARS-CoV-2 RNA in the hospital emerged from the fecal-derived aerosols. Thus, bioaerosols generated by toilet flushing could potentially contribute to the environmental transmission of SARS-CoV-2. Though there does not exist any specific research on the toilet-associated generation of infectious bioaerosols containing SARS-CoV-2, it would be judicious to use precautions to prevent this transmission route [9]. Closing the lid on the toilet before flushing and cleaning the toilet seat before using it has been recommended [17,19]. Environmental disinfection of toilet areas should be imposed in healthcare facilities and public toilets [20]. Toilets are generally indoor, having limited potential for aerosolized virus dilution as compared with the outdoor settings (wastewater treatment facilities and sprinkler irrigation systems). It has also been recommended to ensure sufficient and effective ventilation, possibly enhanced by air filtration and disinfection [22]. In addition, modification in toilet designs should also be considered [19]. Raising public awareness is crucial to impede the toilet-associated transmission of SARS-CoV-2.

2.2. Bioaerosol Produced during Wastewater Treatment

Wastewater contains a high number of pathogens such as viruses, bacteria, and parasites. Processes involved in wastewater treatment may lead to the aerosolization of these pathogens [23,24]. Therefore, bioaerosols generated at cooling towers and wastewater treatment plants (WWTPs) have been widely considered as a potential health hazard for sewage workers and nearby communities [24,25]. For example, Masclaux et al. [23] found adenovirus RNA in 100% of summer air samples of WWTP and 97% of winter samples. They detected norovirus in only 3 of the 123 air samples, but no sample contained the hepatitis E virus. Courault et al. [25] detected hepatitis E virus and norovirus RNA in the aerosol produced from active sludge basins of WWTPs and in that of plots irrigated with the treated water. Indeed, risks of SARS-CoV-2 aerosolization can be particularly high in uncovered aerobic wastewater treatment facilities like an aerobic tank [26] and activated sludge process [23]. The potential for the coronaviruses to become aerosolized increases with the transport in water [27], particularly during the pumping of wastewater, during its discharge and subsequent flow through the drainage network [28]. Substantial load of SARS-CoV-2 arriving at WWTPs should raise concerns for its aerosolization and ultimate implications in public health. Due to the significant concentrations of SARS-CoV-2 RNA in wastewater, WWTPs are gaining attention for early tracking and removal of this virus [29]. The viral loads can be decreased in WWTPs by wastewater disinfection and filtration [3,30]. However, the effects of the disinfection process on the microbial community in wastewater, antimicrobial resistance, and associated environmental risks should also be considered [31]. Safety practices should be particularly ensured to protect the health of sanitation workers.

2.3. Bioaerosol Produced during Irrigation

The use of sprinkler/spray irrigation is prevalent in many countries to apply treated wastewater in urban green spaces and agricultural soils. Sprinkler irrigation can potentially aerosolize the pathogens if present in the wastewater [32]. It can provide a respiratory route for exposure to the irrigators and community members in the vicinity [25,33]. In addition to the irrigation method, viruses and other pathogens transported to the soil by irrigation water can also be aerosolized later during windy spells [34]. Indeed, 1–15% of viruses transported to the soil with irrigation water were aerosolized, of which 11–89% were aerosolized within the first 30 min [34]. Risk assessment of airborne enteric viruses, released from wastewater used for irrigation in France, revealed that an increase in wind speed and a decrease in distance from the pathogen source can significantly increase the probability of infection [25]. Similarly, pathogens can be aerosolized in wastewater canals [35]. This can pose disproportionate risks to developing communities having poor sewage infrastructure. As reported in the 2017 World Water Development Report of the United Nations, 80% of wastewater worldwide (>95% in some developing countries) is discharged into the environment without suitable treatment.

3. Concluding Remarks and Perspectives

The COVID-19 patients discharge infectious SARS-CoV-2 virus in their feces [36,37], highlighting the potential of transmission through fecal–oral route. However, the existing data on SARS-CoV-2 detection in wastewater were obtained by PCR and not by cell infection. Therefore, there exists a clear knowledge gap regarding the detection of infectious SARS-CoV-2 in wastewater. Moreover, further research is needed to assess the survival of the infectious viruses in wastewater. The evidence from surrogate corona viruses such as murine hepatitis virus (MHV) and transmissible gastroenteritis virus (TGEV) showed that it takes at least 10 days for 99% inactivation of these viruses in lake water [38]. That may further highlight the transmission potential of corona viruses such as SARS-COV2 through wastewater.

Though there is no concrete evidence for the spread of SARS-CoV-2 through aerosolized wastewater, its role in the environmental transmission of COVID-19 cannot be ruled out. Protection of workers in these fields needs immediate attention. For that, authorities should ensure that workers, particularly those facing greater exposure risk, have access to appropriate protective equipment, adequate training in infection control, high testing rates, and paid sick leave. The situation is particularly critical in developing communities due to poor sewage infrastructures and their lack of access to adequate water and hygiene facilities. Substantial viral load within aerosolized wastewater calls for effective disinfection of wastewater to prevent the formation of pathogen-laden aerosols. Expertise in wastewater science and technology would be required, whereas transition to the aerosol phase calls for treatment according to aerosol standards. Research is also needed for the rapid characterization of aerosols, measurement of SARS-CoV-2 in wastewater aerosols, and their risk assessment.

Author Contributions: M.U.: Conceptualization, Writing—Review and Editing, M.F. (1): Writing—Review and Editing, M.F. (2): Writing—Review and Editing, I.A.: Writing—Review and Editing. All authors have read and agreed to the published version of the manuscript.

Funding: This research was funded by a research grant to M. Usman from Madayn—Public Establishment for Industrial Estates, Oman (CHAIR/DVC/MADAYN/20/02).

Institutional Review Board Statement: Not applicable.

Informed Consent Statement: Not applicable.

Data Availability Statement: No new data were created or analyzed in this study. Data sharing is not applicable to this article.

Conflicts of Interest: The authors declare no conflict of interest.

References

1. Bivins, A.; North, D.; Ahmad, A.; Ahmed, W.; Alm, E.; Been, F.; Bhattacharya, P.; Bijlsma, L.; Boehm, A.B.; Brown, J.; et al. Wastewater-Based Epidemiology: Global Collaborative to Maximize Contributions in the Fight Against COVID-19. *Environ. Sci. Technol.* **2020**, *54*, 7754–7757. [CrossRef] [PubMed]
2. Lodder, W.; de Roda Husman, A.M. SARS-CoV-2 in wastewater: Potential health risk, but also data source. *Lancet Gastroenterol. Hepatol.* **2020**, *5*, 533–534. [CrossRef]

3. Usman, M.; Farooq, M.; Hanna, K. Existence of SARS-CoV-2 in Wastewater: Implications for Its Environmental Transmission in Developing Communities. *Environ. Sci. Technol.* **2020**, *54*, 7758–7759. [CrossRef]

4. Franklin, A.B.; Bevins, S.N. Spillover of SARS-CoV-2 into novel wild hosts in North America: A conceptual model for perpetuation of the pathogen. *Sci. Total Environ.* **2020**, *733*, 139358. [CrossRef]

5. Collivignarelli, M.C.; Collivignarelli, C.; Carnevale Miino, M.; Abbà, A.; Pedrazzani, R.; Bertanza, G. SARS-CoV-2 in sewer systems and connected facilities. *Process Saf. Environ. Prot.* **2020**, *143*, 196–203. [CrossRef]

6. SanJuan-Reyes, S.; Gómez-Oliván, L.M.; Islas-Flores, H. COVID-19 in the environment. *Chemosphere* **2021**, *263*, 127973. [CrossRef]

7. Bogler, A.; Packman, A.; Furman, A.; Gross, A.; Kushmaro, A.; Ronen, A.; Dagot, C.; Hill, C.; Vaizel-Ohayon, D.; Morgenroth, E.; et al. Rethinking wastewater risks and monitoring in light of the COVID-19 pandemic. *Nat. Sustain.* **2020**, *3*, 981–990. [CrossRef]

8. Knowlton, S.D.; Boles, C.L.; Perencevich, E.N.; Diekema, D.J.; Nonnenmann, M.W.; Program, C.D.C.E. Bioaerosol concentrations generated from toilet flushing in a hospital-based patient care setting. *Antimicrob. Resist. Infect. Control* **2018**, *7*, 16. [CrossRef] [PubMed]

9. Gormley, M.; Aspray, T.J.; Kelly, D.A. COVID-19: Mitigating transmission via wastewater plumbing systems. *Lancet Glob. Health* **2020**, *8*, e643. [CrossRef]

10. Anderson, E.L.; Turnham, P.; Griffin, J.R.; Clarke, C.C. Consideration of the Aerosol Transmission for COVID-19 and Public Health. *Risk Anal.* **2020**, *40*, 902–907. [CrossRef]

11. WHO. Transmission of SARS-CoV-2: Implications for Infection Prevention Precautions. 2020. Available online: https://www.who.int/news-room/commentaries/detail/transmission-of-sars-cov-2-implications-for-infection-prevention-precautions (accessed on 15 December 2020).

12. Fears, A.; Klimstra, W.; Duprex, P.; Hartman, A.; Weaver, S.; Plante, K.; Mirchandani, D.; Plante, J.A.; Aguilar, P.; Fernández, D.; et al. Persistence of Severe Acute Respiratory Syndrome Coronavirus 2 in Aerosol Suspensions. *Emerg. Infect. Dis.* **2020**, *26*. [CrossRef] [PubMed]

13. Van Doremalen, N.; Bushmaker, T.; Morris, D.H.; Holbrook, M.G.; Gamble, A.; Williamson, B.N.; Tamin, A.; Harcourt, J.L.; Thornburg, N.J.; Gerber, S.I.; et al. Aerosol and Surface Stability of SARS-CoV-2 as Compared with SARS-CoV-1. *N. Engl. J. Med.* **2020**, *382*, 1564–1567. [CrossRef] [PubMed]

14. Yu, I.T.S.; Li, Y.; Wong, T.W.; Tam, W.; Chan, A.T.; Lee, J.H.W.; Leung, D.Y.C.; Ho, T. Evidence of Airborne Transmission of the Severe Acute Respiratory Syndrome Virus. *N. Engl. J. Med.* **2004**, *350*, 1731–1739. [CrossRef]

15. Wilson, G.M.; Jackson, V.B.; Boyken, L.D.; Schweizer, M.L.; Diekema, D.J.; Petersen, C.A.; Breheny, P.J.; Nonnenmann, M.W.; Perencevich, E.N. Bioaerosols generated from toilet flushing in rooms of patients with Clostridioides difficile infection. *Infect. Control Hosp. Epidemiol.* **2020**, *41*, 517–521. [CrossRef] [PubMed]

16. Johnson, D.; Lynch, R.; Marshall, C.; Mead, K.; Hirst, D. Aerosol Generation by Modern Flush Toilets. *Aerosol Sci. Technol.* **2013**, *47*, 1047–1057. [CrossRef] [PubMed]

17. McDermott, C.V.; Alicic, R.Z.; Harden, N.; Cox, E.J.; Scanlan, J.M. Put a lid on it: Are faecal bio-aerosols a route of transmission for SARS-CoV-2? *J. Hosp. Infect.* **2020**, *105*, 397–398. [CrossRef]

18. Gormley, M.; Aspray, T.J.; Kelly, D.A.; Rodriguez-Gil, C. Pathogen cross-transmission via building sanitary plumbing systems in a full scale pilot test-rig. *PLoS ONE* **2017**, *12*, e0171556. [CrossRef]

19. Li, Y.-Y.; Wang, J.-X.; Chen, X. Can a toilet promote virus transmission? From a fluid dynamics perspective. *Phys. Fluids* **2020**, *32*, 065107. [CrossRef]

20. Liu, Y.; Ning, Z.; Chen, Y.; Guo, M.; Liu, Y.; Gali, N.K.; Sun, L.; Duan, Y.; Cai, J.; Westerdahl, D.; et al. Aerodynamic analysis of SARS-CoV-2 in two Wuhan hospitals. *Nature* **2020**, *582*, 557–560. [CrossRef]

21. Ding, Z.; Qian, H.; Xu, B.; Huang, Y.; Miao, T.; Yen, H.-L.; Xiao, S.; Cui, L.; Wu, X.; Shao, W.; et al. Toilets dominate environmental detection of severe acute respiratory syndrome coronavirus 2 in a hospital. *Sci. Total Environ.* **2021**, *753*, 141710. [CrossRef]

22. Tang, J.W.; Bahnfleth, W.P.; Bluyssen, P.M.; Buonanno, G.; Jimenez, J.L.; Kurnitski, J.; Li, Y.; Miller, S.; Sekhar, C.; Morawska, L.; et al. Dismantling myths on the airborne transmission of severe acute respiratory syndrome coronavirus (SARS-CoV-2). *J. Hosp. Infect.* **2021**. [CrossRef] [PubMed]

23. Masclaux, F.G.; Hotz, P.; Gashi, D.; Savova-Bianchi, D.; Oppliger, A. Assessment of airborne virus contamination in wastewater treatment plants. *Environ. Res.* **2014**, *133*, 260–265. [CrossRef]

24. Tian, J.-H.; Yan, C.; Nasir, Z.A.; Alcega, S.G.; Tyrrel, S.; Coulon, F. Real time detection and characterisation of bioaerosol emissions from wastewater treatment plants. *Sci. Total Environ.* **2020**, *721*, 137629. [CrossRef]

25. Courault, D.; Albert, I.; Perelle, S.; Fraisse, A.; Renault, P.; Salemkour, A.; Amato, P. Assessment and risk modeling of airborne enteric viruses emitted from wastewater reused for irrigation. *Sci. Total Environ.* **2017**, *592*, 512–526. [CrossRef]

26. Han, Y.; Yang, K.; Yang, T.; Zhang, M.; Li, L. Bioaerosols emission and exposure risk of a wastewater treatment plant with A2O treatment process. *Ecotoxicol. Environ. Saf.* **2019**, *169*, 161–168. [CrossRef] [PubMed]

27. Casanova, L.; Rutala, W.A.; Weber, D.J.; Sobsey, M.D. Survival of surrogate coronaviruses in water. *Water Res.* **2009**, *43*, 1893–1898. [CrossRef] [PubMed]

28. Quilliam, R.S.; Weidmann, M.; Moresco, V.; Purshouse, H.; O'Hara, Z.; Oliver, D.M. COVID-19: The environmental implications of shedding SARS-CoV-2 in human faeces. *Environ. Int.* **2020**, *140*, 105790. [CrossRef]

29. Lesimple, A.; Jasim, S.Y.; Johnson, D.J.; Hilal, N. The role of wastewater treatment plants as tools for SARS-CoV-2 early detection and removal. *J. Water Process Eng.* **2020**, *38*, 101544. [CrossRef]

30. Naddeo, V.; Liu, H. Editorial Perspectives: 2019 novel coronavirus (SARS-CoV-2): What is its fate in urban water cycle and how can the water research community respond? *Environ. Sci. Water Res. Technol.* **2020**, *6*, 1213–1216. [CrossRef]

31. Usman, M.; Farooq, M.; Hanna, K. Environmental side effects of the injudicious use of antimicrobials in the era of COVID-19. *Sci. Total Environ.* **2020**, *745*, 141053. [CrossRef]

32. Pepper, I.L.; Gerba, C.P. Risk of infection from Legionella associated with spray irrigation of reclaimed water. *Water Res.* **2018**, *139*, 101–107. [CrossRef]

33. Hamilton, K.A.; Hamilton, M.T.; Johnson, W.; Jjemba, P.; Bukhari, Z.; LeChevallier, M.; Haas, C.N. Health risks from exposure to Legionella in reclaimed water aerosols: Toilet flushing, spray irrigation, and cooling towers. *Water Res.* **2018**, *134*, 261–279. [CrossRef]

34. Girardin, G.; Renault, P.; Bon, F.; Capowiez, L.; Chadoeuf, J.; Krawczyk, C.; Courault, D. Viruses carried to soil by irrigation can be aerosolized later during windy spells. *Agron. Sustain. Dev.* **2016**, *36*, 59. [CrossRef]

35. Dickin, S.K.; Schuster-Wallace Corinne, J.; Qadir, M.; Pizzacalla, K. A Review of Health Risks and Pathways for Exposure to Wastewater Use in Agriculture. *Environ. Health Perspect.* **2016**, *124*, 900–909. [CrossRef]

36. Wang, W.; Xu, Y.; Gao, R.; Lu, R.; Han, K.; Wu, G.; Tan, W. Detection of SARS-CoV-2 in Different Types of Clinical Specimens. *JAMA* **2020**, *323*, 1843–1844. [CrossRef] [PubMed]

37. Xiao, F.; Sun, J.; Xu, Y.; Li, F.; Huang, X.; Li, H.; Zhao, J.; Huang, J.; Zhao, J. Infectious SARS-CoV-2 in Feces of Patient with Severe COVID-19. *Emerg. Infect. Dis. J.* **2020**, *26*, 1920. [CrossRef] [PubMed]

38. Tran, H.N.; Le, G.T.; Nguyen, D.T.; Juang, R.-S.; Rinklebe, J.; Bhatnagar, A.; Lima, E.C.; Iqbal, H.M.N.; Sarmah, A.K.; Chao, H.-P. SARS-CoV-2 coronavirus in water and wastewater: A critical review about presence and concern. *Environ. Res.* **2021**, *193*, 110265. [CrossRef]

Review

Nexus between Water Security Framework and Public Health: A Comprehensive Scientific Review

Sushila Paudel [1], Pankaj Kumar [2,*], Rajarshi Dasgupta [2], Brian Alan Johnson [2], Ram Avtar [3], Rajib Shaw [4], Binaya Kumar Mishra [5] and Sakiko Kanbara [6]

[1] Graduate School of Nursing, University of Kochi, 2751-1 Ike, Kochi 781-8515, Japan; s17G403w@st.u-kochi.ac.jp

[2] Natural Resources and Ecosystem Services, Institute for Global Environmental Strategies, Hayama, Kanagawa 240-0115, Japan; dasgupta@iges.or.jp (R.D.); johnson@iges.or.jp (B.A.J.)

[3] Faculty of Environmental Earth Science, Hokkaido University, Sapporo 060-0810, Japan; ram@ees.hokudai.ac.jp

[4] Graduate School of Media and Governance, Keio University, Fujisawa 252-0882, Japan; shaw@sfc.keio.ac.jp

[5] School of Engineering, Pokhara University, Pokhara 33700, Nepal; bkmishra@pu.edu.np

[6] Faculty of Nursing, University of Kochi, 2751-1 Ike, Kochi 781-8515, Japan; kanbara@cc.u-kochi.ac.jp

* Correspondence: kumar@iges.or.jp

Citation: Paudel, S.; Kumar, P.; Dasgupta, R.; Johnson, B.A.; Avtar, R.; Shaw, R.; Mishra, B.K.; Kanbara, S. Nexus between Water Security Framework and Public Health: A Comprehensive Scientific Review. *Water* **2021**, *13*, 1365. https://doi.org/10.3390/w13101365

Academic Editor: Luís Filipe Sanches Fernandes

Received: 2 April 2021
Accepted: 11 May 2021
Published: 14 May 2021

Publisher's Note: MDPI stays neutral with regard to jurisdictional claims in published maps and institutional affiliations.

Abstract: Water scarcity, together with the projected impacts of water stress worldwide, has led to a rapid increase in research on measuring water security. However, water security has been conceptualized under different perspectives, including various aspects and dimensions. Since public health is also an integral part of water security, it is necessary to understand how health has been incorporated as a dimension in the existing water security frameworks. While supply–demand and governance narratives dominated several popular water security frameworks, studies that are specifically designed for public health purposes are generally lacking. This research aims to address this gap, firstly by assessing the multiple thematic dimensions of water security frameworks in scientific disclosure; and secondly by looking into the public health dimensions and evaluating their importance and integration in the existing water security frameworks. For this, a systematic review of the Scopus database was undertaken using Preferred Reporting Items for Systematic Reviews and Meta-Analyses (PRISMA) guidelines. A detailed review analysis of 77 relevant papers was performed. The result shows that 11 distinct dimensions have been used to design the existing water security framework. Although public health aspects were mentioned in 51% of the papers, direct health impacts were considered only by 18%, and indirect health impacts or mediators were considered by 33% of the papers. Among direct health impacts, diarrhea is the most prevalent one considered for developing a water security framework. Among different indirect or mediating factors, poor accessibility and availability of water resources in terms of time and distance is a big determinant for causing mental illnesses, such as stress or anxiety, which are being considered when framing water security framework, particularly in developing nations. Water quantity is more of a common issue for both developed and developing countries, water quality and mismanagement of water supply-related infrastructure is the main concern for developing nations, which proved to be the biggest hurdle for achieving water security. It is also necessary to consider how people treat and consume the water available to them. The result of this study sheds light on existing gaps for different water security frameworks and provides policy-relevant guidelines for its betterment. Also, it stressed that a more wide and holistic approach must be considered when framing a water security framework to result in sustainable water management and human well-being.

Keywords: water security; water insecurity; water scarcity; water security framework; public health; primary health care; COVID-19

1. Introduction

Water security is a concept that has recently gained widespread global attention. With the increase in population, rapid urbanization, overexploitation of natural resources, encroachment on natural forests, declining soil fertility, lack of capacity to adapt to climate change, and inadequate capacity of institutions for water management, achieving water security has become one of the emerging global challenges [1–3]. Nearly 1.8 billion people worldwide are already living in countries experiencing high water stress or scarcity [4], and by 2025 more than 2.8 billion people in 48 countries may face water stress [5]. Furthermore, the number of people exposed to water stress could double by 2050, when compared to 2010 [5]. All stakeholders must act quickly to address these challenges of rapid population growth, natural resource degradation, and climate change for better adaptation and achieving water security in a timely manner [3].

Over the last few decades, the concept of security has moved beyond a narrow emphasis on military threats and conflicts to wider concepts of human security, wherein water serves as a central link between health, economic, political, personal, food, energy, and environmental aspects of human security [6]. While the term 'water security' has been conceptualized with a variety of meanings by different scholars, managers, planners, and stakeholders to fit in their specific contexts [7], the United Nations task force on water security has holistically defined water security as *"the capacity of a population to safeguard sustainable access to adequate quantities of acceptable quality water for sustaining livelihoods, human well-being, and socio-economic development, for ensuring protection against water-borne pollution and water-related disasters, and for preserving ecosystems in a climate of peace and political stability"* [4].

Water security is a key factor affecting public health. According to the Sustainable Development Goal (SDG) 6 synthesis study [4], 2.1 billion people lack access to clean drinking water, 4.5 billion people lack access to safely managed sanitation, 0.9 billion people still practice open defecation, and in the least developed nations only 27% of the population has access to soap and water for hand sanitation. In particular, vulnerable groups (e.g., ethnic minorities, children, the urban poor) are those who typically suffer most. Inadequate accessibility to clean water is also often linked with gender inequality. Every day, women and girls are estimated to spend 200 million hours hauling water around the world. In rural Africa, the average woman walks 6 km a day to carry 40 pounds of water [8]. These factors can negatively affect women's opportunities for pursuing education and careers. When there is an absence, insufficient or poorly regulated water and sanitation services, individuals become vulnerable to different preventable health risks [9]. From an environmental point of view, water-related diseases can be waterborne diseases caused by ingestion of contaminated water (e.g., Diarrhea, typhoid); water-washed diseases caused by poor personal hygiene (ex. lice, skin rashes), water-based diseases caused by parasites living in the water (e.g., some helminths), and diseases transmitted by water-associated insect vectors that breed in water (e.g., dengue, malaria) [10].

The incidence of outbreaks depends on the level of scarcity, population density, economic growth, extreme weather events [11]. When people lack even a basic drinking water service, they depend on surface water and/or wastewater that is not safe. Already, at least two billion people around the world are using drinking water sources that are contaminated with feces [9]. Water-borne diseases also occur through leakage of contaminated run-off water, or within the distribution of pipe systems [12].

Despite advancements in science and technology and water security measures, waterborne diseases kill 2195 children every day which is more than AIDS, malaria, and measles combined. This accounts for 1 in 9 child deaths worldwide, making waterborne diseases the second leading cause of death among children under the age of five, even in the 21st century [13]. Diarrheal diseases, the most common type of waterborne diseases, are particularly serious for children and vulnerable people in low-income countries [9]. On top of that, occasional climate-related hazards such as floods and droughts can further increase the pathogen load making water unsafe to drink. Flood can damage water infrastructures, sanitation facilities, reduces water quality, and can mix up drinking water with industrial and agricultural waste, increasing the risk of waterborne diseases [14] while droughts lead to shortages of water and poor water quality [13,15,16].

The risk has been even greater due to the emergence of SARS, MERS, and now COVID-19 [17]. Furthermore, water scarcity, poor quality, and poor accessibility to clean water can lead to additional mental health conditions such as persistent psychological stress, social alienation, intra-community disputes, despair, hopelessness, depression, and anxiety especially in developing countries [18,19]. Finally, due to rapid global changes including climate change, land-use change, and population growth, the risks of waterborne diseases are expected to further increase [2].

There are several frameworks or assessments developed to measure water security with various scales, aspects, and dimensions [7,20–22]. Considering the above-mentioned facts, health-related issues cannot be denied as a focal point or dimension to designing the water security framework. Thus, a comprehensive understanding of how health has been incorporated as a dimension in water security frameworks in different contexts around the world is needed. This research aims to address this need through a systematic analysis of existing water security frameworks, seeking to answer the following research questions:

1. How have the different dimensions of water security been reflected in scientific disclosures? What is the geographical and rural/urban focus of studies on water security and methods used for water security analysis?
2. What are the health impacts mentioned in the papers?
3. What is the importance of health-related issues in framing any water security framework?

2. Materials and Methods

A systematic literature review was conducted, using the Scopus database (http://www.scopus.com/, accessed on 5 December 2020) to collect existing literature related to water security frameworks. For this analysis, we followed the Preferred Reporting Items for Systematic Reviews and Meta-Analyses (PRISMA) guidelines to document our literature review process [23]. As the search query, we used the Boolean string: TITLE-ABS-KEY (water security) AND (TITLE-ABS-KEY (framework) OR TITLE-ABS-KEY (assessment) OR TITLE-ABS-KEY (health AND hazard)). Publication dates for the search were limited from the year 2000 to 2020, and all searches were limited to the English language texts. The full-texts of all articles retrieved from this search were downloaded and manually screened, and articles meeting any of the following inclusion criteria were used for further analysis: (1) articles discussing water security frameworks or assessments with or without an associated hazard; (2) articles reporting dimensions, components, or indicators of water security; or (3) articles including the term 'health'. Finally, papers matching our inclusion criteria were reviewed and the following information extracted: (a) Spatio-temporal information (i.e., the publication date and location of the studies); (b) the context in which the term water security was used (e.g., in relation to hazard-prone areas, lack of water infrastructure, water pollution, etc.); (c) the key components of the water security framework described in the study; (d) direct and indirect health impacts mentioned, and the importance given to health-related issues in framing a water security framework.

3. Results

The results of our literature search using the PRISMA guidelines are shown in Figure 1. A total of 933 papers were initially retrieved from Scopus using our defined search query. After a manual screening of these 933 articles, a total of 77 articles were found to be matching our defined inclusion criteria. All other journal articles not matching our inclusion criteria, as well as book chapters, conference papers, and other non-peer reviewed reports were excluded from further analysis.

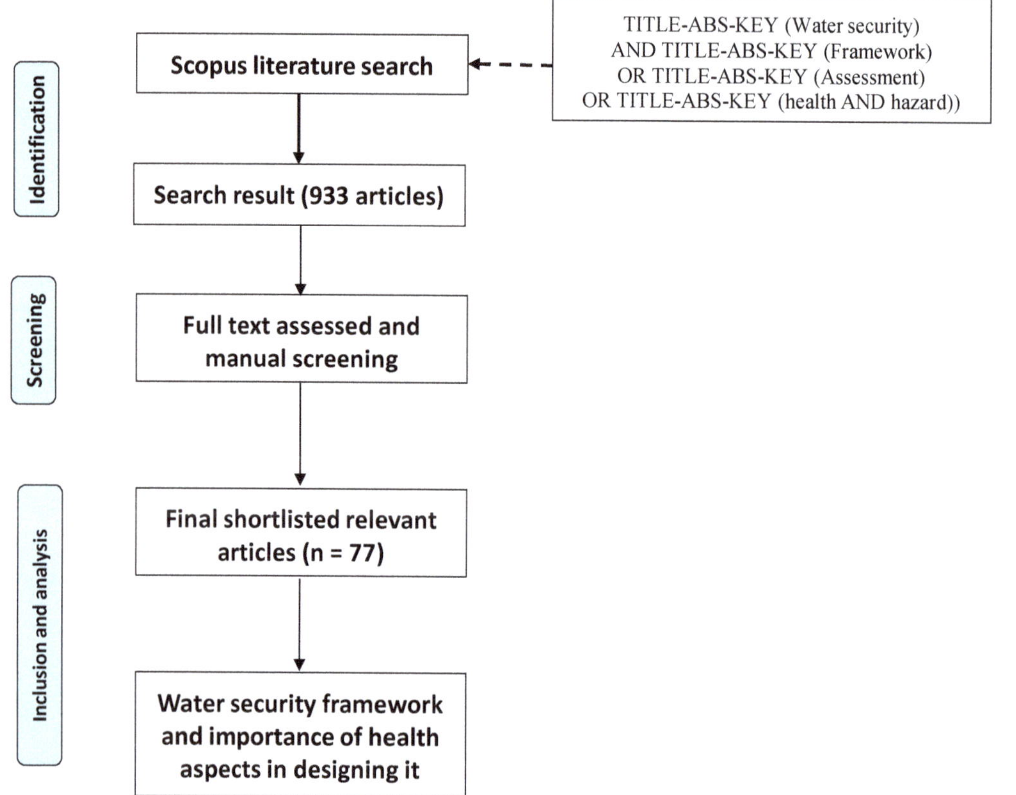

Figure 1. PRISMA flowchart of review results.

3.1. Spatio-Temporal Identity of the Available Literature on Water Security

As shown in Figure 2, all papers matching our inclusion criteria were published between the years 2008 and 2021. The number of articles published per year showed an increasing trend starting from 2015, with almost 79 percent (61 out of 77) of papers on water security being published since 2015. This suggests increasing global attention toward the issue of water security in recent years, possibly as a result of the SDGs (which were agreed upon in 2015), as one of these global goals (i.e., SDG 6: "Clean water and sanitation for all") is directly linked to water security. Indeed, the SDGs have pushed countries around the world to think more holistically about achieving sustainable water development through attaining water security.

With regards to the geographic focus of the articles, 81 percent ($n = 62$) focused on water security in a particular continent, while the remaining 31 percent ($n = 24$) did not mention any specific region. The spatial distribution of the study locations is plotted on a world map (Figure 3). It is found that the majority of the studies related to water security frameworks and health were focused on the Asian continent ($n = 29$), followed by North America ($n = 12$). Conversely, relatively few studies focused on Africa, Australia, Europe, and South America. In terms of the countries that were the focus of most studies, China was the most common ($n = 10$), followed by Canada ($n = 6$). Only one study (1 of 77) covered Europe. However, this study found 24 papers, which did not have any geographical area of focus, because they were either conceptual papers describing water security frameworks or literature review papers such as Bichai et al. [24], UN water [4]. The plotted map does not include studies done in multiple regions: North America and South

America [25]; Artic region with seven countries [26]; seven urban case studies selected from Asia, Europe, North America, and South America [27]; and 27 sites in 21 low-and middle-income countries across Africa, Asia, the Middle East, and the Americas [28].

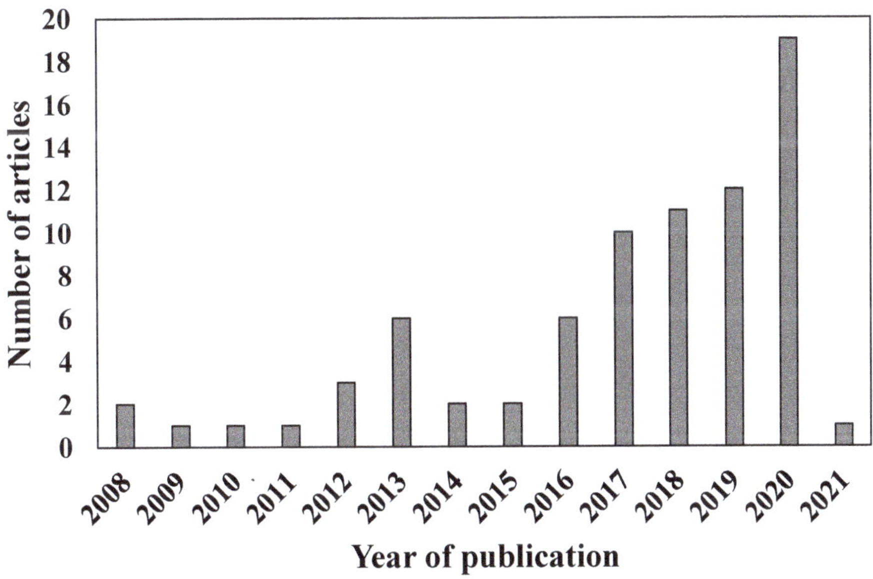

Figure 2. Systematic review by publication date.

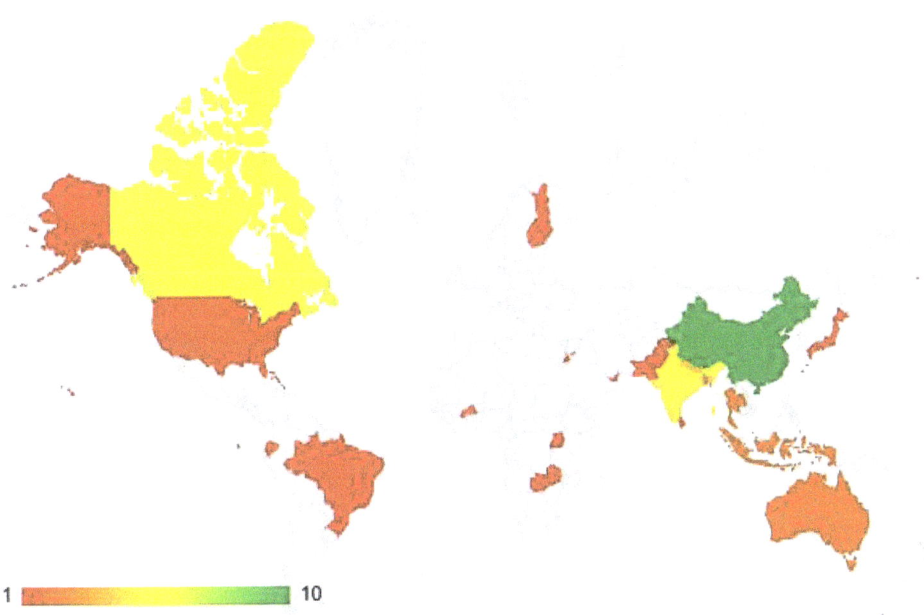

Figure 3. Systematic review by study locations (countries).

3.2. Urban vs. Rural Focus

Here, all the reviewed papers are further categorized based on their focus on rural or urban areas. 49 percent (n = 38) of the studies had a focus on rural areas, urban areas, or both rural and urban areas. The majority focused on urban areas (n = 22), followed by rural (n = 13), and finally papers focusing on both urban and rural areas (n = 3) (Figure 4). Papers dealing with rural areas mainly emphasized accessibility, availability, pollution, lack of awareness, poor governance, etc., as key hurdles for achieving water security. On the other hand, the papers focusing on urban areas mainly emphasized water infrastructure-related issues such as non-revenue loss and leakage from supply pipes, pollution, poor governance, flooding, climate change, etc., as key drivers for water insecurity. Finally, papers falling under the mixed category, deal with the coastal areas or river basins including both rural and urban sites. On the other hand, 51% of the papers did not have any focus area, they were either literature review papers or conceptual papers.

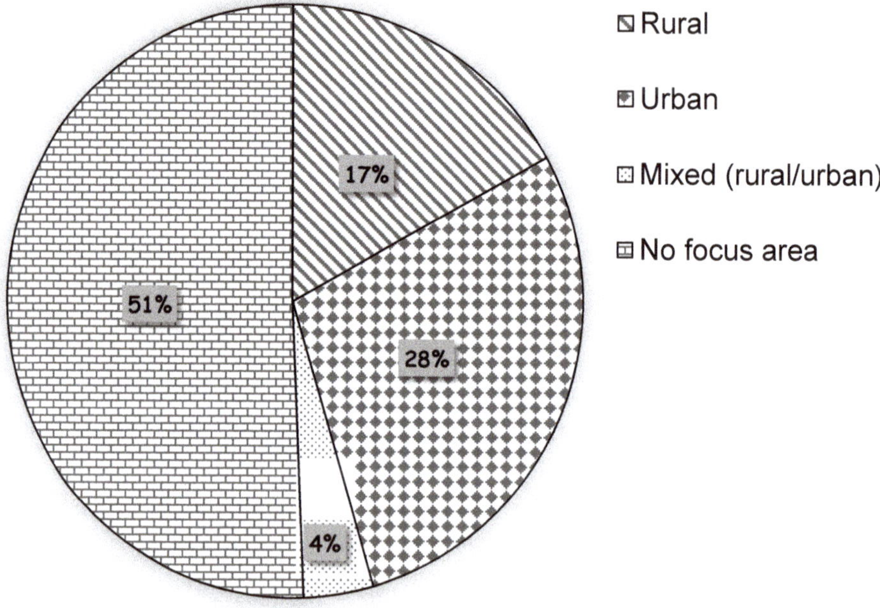

Figure 4. Systematic review by the focus of the study.

3.3. Analytical Approach

To understand the types of analytical approaches used in relation to water security frameworks, we grouped the reviewed studies based on their methodology used for analysis. From this, we found that all the analytical approaches could be classified into four major categories: big data analysis, literature review, qualitative analysis, and quantitative analysis (Figure 5). Around 16% of studies (6 of 77) used big data analysis, e.g., analyzing data from the national census or the global water data portal; 18% of the papers used qualitative analysis to analyze data from household surveys and key informants' interviews. Studies using a literature review as the principle analysis method accounted for 26% of all papers that contained the study of national and international reports, articles, local governmental reports, and case studies. Finally, 40% (the largest percentage) of the papers used quantitative analysis such as numerical simulations, statistical analysis of water quality data, machine-learning (e.g., neural network) analysis, remote sensing, and GIS-based analysis. The reason for quantitative analysis being the most dominant one is possibly due to the advancement of different tools and technologies for analyzing water quality and quantity. Also, many countries have become committed to achieving SDGs, and for this they are enhancing the monitoring activities of their precious water resources. However, for many developing countries, in absence of any past data, it is mandatory to do the baseline studies to get a clear picture about the current status and hence helpful take any timely measures.

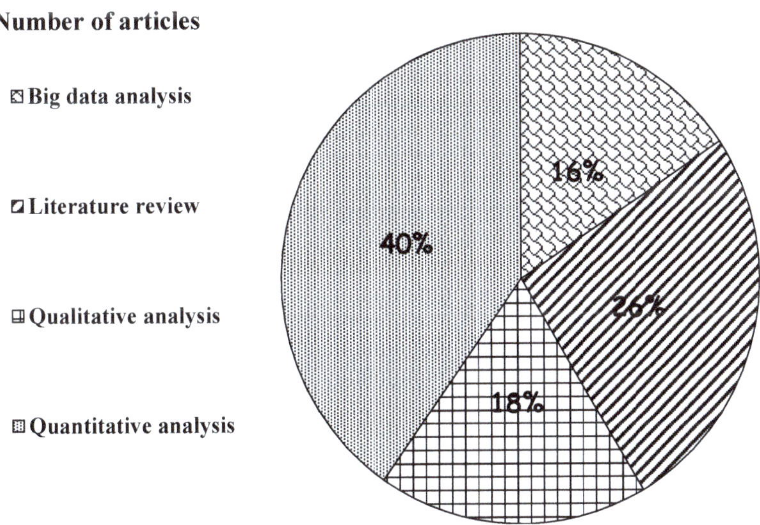

Figure 5. Systematic review by study methods/approach.

3.4. Water Security Dimensions

All of the reviewed papers used multiple indicators or variables to measure water security under a different framework. All the variables and indicators were categorized under 11 dimensions in this study, and the result is shown in Figure 6. According to the number of appearances, the order of different dimensions considered in these papers are, namely public health > water quality > availability > policy and governance > ecosystem > socio-economic > water quantity = accessibility > risk/hazards > infrastructure and technology > sanitation and hygiene. Also, a summary of the list of reviewed papers and their association with different water security dimensions is presented in Table S1. Although health appeared as the most common dimension in these reviewed papers (51 percent of studies, *n* = 39), number of papers measuring direct health impacts such as incidence rates of water borne diseases, were still quite low (n = 14), which highlights the lack of priority being given to health-related issues when framing water security frameworks. Moreover, only a few papers discussed indirect health effects such as anxiety and stress, where water scarcity plays a mediating effect [18,29,30]. Most of the papers discussed direct health effects, which included issues such as the presence of different contaminants and pathogens which cause adverse health effects. The next dimension is water quality followed by availability with 47% and 34% of the share among the papers reviewed in this study [25,31,32]. Here, for the water quality component, most of the papers discussed different contaminants with grave health concerns and different approaches to monitor and manage them to achieve water security. A majority of these papers also discussed health aspects, hence they both share a large number of common papers in this graph. When talking about availability, this is more about water scarcity due to geographical or climatic conditions such as arid regions and highlighting the different possible options for water resource management. The next dominant factor is policy and governance with 26% of papers covering it. It is evident especially from developing countries that poor policy and governance leads to water insecurity whether it is related to wastewater management in a city [2], or that poor policy related to water security based on time and distance to fetch water, often lead to stress and anxiety [33]. The fifth dimension is the ecosystem which appeared in 25% of the reviewed papers. The main idea to bring ecosystem in the discussion here is that an ecosystem-based approach must be adopted to manage not only water security but human well-being in any particular region [34]. The sixth dimension found in this study is socio-economic with an appearance in 22% of the reviewed papers. Here, the socio-economic dimension reflects on water inadequacy for people living in densely populated areas or informal settlements due to poor water systems or a poor capacity to access the available water [35–37]. The next few dimensions from seventh to the tenth rank, such as accessibility, water quantity, risk/hazard, infrastructure, and technology carries approximately similar occurrences in the reviewed papers [38–41]. These dimensions mainly discuss the effect of rapid global changes viz. climate change led to extreme weather conditions such as flooding or drought, on different water-related infrastructure, water accessibility, and water quantity. The last dimension i.e., sanitation and hygiene appeared in about 10% of the reviewed papers. The paper discussing sanitation and hygiene under a water security framework highlights the relation between poor water availability and sanitation and ultimately its impact on health whether its health of people or the ecosystem [42]. The main reason behind these low occurrences despite being a critical element of water security is that sanitation and hygiene are separately being discussed under the topic of Water, Sanitation, and Health (WASH). However, our goal here is to relate these dimensions under the umbrella of the water security framework.

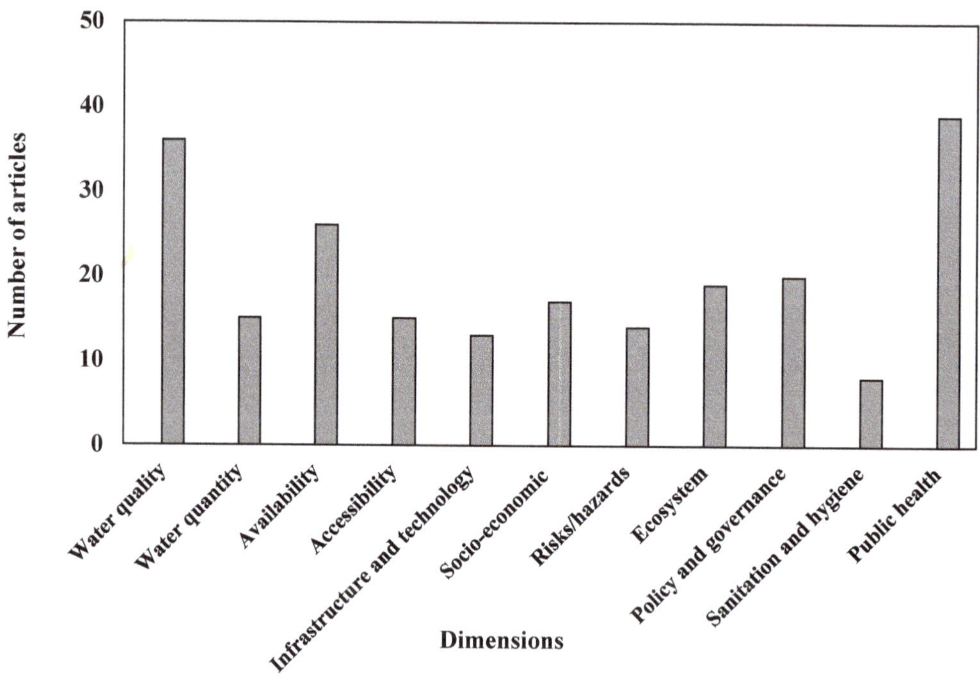

Figure 6. Summary of different water security dimensions.

Among different frameworks reviewed in this study, the water security framework reported by [31], which is also called the DECS framework (Drinking water, Ecosystem, Climate change, and water-related hazards, and Socioeconomic aspects) adopted from the United Nations task force on water security, covered almost all dimensions, followed by Urban Water Security (UWS) framework by Romero-Lankao and Gantz and Urban Water Security Index (UWSI) [43]. On the other hand, frameworks that have widely covered the public health perspectives were the biocultural model of household water insecurity [28]; DECS UWS Framework [31]; UWSI [43]; HWIAS [44].

3.5. Public Health

This category represents both direct and indirect health impacts. According to the World Health Organization [45], health is defined as the state of complete physical, mental and social well-being and not merely the absence of diseases or infirmity. In this review, direct health impacts were considered to be the state free from water-borne diseases or the health impacts caused by the intake of water. Indirect health impacts were considered to be the mediating factors such as the accessibility to adequate quantities of acceptable quality water for physical and mental well-being. As mentioned, a total of 39 papers dealt with public health (both direct and indirect) issues. The result of direct health impacts is shown in Figure 7.

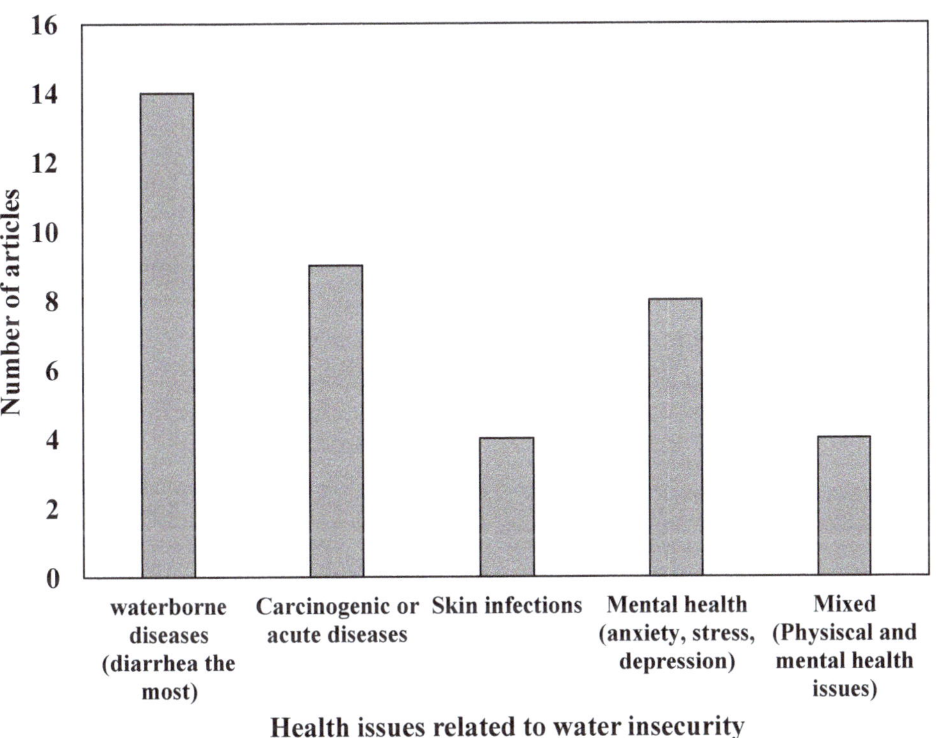

Figure 7. Summary of direct health issues related to water insecurity found in reviewed papers.

Among direct health impacts, physical and mental health were considered. In physical health, water-borne diseases (diarrhea) were most common, considered by 14 papers. However, only 5 among the reviewed frameworks mentioned the inclusion of the incidence rate of diseases as a crucial variable to measure water security [31,46–50]. This was followed by carcinogenic diseases caused by the presence of different pollutants such as heavy metals, including Cd, Pb, etc. [51] and finally skin diseases because of the presence of Arsenic [52]. On the other hand, mental health impacts such as anxiety and water stress appeared in 12 papers, i.e., 16% of the total reviewed papers. Experienced anxieties from water issues are strongly linked to household water needs not being met, time investments required for fetching water, suspected waterborne sickness, and household size [18,29].

Indirect health impacts or the mediating factors were considered by 25 papers. Aboelnga et al. [31] presented the DECS framework which has one of the dimensions as "drinking water and human wellbeing", which includes the sub-indicators such as water availability; diversity of water and energy source; consumption; reliability, water quality, accessibility, adequacy, equity, water bodies' and dependency ratio [47]. Similarly, the water-energy-food nexus framework by Marttunen et al. [20] has one of the dimensions as "human health and well-being" that includes the indicators such as quality and quantity of drinking water, sanitation, and hygiene, recreational opportunities. Similarly, water quality parameters such as heavy metals and microorganisms possess health threats. High Mg intake can cause hypermagnesemia; high Ca intake can cause cholesterol, muscle cramps, and kidney stones; high hard water intake can cause eczema, intestinal problems, and loss of fertility; high fluoride intake can cause fluorosis and osteosarcoma; and high nitrate intake can cause methemoglobinemia [39,53–55] and diseases related to the presence of microorganisms such as E. coli and fecal coliform can cause serious waterborne infectious diseases [44,46,56].

Table 1 shows how we use the term 'human health' while measuring water security as a whole. On one hand, water security measurement is limited to the provision of safe, adequate water supply and sanitation facilities, and on the other hand, the waterborne morbidity and mortality rates are also taken into consideration to achieve water security. The incident rate of waterborne diseases (particularly diarrhea), which is measured in cases per 100,000 population per year, is commonly considered to be one of the indicators of water security [26,31,46–48]. Carcinogenic diseases are measured by checking the prevalence of diseases such as skin malignancy [57]; carcinogenic risks are predicted through hydrochemical evaluation [51]; and mental health issues are often measured through self-reported items and behavioral manifestations [29,44]. Measuring the public's health is an important step in focusing attention and resources on improving health, and slowing the nation's declining quality of life, which is threatening the country's future [58].

Table 1. Public health measurement in water security.

Direct Health Measurements	Indirect Health Measurements
- Waterborne diseases (diarrhea) - Carcinogenic risks/diseases - Mental health issues (anxiety, depression) - Skin infections	- Drinking water and well-being (water quality, availability, accessibility, water adequacy, sanitation, hygiene, etc.) - Human health and well-being (quality, sanitation, hygiene, recreational opportunities) - Water quality parameters (Heavy metals, micro-organisms)

Having clean water provision may not define good human health and having contaminated water may not always mean bad health of a population. To support this statement, the public health dimension (water-borne diseases) had a poor score (1) despite the excellent accessibility to drinking water [31]. It was correlated with poor access to hygiene, and with the intermittent water supply where there is the possibility of microbial regrowth due to static conditions. In contrast, despite the evidence of microbial contamination in the city's drinking water, incidences of waterborne diseases were found to be the lowest [46], attributed to the good hygiene practices by residents and public awareness regarding disinfection of water before drinking.

4. Discussion

Water is a basic need and is fundamental to life and health. Drinking water supply and sanitation are among the essential components of primary health care [59,60]. All people must have access to at least a satisfactory level of water in terms of adequacy, safety, and accessibility because just improving access to clean drinking water reduces a major burden on human health [12]. Depending upon climate, a person's physiology, social culture, and norms, the Sphere Standards suggest at least 15 L of water per day per person for basic survival [61]. It further elaborates that the maximum distance between any household and the nearest water point is 500 m; queueing time at a water source is no more than 15 min, and filling a 20-L container takes no more than three minutes [61]. Yet, every day, women and girls are estimated to spend 200 million hours hauling water around the world [8].

During the Covid19 pandemic, where handwashing is crucial to reduce infection, a Household water Insecurity Scale (HWIS) study showed that many households in low- and middle-income countries were not able to wash hands due to lack of basic handwashing facilities [17,62]. This reveals that significant investments in water infrastructure, water governance, promotion of knowledge, and behavioral changes are crucial [62,63].

Apart from the existing conditions, there are occasional conditions such as disasters, natural or man-made that affect safe water availability; accessibility; reliability; and adequacy, consequently worsening physical and mental health. A study by Rosinger et al. [64] after a historic flood found a higher dehydration prevalence among children in households with high water insecurity. Dehydration, on the other hand, has mental health impacts such as anxiety and depression [65]; mood effects [66]; tension and fatigue [67]; tension, depression, and confusion [30]; and short sleep duration [64]. A case from Haiti backs up theories that suggest household water insecurity plays a central, influential role as a potential driver of common mental illness in households through direct and indirect pathways such as food insecurity and sanitation [68].

Water-borne outbreaks could be minimized if the water is treated before drinking. A study by Carlton et al. [69] had an interesting finding that good sanitation, hygiene, and social cohesion did not modify the relationship between heavy rainfall events and diarrhea, instead of drinking water treatment by households was the factor to reduce the diarrheal events. Thus, they emphasized adopting water treatment behavior as a climatic adaptation to reduce climate vulnerability. Furthermore, even if there is a good WASH intervention, if other factors such as climatic variability, ecosystem, and other factors are not considered, there are very few chances to prevent frequent occurrences of water-borne diseases [69]. This evidence clearly shows that just having good access to drinking water or with good WASH services does not always mean good health. Therefore, measuring the direct physical and mental health impacts caused by the intake of water distributed would give justice to dimensions of water security for having a sustainable healthy livelihood.

This study addresses a gap in how we perceive health while measuring water security. Health, in medical discipline especially by doctors and nurses, is defined as the absence of diseases or infirmity, while health in water disciplines can be ecological health and the mediating factors for health. On one hand, health depends on hygienic behavior, and on the other hand, it depends on environmental conditions. Under environmental pressures such as climate change and other factors, the risks of waterborne diseases tend to increase, so much effort is being intervened to supply adequate safe water, which then is known as water security, while health professionals look after the morbidity and mortality rates, and conclude that water security is not achieved despite advancements in science and technology and water security measures. The biohazard 'Covid19' is another example that reflects the urgent call for water security. Studies also concluded that the household culture, norms, and how people treat water before consumption affects the value measured. This gap should be bridged by emphasizing more local and household surveys with the inclusion of health indicators such as households with water-borne diseases, hygiene, and sanitation both in routine and disaster periods for sustainable human security. Also, a specific health-specific water security framework is imperative for dealing with issues arising from public health.

5. Conclusions

This study reports the water security frameworks and the place for health in them. Asia is the main hotspot for water security issues owing to rapid population growth, urbanization, and extreme weather conditions due to climate change. Mental health issues such as stress caused by poor accessibility and availability (in terms of time and distance) is a big determinant when framing water security frameworks, particularly in developing nations. Water-borne diseases due to poor water quality and sanitation play a major role in determining the water security framework for countries with poor governance on water and sanitation. Among different health issues, diarrhea is the most prevalent one considered for developing a water security framework because of poor water quality, especially the presence of biological pollutants. Although water security is a concept with several aspects and dimensions, public health is often not integrated into water security frameworks. More local and household surveys should be emphasized, with the inclusion of health indicators such as the percentage of households with water-borne diseases, hygiene, and sanitation both in routine and disaster periods for sustainable human security. As an important implication, this study provides knowledge and information of public health indicators in the available water security frameworks. This information is important for scientific communities, managers, policy planners, and stakeholders not only to conceptualize water security in a variety of meanings but also to measure water-related health issues for a population and its relation to other socio-environmental, provisional, and technical issues commonly prevailing in the human society. For the future, this study emphasizes developing comprehensive ways to refine the public health indicators at a ground level, to measure the significant burden that water insecurity places on human health. Similarly, a holistic approach with regular monitoring and future prediction of water resources and designing management measures on a timely basis are very much needed. Nexus approaches are needed, considering aspects of water-food-health-energy [20] or socio-hydrology [18] to achieve this very complex issue of water security and human well-being.

Supplementary Materials: The following are available online at https://www.mdpi.com/article/10.3390/w13101365/s1, Table S1: List of reviewed papers (from the Scopus search) associated with different water security dimensions.

Author Contributions: Conceptualization: P.K., S.P.; methodology: P.K., S.P., R.D.; formal analysis: P.K., S.P.; investigation: P.K., S.P.; writing—original draft preparation: P.K., S.P.; writing—review and editing: P.K., R.D., B.A.J., R.S., R.A., B.K.M., S.K. All authors have read and agreed to the published version of the manuscript.

Funding: This research received no external funding.

Institutional Review Board Statement: Not applicable.

Informed Consent Statement: Not applicable.

Data Availability Statement: Not applicable.

Acknowledgments: Authors also want to acknowledge the support from the Asia-Pacific Network for Global Change Research (APN) under Collaborative Regional Research Programme (CRRP) project with project reference number CRRP2019-01MY-Kumar, to accomplish this research work.

Conflicts of Interest: The authors declare no conflict of interest.

References

1. Pandey, C.L.; Maskey, G.; Devkota, K.; Ojha, H. Investigating the Institutional Landscape for Urban Water Security in Nepal. *Sustain. J. Rec.* **2019**, *12*, 173–181. [CrossRef]
2. Kumar, P. Numerical quantification of current status quo and future prediction of water quality in eight Asian megacities: Challenges and opportunities for sustainable water management. *Environ. Monit. Assess.* **2019**, *191*, 319. [CrossRef] [PubMed]

3. Maja, M.M.; Ayano, S.F. The Impact of Population Growth on Natural Resources and Farmers' Capacity to Adapt to Climate Change in Low-Income Countries. *Earth Syst. Environ.* **2021**, 1–13. [CrossRef]
4. UN Water. *Water Security and the Global Water Agenda*; UNU-INWEH: Hamilton, ON, Canada, 2013; p. 38. ISBN 9789280860382.
5. UNESCO. *The United Nations World Water Development Report 2017*; Wastewater: The Untapped Resource; UNESCO: Paris, France, 2017.
6. UNESCO and UNESCO International Centre for Water Security and Sustainable Management (UNESCO and UNESCO ICWSSM). *Water Security and Sustainable Development Goals. Global Water Security Issues Series*; I-WSSM; UNESCO: Paris, France; Daejeon, Korea, 2019; p. 210. ISBN 9789231003233.
7. Gerlak, A.K.; House-Peters, L.; Varady, R.G.; Albrecht, T.; Zúñiga-Terán, A.; De Grenade, R.R.; Cook, C.; Scott, C.A. Water security: A review of place-based research. *Environ. Sci. Policy* **2018**, *82*, 79–89. [CrossRef]
8. UNICEF. *UNICEF Annual Report 2016*; UNICEF: New York, NY, USA, 2017; p. 84.
9. WHO. Drinking-Water. 2019. Available online: https://www.who.int/news-room/fact-sheets/detail/drinking-water (accessed on 15 February 2021).
10. White, G.F.; Bradley, D.J.; White, A.U. *Drawers of Water: Domestic Water Use in East Africa*; University of Chicago Press: Chicago, IL, USA, 1972; p. 306.
11. Jofre, J.; Blanch, A.R.; Lucena, F. Water-Borne Infectious Disease Outbreaks Associated with Water Scarcity and Rainfall Events. In *The Handbook of Environmental Chemistry*; Springer: Berlin, Heidelberg, 2009; Volume 8, pp. 147–159. [CrossRef]
12. WHO. *Water Safety in Distribution Systems. Public Health, Environmental and Social Determinants Water, Sanitation, Hygiene and Health*; WHO: Geneva, Switzerland, 2014; p. 157. Available online: https://www.who.int/water_sanitation_health/publications/Water_safety_distribution_systems_2014v1.pdf?ua=1 (accessed on 1 April 2021).
13. CDC. Health Implications of Drought. 2021. Available online: https://www.cdc.gov/nceh/drought/implications.htm (accessed on 28 March 2021).
14. Talbot, C.J.; Bennett, E.M.; Cassell, K.; Hanes, D.M.; Minor, E.C.; Paerl, H.; Raymond, P.A.; Vargas, R.; Vidon, P.G.; Wollheim, W.; et al. The impact of flooding on aquatic ecosystem services. *Biogeochemistry* **2018**, *141*, 439–461. [CrossRef]
15. EEA. Water Scarcity and Drought. 2018. Retrieved from European Environment Agency. Available online: https://www.eea.europa.eu/archived/archived-content-water-topic/water-resources/water-scarcity-and-drought (accessed on 28 March 2021).
16. Sadoff, C.W.; Hall, J.W.; Grey, D.; Aerts, J.C.J.H.; Ait-Kadi, M.; Brown, C.; Cox, A.; Dadson, S.; Garrick, D.; Kelman, J.; et al. *Securing Water, Sustaining Growth: Report of the GWP/OECD Task Force on Water Security and Sustainable Growth*; University of Oxford: Oxford, UK, 2015; p. 180.
17. Staddon, C.; Everard, M.; Mytton, J.; Octavianti, T.; Powell, W.; Quinn, N.; Uddin, S.M.N.; Young, S.L.; Miller, J.D.; Budds, J.; et al. Water insecurity compounds the global corona-virus crisis. *Water Int.* **2020**, *45*, 416–422. [CrossRef]
18. Kumar, P.; Avtar, R.; Dasgupta, R.; Johnson, B.A.; Mukherjee, A.; Ahsan, N.; Nguyen, D.C.H.; Nguyen, H.Q.; Shaw, R.; Mishra, B.K. Socio-hydrology: A key approach for adaptation to water scarcity and achieving human well-being in large riverine islands. *Prog. Disaster Sci.* **2020**, *8*, 100134. [CrossRef]
19. Stanwell-Smith, R. Classification of Water-Related Disease. In *Biological, Physiological and Health Sciences, Encyclopedia of Life Support Systems (EOLSS), Developed under the Auspices of the UNESCO, 2009*; Grabow, W.O.K., Ed.; Eolss Publishers: Paris, France, 2009.
20. Marttunen, M.; Mustajoki, J.; Sojamo, S.; Ahopelto, L.; Keskinen, M. A Framework for Assessing Water Security and the Water–Energy–Food Nexus—The Case of Finland. *Sustainbility* **2019**, *11*, 2900. [CrossRef]
21. Gain, A.K.; Giupponi, C.; Wada, Y. Measuring global water security towards sustainable development goals. *Environ. Res. Lett.* **2016**, *11*, 124015. [CrossRef]
22. Allan, J.V.; Kenway, S.J.; Head, B.W. Urban water security—What does it mean? *Urban Water J.* **2018**, *15*, 899–910. [CrossRef]
23. Moher, D.; Liberati, A.; Tetzlaff, J.; Altman, D.G. The PRISMA Group. Preferred reporting items for systematic reviews and meta-analyses: The PRISMA statement. *PLoS Med.* **2009**, *6*, 336–341. [CrossRef] [PubMed]
24. Bichai, F.; Smeets, P.W.M.H. Integrating Water Quality into Urban Water Management and Planning While Addressing the Challenge of Water Security. In *Understanding and Managing Urban Water in Transition*; Springer: Dordrecht, The Netherlands, 2015; pp. 135–154. [CrossRef]
25. Díaz-Caravantes, R.E.; Zuniga-Teran, A.; Martín, F.; Bernabeu, M.; Stoker, P.; Scott, C. Urban water security: A comparative study of cities in the arid Americas. *Environ. Urban.* **2020**, *32*, 275–294. [CrossRef]
26. Nilsson, L.M.; Destouni, G.; Berner, J.; Dudarev, A.A.; Mulvad, G.; Odland, J.Ø.; Parkinson, A.; Tikhonov, C.; Rautio, A.; Evengård, B. A Call for Urgent Monitoring of Food and Water Security Based on Relevant Indicators for the Arctic. *Ambio* **2013**, *42*, 816–822. [CrossRef] [PubMed]
27. Krueger, E.; Rao, P.S.C.; Borchardt, D. Quantifying urban water supply security under global change. *Glob. Environ. Chang.* **2019**, *56*, 66–74. [CrossRef]
28. Brewis, A.; Workman, C.; Wutich, A.; Jepson, W.; Young, S.; Adams, E.; Ahmed, J.F.; Alexander, M.; Balogun, M.; Boivin, M.; et al. Household water insecurity is strongly associated with food insecurity: Evidence from 27 sites in low- and middle-income countries. *Am. J. Hum. Biol.* **2020**, *32*, e23309. [CrossRef]
29. Marcantonio, R.A. Water, anxiety, and the human niche: A study in Southern Province, Zambia. *Clim. Dev.* **2019**, *12*, 310–322. [CrossRef]

30. Muñoz, C.X.; Johnson, E.C.; McKenzie, A.L.; Guelinckx, I.; Graverholt, G.; Casa, D.J.; Maresh, C.M.; Armstrong, L.E. Habitual total water intake and dimensions of mood in healthy young women. *Appetite* **2015**, *92*, 81–86. [CrossRef]
31. Aboelnga, H.T.; El-Naser, H.; Ribbe, L.; Frechen, F.-B. Assessing Water Security in Water-Scarce Cities: Applying the Integrated Urban Water Security Index (IUWSI) in Madaba, Jordan. *Water* **2020**, *12*, 1299. [CrossRef]
32. Bichai, F.; Grindle, A.K.; Murthy, S.L. Addressing barriers in the water-recycling innovation system to reach water security in arid countries. *J. Clean. Prod.* **2018**, *171*, S97–S109. [CrossRef]
33. De Miguel, A.; Froebrich, J.; Jaouani, A.; Souissi, Y.; Elmahdi, A.; Mateo-Sagasta, J.; Al-Hamdi, M.; Frascari, D. Innovative Research Approaches to Cope with Water Security in Africa. *Integr. Environ. Assess. Manag.* **2020**, *16*, 853–855. [CrossRef]
34. Qin, K.; Liu, J.; Yan, L.; Huang, H. Integrating ecosystem services flows into water security simulations in water scarce areas: Present and future. *Sci. Total Environ.* **2019**, *670*, 1037–1048. [CrossRef] [PubMed]
35. Smit, S.; Musango, J.K.; Kovacic, Z.; Brent, A.C. Towards Measuring the Informal City: A Societal Metabolism Approach. *J. Ind. Ecol.* **2018**, *23*, 674–685. [CrossRef]
36. Crow, B.; Odaba, E. Access to water in a Nairobi slum: Women's work and institutional learning. *Water Int.* **2010**, *35*, 733–747. [CrossRef]
37. Deshpande, T.; Michael, K.; Bhaskara, K. Barriers and enablers of local adaptive measures: A case study of Bengaluru's informal settlement dwellers. *Int. J. Justice Sustain.* **2018**, *24*, 167–179. [CrossRef]
38. Nepal, S.; Neupane, N.; Belbase, D.; Pandey, V.P.; Mukherji, A. Achieving water security in Nepal through unravelling the water-energy-agriculture nexus. *Int. J. Water Resour. Dev.* **2021**, *37*, 67–93. [CrossRef]
39. Saidmamatov, O.; Rudenko, I.; Pfister, S.; Koziel, J. Water–Energy–Food Nexus Framework for Promoting Regional Integration in Central Asia. *Water* **2020**, *12*, 1896. [CrossRef]
40. Islam, A.R.M.T.; Siddiqua, M.T.; Zahid, A.; Tasnim, S.S.; Rahman, M. Drinking appraisal of coastal groundwater in Bangladesh: An approach of multi-hazards towards water security and health safety. *Chemosphere* **2020**, *255*, 126933. [CrossRef]
41. Islam, M.S.; Sadiq, R.; Rodriguez, M.J.; Najjaran, H.; Hoorfar, M. Reliability Assessment for Water Supply Systems under Uncertainties. *J. Water Resour. Plan. Manag.* **2014**, *140*, 468–479. [CrossRef]
42. Bradley, D.J.; Bartram, J.K. Domestic water and sanitation as water security: Monitoring, concepts and strategy. *Philos. Trans. R. Soc. A Math. Phys. Eng. Sci.* **2013**, *371*, 20120420. [CrossRef]
43. Romero-Lankao, P.; Gnatz, D.M. Conceptualizing urban water security in an urbanizing world. *Curr. Opin. Environ. Sustain.* **2016**, *21*, 45–51. [CrossRef]
44. Young, S.L.; Boateng, G.O.; Jamaluddine, Z.; Miller, J.D.; Frongillo, E.A.; Neilands, T.B.; Collins, S.M.; Wutich, A.; Jepson, W.E.; Stoler, J. The Household Water InSecurity Experiences (HWISE) Scale: Development and validation of a household water insecurity measure for low-income and middle-income countries. *BMJ Glob. Health* **2019**, *4*, e001750. [CrossRef]
45. WHO. *Guidelines for Drinking Water Quality. Recommendation Edn*; World Health Organization: Geneva, Switzerland, 1993; Volume 1–2, p. 672.
46. Babel, M.S.; Shinde, V.R.; Sharma, D.; Dang, N.M. Measuring water security: A vital step for climate change adaptation. *Environ. Res.* **2020**, *185*, 109400. [CrossRef] [PubMed]
47. Mushavi, R.C.; Burns, B.F.; Kakuhikire, B.; Owembabazi, M.; Vořechovská, D.; McDonough, A.Q.; Cooper-Vince, C.E.; Baguma, C.; Rasmussen, J.D.; Bangsberg, D.R.; et al. "When you have no water, it means you have no peace": A mixed-methods, whole-population study of water insecurity and depression in rural Uganda. *Soc. Sci. Med.* **2020**, *245*, 112561. [CrossRef] [PubMed]
48. Hair, J.F., Jr.; Black, W.C.; Babin, B.J.; Anderson, R.E. *Multivariate Data Analysis*, 7th ed.; Pearson Education: New York, NY, USA, 2009; p. 816.
49. Khan, S.; Guan, Y.; Khan, F.; Khan, Z. A Comprehensive Index for Measuring Water Security in an Urbanizing World: The Case of Pakistan's Capital. *Water* **2020**, *12*, 166. [CrossRef]
50. Aboelnga, H.T.; Ribbe, L.; Frechen, F.-B.; Saghir, J. Urban Water Security: Definition and Assessment Framework. *Resources* **2019**, *8*, 178. [CrossRef]
51. Jalilov, S.-M.; Kefi, M.; Kumar, P.; Masago, Y.; Mishra, B.K. Sustainable Urban Water Management: Application for Integrated Assessment in Southeast Asia. *Sustainbility* **2018**, *10*, 122. [CrossRef]
52. Chowdhury, S.; Mazumder, M.J.; Al-Attas, O.; Husain, T. Heavy metals in drinking water: Occurrences, implications, and future needs in developing countries. *Sci. Total Environ.* **2016**, *569-570*, 476–488. [CrossRef]
53. Kumar, P.; Kumar, M.; Ramanathan, A.L.; Tsujimura, M. Tracing the factors responsible for arsenic enrichment in groundwater of the middle Gangetic Plain, India: A source identification perspective. *Environ. Geochem. Health* **2009**, *32*, 129–146. [CrossRef]
54. Vieira, L.A.; Alves, R.D.; Menezes-Silva, P.E.; Mendonça, M.A.; Silva, M.L.; Sousa, L.F.; Loram-Lourenço, L.; da Silva, A.A.; Costa, A.C.; Silva, F.G.; et al. Water contamination with atrazine: Is nitric oxide able to improve Pistia stratiotes phytoremediation capacity? *Environ. Pollut.* **2021**, *272*, 115971. [CrossRef]
55. Singhal, T. A Review of Coronavirus Disease-2019 (COVID-19). *Indian J. Pediatr.* **2020**, *87*, 281–286. [CrossRef]
56. Zhou, P.; Yang, X.-L.; Wang, X.-G.; Hu, B.; Zhang, L.; Zhang, W.; Si, H.-R.; Zhu, Y.; Li, B.; Huang, C.-L.; et al. A pneumonia outbreak associated with a new coronavirus of probable bat origin. *Nature* **2020**, *579*, 270–273. [CrossRef] [PubMed]
57. Choudhury, I.M.; Shabnam, N.; Ahsan, T.; Abu Ahsan, S.M.; Kabir, S.; Khan, R.M.; Miah, A.; Uddin, M.K.; Liton, A.R. Cutaneous Malignancy due to Arsenicosis in Bangladesh: 12-Year Study in Tertiary Level Hospital. *BioMed Res. Int.* **2018**, *2018*, 1–9. [CrossRef]

58. Thacker, S.B.; Stroup, D.F.; Carande-Kulis, V.; Marks, J.S.; Roy, K.; Gerberding, J.L. Measuring the Public's Health. *Public Health Rep.* **2006**, *121*, 14–22. [CrossRef] [PubMed]
59. Wilson, N.J.; Harris, L.M.; Joseph-Rear, A.; Beaumont, J.; Satterfield, T. Water is medicine: Reimagining water security through Tr'ondëk Hwëch' in relationship to treated and traditional water sources in Yukon, Canada. *Water* **2019**, *11*, 624. [CrossRef]
60. UNICEF-WHO. Joint Committee on Health Policy. Twenty-fourth Session. 1983. Available online: https://apps.who.int/iris/bitstream/handle/10665/59104/JC24-UNICEF-WHO-83.6-eng.pdf?sequence=1&isAllowed=y (accessed on 29 March 2021).
61. Sphere Association. *The Sphere Handbook: Humanitarian Charter and Minimum Standards in Humanitarian Response*, 4th ed.; Sphere Association: Geneva, Switzerland, 2018; Available online: www.spherestandards.org/handbook (accessed on 3 March 2021).
62. Hannah, D.M.; Lynch, I.; Mao, F.; Miller, J.D.; Young, S.L.; Krause, S. Water and sanitation for all in a pandemic. *Nat. Sustain.* **2020**, *3*, 773–775. [CrossRef]
63. Avtar, R.; Kumar, P.; Supe, H.; Jie, D.; Sahu, N.; Mishra, B.K.; Yunus, A.P. Did the COVID-19 Lockdown-Induced Hydrological Residence Time Intensify the Primary Productivity in Lakes? Observational Results Based on Satellite Remote Sensing. *Water* **2020**, *12*, 2573. [CrossRef]
64. Rosinger, A.Y.; Chang, A.-M.; Buxton, O.M.; Li, J.; Wu, S.; Gao, X. Short sleep duration is associated with inadequate hydration: Cross-cultural evidence from US and Chinese adults. *Sleep* **2018**, *42*, zsy210. [CrossRef]
65. Haghighatdoost, F.; Feizi, A.; Esmaillzadeh, A.; Rashidi-Pourfard, N.; Keshteli, A.H.; Roohafza, H.; Adibi, P. Drinking plain water is associated with decreased risk of depression and anxiety in adults: Results from a large cross-sectional study. *World J. Psychiatry* **2018**, *8*, 88–96. [CrossRef]
66. Pross, N.; Demazières, A.; Girard, N.; Barnouin, R.; Metzger, D.; Klein, A.; Perrier, E.; Guelinckx, I. Effects of Changes in Water Intake on Mood of High and Low Drinkers. *PLoS ONE* **2014**, *9*, e94754. [CrossRef]
67. Ganio, M.S.; Armstrong, L.E.; Casa, D.J.; McDermott, B.P.; Lee, E.C.; Yamamoto, L.M.; Marzano, S.; Lopez, R.M.; Jimenez, L.; Le Bellego, L.; et al. Mild dehydration impairs cognitive performance and mood of men. *Br. J. Nutr.* **2011**, *106*, 1535–1543. [CrossRef]
68. Brewis, A.A.; Piperata, B.; Thompson, A.L.; Wutich, A. Localizing resource insecurities: A biocultural perspective on water and wellbeing. *Wiley Interdiscip. Rev. Water* **2020**, *7*. [CrossRef]
69. Carlton, E.J.; Liang, S.; McDowell, J.Z.; Li, H.; Luoe, W.; Remais, J.V. Regional disparities in the burden of disease attributable to unsafe water and poor sanitation in China. *Bull. World Health Organ.* **2012**, *90*, 578–587. [CrossRef] [PubMed]

Review

Water Security in a Changing Environment: Concept, Challenges and Solutions

Binaya Kumar Mishra [1], Pankaj Kumar [2,*], Chitresh Saraswat [3], Shamik Chakraborty [4] and Arjun Gautam [1]

[1] School of Engineering, Pokhara University, Pokhara 33700, Nepal; bkmishra@pu.edu.np (B.K.M.); arjun.gautam@pu.edu.np (A.G.)
[2] Natural Resources and Ecosystem Services, Institute for Global Environmental Strategies, Hayama 240-0115, Japan
[3] The Fenner School of Environment & Society, Australian National University, Acton, Canberra 2601, Australia; Chitresh.Saraswat@anu.edu.au
[4] Faculty of Sustainability Studies, Hosei University, Tokyo 1-2-8160, Japan; shamik.chakraborty.76@hosei.ac.jp
* Correspondence: kumar@iges.or.jp; Tel.: +81-046-855-3858 (ext. 3058)

Abstract: Water is of vital and critical importance to ecosystems and human societies. The effects of human activities on land and water are now large and extensive. These reflect physical changes to the environment. Global change such as urbanization, population growth, socioeconomic change, evolving energy needs, and climate change have put unprecedented pressure on water resources systems. It is argued that achieving water security throughout the world is the key to sustainable development. Studies on holistic view with persistently changing dimensions is in its infancy. This study focuses on narrative review work for giving a comprehensive insight on the concept of water security, its evolution with recent environmental changes (e.g., urbanization, socioeconomic, etc.) and various implications. Finally, it presents different sustainable solutions to achieve water security. Broadly, water security evolves from ensuring reliable access of enough safe water for every person (at an affordable price where market mechanisms are involved) to lead a healthy and productive life, including that of future generations. The constraints on water availability and water quality threaten secured access to water resources for different uses. Despite recent progress in developing new strategies, practices and technologies for water resource management, their dissemination and implementation has been limited. A comprehensive sustainable approach to address water security challenges requires connecting social, economic, and environmental systems at multiple scales. This paper captures the persistently changing dimensions and new paradigms of water security providing a holistic view including a wide range of sustainable solutions to address the water challenges.

Keywords: water security; water scarcity; climate change; IWRM; socioeconomic changes; sustainable development

Citation: Mishra, B.K.; Kumar, P.; Saraswat, C.; Chakraborty, S.; Gautam, A. Water Security in a Changing Environment: Concept, Challenges and Solutions. *Water* **2021**, *13*, 490. https://doi.org/10.3390/w13040490

Academic Editor: Richard C. Smardon

Received: 7 January 2021
Accepted: 9 February 2021
Published: 14 February 2021

1. Introduction

Water is the foundation of life and necessity for everyone. However, it is becoming an increasingly scarce and degraded natural resource for millions of the world's population. Adequate water, which is necessary for various uses for a rapidly growing population, is one of the major challenges in recent years. It is a critical issue as the increase in food production to meet the future population will have to be achieved with the same water resources. Growing populations and climate change are added burdens to the global water crisis. More than 1.1 billion people have inadequate access to clean drinking water globally, and approximately 2.6 billion people lack basic sanitation facilities [1,2]. Water stress is increasing rapidly especially in developing nations of the world. According to the United Nations [3], global water use over the last century has been growing at twice the rate of population increase. As a result, approximately 1.2 billion people live in areas of physical water scarcity, where supply of water is not enough to meet the demand [4]. Apart from

181

physical scarcity, economic scarcity is another major issue for water insecurity. As of now, about 1.6 billion people face economic water scarcity, where people do not have enough financial means to access existing water sources. Lack of safe water access is gradually becoming a crisis for millions of people around the world that is responsible for poor health, destruction of livelihoods, and unnecessary suffering for the poor [5].

Water shortages are the most pressing challenge for socioeconomic and human development in general. Water shortages can lead to ecosystem degradation, worsening health and destruction of livelihood [6]. Increasing human pressure threatens the ability to provide adequate water resources and functioning of ecosystem services in the arid and semiarid regions and are particularly vulnerable to climate and land changes. As the human population grows and economic activity grows, water degradation has become a global concern [7]. Poor water quality, which makes water unfit for use, has multiple health and environmental consequences, and further reduces water availability. Contamination of the surface and groundwater is becoming one of the biggest threats to the available fresh water.

Today, more than half of the global population reside in urban areas [3]. With change in demography characterized by massive migration into cities, it is projected to further increase in percentage of urban population. Urban water supply systems for various applications as well as wastewater management for the increasing population is a serious threat, especially in developing countries. These cities have not been able to provide minimum water services to their growing population. Considering the urbanization pattern, it is urgent to improve water supply and wastewater treatment systems. These components should be seriously incorporated into urban planning.

A brief description of population and share of freshwater availability across different continents is shown in Figure 1 to highlight the issue of water scarcity.

Increase in water-related disasters is another important issue in the context of global change [8]. The number of deaths and economic damage caused by water-related disasters such as floods, droughts, landslides, and land subsidence has increased dramatically. Climate change and variability, land use changes, urbanization, migration patterns, energy problems, and food production caused by demographic change and economic development can exacerbate more uncertain risks.

Flooding is one of the most damaging natural disasters in the world. Floods are often seen as a natural phenomenon due to extreme weather events. However, in practice, human activity changes the environment in multiple ways, changes the water cycle, and hence more flooding. Therefore, flood risk and flood risk management are closely related to human activities. The impact of land-use change includes not only urbanization, but also the development and consolidation of agriculture [9]. Land-use policy has been a key factor in permitting urban developments in areas of risk. Flood risk results from extreme events, changes in the natural environment, and poor disaster management institutions which have the ability to reduce and manage risks.

Similarly, drought has multiple physical and social aspects. Lack of precipitation alters water resources and agriculture systems, and the impact can be severe, depending on the resilience of local communities and populations. Tensions between competing water use further worsen due to conflict between human use and environmental flow requirements [10]. Drought can constrain the multiple societal uses of water, including energy production, at local and regional levels. Many people rely on groundwater for drinking water, food security, and sustainable living. Groundwater use has increased significantly over the past 50 years due to its better quality and easy availability even during droughts [11].

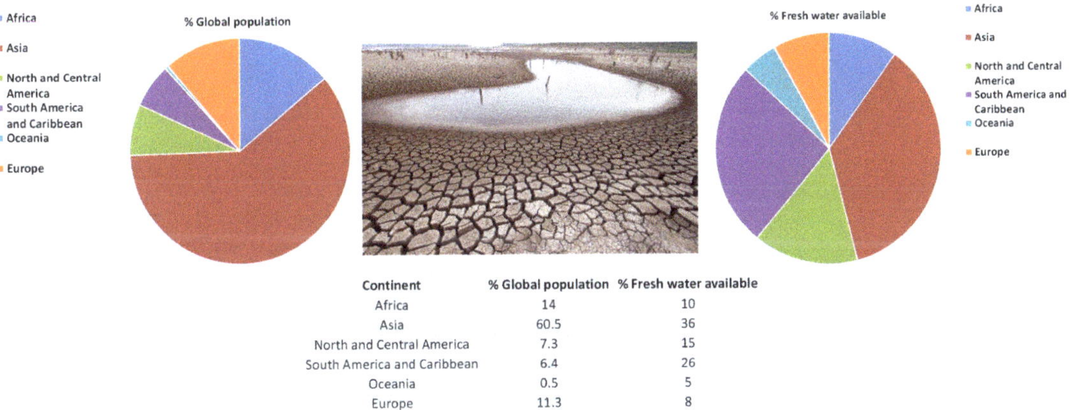

Continent	% Global population	% Fresh water available
Africa	14	10
Asia	60.5	36
North and Central America	7.3	15
South America and Caribbean	6.4	26
Oceania	0.5	5
Europe	11.3	8

Figure 1. A brief description of population distribution and water availability around the world [1,3,8,12,13].

These challenges require further research, implementation of new science-based methodologies, and endorsement of principles of integrated water resources management which can sustainably address various water-related issues. It is important to understand and manage water quantity and quality worldwide, and especially in the developing world. Overcoming the water crisis remains one of the most critical challenges our generation is facing [12] and developing clean potable water, managing wastewater efficiently and providing basic sanitation facilities for sustainability and human progress [13,14]. The United Nations (UN) Sustainable Development Goals 2030 will not be possible without achieving a water secure world first.

While resource allocation and competition requirements are to represent the first set of water security challenges, the second major focus for water security is on extreme events. It is very important to design a wide range of sustainable solutions, which addresses various water problems. With increased concerns on the sustainable use of limited water resources, in recent years, policy makers, organizations, funding agencies, and individuals largely use the 'water security' term to express their opinions for solving various water-related problems. However, there is no consensus on framing of water security concepts for solving water problems in a sustainable manner. Water security employing different disciplines is proposed as a starting point for solving various water related problems.

In the past few years, constructive efforts have been made to improve water security issues. Water security includes the sustainable use and protection of water systems, protection against water-related hazards, sustainable development of water resources, and protection of water functions and services for humans and the environment [15].

There is a need for holistic approaches to address the water security challenges and inclusion of the social, economic, and environmental dimensions at multiple scales. The holistic approaches can act as a catalyst for progress in many sectors such as public health, energy security, climate resilience, poverty reduction, and accelerate the pace of achieving the Sustainable Development Goals (SDGs). Water security is important to understand and manage various challenges in the context of environmental change. The interface between communities of researchers, practitioners, and stakeholders is seen as increasingly important for the recognition and management of water security.

Lack of infrastructure and capacity development; weak and rigid governance systems, and lack of interdisciplinary approach are among the major reasons for the increase in solving various water-related problems. Framing the challenges of water security goes beyond single-issue indicators such as water stress. In a rapidly changing world, we are solving new problems with old solutions. The status quo is no longer enough, and conventional models will not be able to achieve water security issues. It is crucial to shift from ad hoc and isolated water solutions to integrated water resources management approaches, which yield more sustainable and resilient systems. A new paradigm is therefore needed that considers alternative solutions for achieving water security. The paper highlights water security issues and sustainable solutions at different scales and presents good practices to address water security challenges. This paper aims to review various water related problems and challenges in the global change context, explore emerging paradigms of water security assessment, and finally seek sustainable solutions towards achieving water security.

Therefore, this paper is aimed to address key questions on water security as: (i) what is water security? (ii) What are various dimensions of water security? (iii) How can we assess water security? and (iv) What are different sustainable solutions to improve water security? Answers to these questions will be of great value to policy-makers, who are responsible for making well-informed decisions and investment in the field water management.

2. Methodology

For this study we carried out a narrative literature review. To carry out the literature review, we searched through keywords water security, water scarcity, climate change, integrated water resource management, and sustainable development to focus our study within the review questions (see Section 1). The Claivariate Analytics's Web of Science (WoS) was used to search for scientific literature and articles pertaining to the combination of keywords mentioned above from year 2000 to 2020. Web of Science was chosen as it is one of the world's largest scientific citation search platforms, has an extensive coverage of a wide spectrum of studies related to a topic of interest, and a frequent source for literature review based analysis [16]. A variety of articles were searched to ensure that the majority of the relevant articles and the arguments they capture have been identified. After completing the search, abstracts were read and articles that were common (duplicates) in the searched databases were discarded. All the remaining abstracts were then read to find an answer to our review questions. We did not include all the articles of the abstracts we reviewed because of the relevancy of the arguments in the abstract and articles they represent. We then read all the articles thoroughly, this included a total of 71 articles, which show a diverse line of arguments on the review question. We then synthesized the findings from this literature review into a narrative and integrated into two main types of streams of argument—problems and solutions of water security under changing environmental and socioeconomic context. A list of reviewed papers used to identify different dimensions for water security is shown in Appendix A. The whole study is described in the flowchart shown in Figure 2.

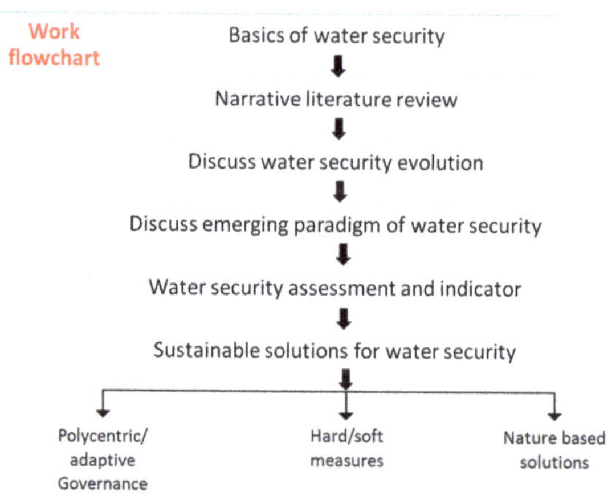

Figure 2. Flow chart explaining the work done in this study.

3. Water Security: Concept and Evolution

Water security is critical to achieving sustainable and comprehensive growth. A water-secure world uses the productive power of water and reduces its destructive impacts. It is a world where everyone has safe, affordable, clean water to live a healthy and productive life. It is a world where communities are protected from floods, droughts, landslides, erosion, and water-borne diseases. Water security promotes environmental protection and social justice and deals with the consequences of poor water management. In general, several definitions have been proposed for water security, reflecting the desire to manage water resources. The 'UN agency on Water' defines the water security as "the capacity of a population to safeguard sustainable access to adequate quantities of water with acceptable quality necessary for sustaining livelihoods, human well-being, socio-economic development, ensuring protection against water- related disasters, and for preserving ecosystems in a climate of peace and political stability" [13]. Increasing water security lies in: (i) ensuring the availability of adequate and reliable water resources of acceptable quality to provide water services for all social and economic activity in a manner that is environmentally sustainable; (ii) mitigating water-related risks such as floods, droughts, and pollution; (iii) addressing the conflicts that may arise from disputes over shared waters, especially in situations of growing stress, and turning them into win–win solutions. Water security is emerging as a possible unified concept for water managers. Operationalizing the concept of water security means identifying its various aspects, setting goals, and exploring actions to achieve these goals.

In recent years, the focus of water security has become more diverse, expanding from water scarcity to clean water, ecosystem services and overall human wellbeing [17]. Conventional approaches focused on a single aspect and narrow perspective such as development studies tend to focus on national scales only, economists' main concerns are about the economic aspect of the problem, hydrologists often only focus on watershed scales, and the social scientists focused on research work around community only [17]. The new dimensions pledged to attain water security in a sustainable way with long-term vision, which proposes a range of sustainable solutions based on economic, environmental, and societal aspects [18].

The spatial and temporal considerations are important determinants for water security. Rijsberman [19] explained the supply problem when water is truly scarce in the physical sense, and demand problem when enough water is available but not used in a better or sustainable way. Hydrological extreme conditions such as floods and droughts can occur at the same location within the same year. Therefore, use of average available water cannot represent a measure of water scarcity or excess. In the monsoon season, many regions of Asia suffer from water scarcity while the average annual water resource availability appears to be abundant. Large countries like India can have water scarcity and flooding at the same time, but in different parts of the country. The situation of uneven distribution of water resources is going to be aggravated due to climate change. The understanding of water scarcity is vital as it affects the views on the most effective policies to deal with water crisis. The Intergovernmental Panel on Climate Change (IPCC) reports indicated that more than 87% of the climate change impact will be on water related infrastructure and increasing negative impacts of global warming is likely to increase the variations, frequency, and severity of weather such as extreme droughts and floods occurring [20]. On the global scale, a current water withdrawal is well below renewable water resources limit, but the concern is the high unpredictability of water resources in coming years [21]. Virtual water trade can be one of the effective options to address water scarcity, which is estimated around 1000 km^3/year internationally although only a very small part of virtual water trade is utilized to compensate for water shortages. Another major challenge is pollution and deterioration of surface and ground water bodies in rapidly growing cities and elsewhere in the developing world. Understanding of causes and measures that lead to improvement of water quality degradation is limited. In order to address water quality degradation, several soft and hard measures need to be formulated at individual, community, and institution level. It needs balance between technical and social approaches. Water quality is another major variable and its analysis is beneficial for security of the environment, human use, and understanding the water stress. Historical evolution of water security technologies is important to understand the present and future concept of water security (Figure 3). Urban regions required new strategies to cope with the transitioning phases of water requirements with the development of urban areas. Water supply for demand management was a concept in the water supply system in cities, which evolved and moved in the direction of water sensitive cities due to various factors. Various drivers are responsible for this transitioning phase of water such as rapid population growth and consumption patterns that use more water. To solve the emerging problem of water security, it is important to change the management response that appears in response to social and political factors [22].

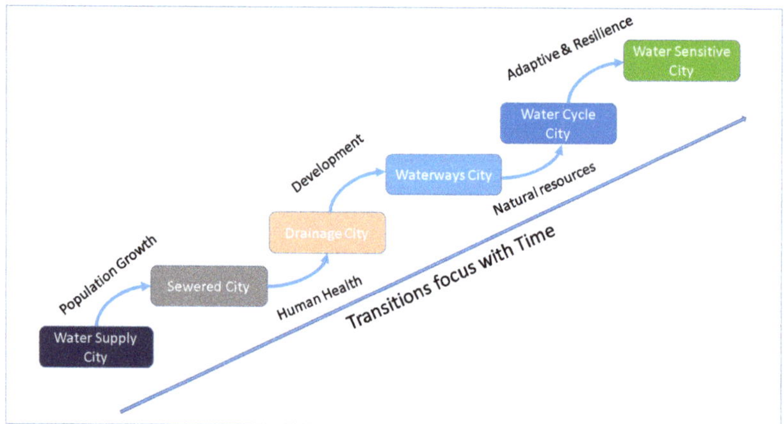

Figure 3. Changing paradigms of Water Security and Transition Phase for cities (adapted from brown et al. 2008 [22]).

4. Emerging Paradigms of Water Security

Water Security embodies a complex, multidimensional and interdependent set of issues. With increasing pressures on water resources, there is heightened competition for the water uses at local, regional, and international scales, both between diverse sectors of the economy and jurisdictions. This comprises basic societal needs such as drinking water supply, irrigation, hydropower, and industrial uses. Achieving water security is also dependent on the water quality, which is a further key dimension of water use. Rivers are used to receive, transport, and dilute wastes, and intensification of human activities is putting increasing pressure on the quality of both surface waters and groundwater, with consequences for water resource availability for various uses. Water-related ecosystem functions, with their dependence on water quantity (and its temporal variability) and water quality is another critical dimension of water security.

Securing water for people and the environment is a necessary condition for sustainable growth, ending poverty and hunger, and achieving the SDGs. The dominant threats to water availability, quality, and supply vary geographically and over time. This entails achieving water security is not a dynamic goal, which evolves with changing dynamics. It is a dynamic process affected by changing climate, political set up, economic growth, and resource degradation. Moreover, as social, cultural, political, economic priorities, and values change, the goal for achieving water security evolves with them. In this background, the study argued for evaluating water security challenges under emerging paradigms and seeking the solutions.

Paradigm is a concept of envisioning the future of societal interaction with the environment based on the shared assumptions and values [23]. The emerging paradigms are significant in the context of continuous shifts in societal interaction with natural resources such as water. In order to bring holistic solutions to water security, the nature and characteristics of paradigm changes are necessary. Historically, paradigm shifts are seen as a continuous and necessary part of policy making. The important question that arises here is why there is a need to rethink water security under emerging or new paradigms. Societal interactions and values around water are continuously changing. For example, the shift from Millennium Development Goals (MDGs) to the SDGs, the focus of the water security concept has shifted from only water supply and demand in cities towards the perception of water as an economic resource shared between countries. The paradigm shift in water emphasizes on how societies are valuing water as a resource and support this transition. This shift comprises water governance, which is the capacity of governments to manage water equitably and efficiently, including across borders [24]. Due to increased economic development and climate change the uncertainty is increased which impacted the interactions of society with water resources. In the traditional paradigms, the water crisis is generally considered as technology related problems but under emerging paradigms focus is shifted towards recycle and reuse of water, considering waste and stormwater as resources, managing demand effectively, promoting green infrastructure, increasing community and stakeholder participation, effective governance, and multidisciplinary approach to achieve water security.

The four different paradigms shifts are identified in this paper to evaluate the water security and seek the solutions to achieve the goal. In changing context and challenges, the new approaches to govern the water is suggested by Huitema et al. [25], are the adaptive water governance and polycentric approaches to govern. Another perspective of Integrated water resource management (IWRM) is popular among the developing nations, how the IWRM evolves in current perspective. The nature-based solutions perspective is gaining popularity worldwide to achieve water security and still many scholars focus on the combination of hard/soft approaches to achieve water security (see Section 5). The paradigm shift in the water sector constitutes how water is managed in the current context under different challenges and transitioning the way of managing water towards diversifying water resources and building resilient infrastructure. Shift in financing the water sector is significant in emerging paradigms where engaging the private sector partnerships and reducing the nonrevenue water is in focus. Collaboration at different scales through regional and national cooperation and agreements are helpful in shifting towards system thinking approaches. This study envisioned the challenges and solutions to achieve water security under emerging paradigms in a changing environment.

5. Water Security Assessment and Indicators

Water security captures the complex concept of overall water management with resource conservation and its use [26]. Water security should be represented by indicators to promote its quantification. There are different kinds of water scarcity indicators proposed by various studies and explained in literature such as water stress indicator or Falkenmark indicator, water resources vulnerability index, economic water scarcity index, and water poverty index, among many others. There are several advantages of translating water security into numerical terms to encourage clarity and establish a common understanding of a concept around which there currently exists substantial ambiguity. This helps to facilitate discussion on the presence or absence of water security, or the scale and threshold for evaluating the degree of water security. This also assesses the extent to which water security on the ground has been achieved in various locations.

The treatment of wastewater is a very important aspect of water security. Historically, wastewater was considered as waste only, although it could be a valuable resource. If treated properly, wastewater can be reused and contribute to lesser pollution in water bodies. The recycling of water is the concept of reusing the treated wastewater or stormwater for many useful purposes, such as irrigation, toilet flushing, needs of industries, and in some aspect for ground water recharge [27,28]. Recycled water or water reuse offers many kinds of resources and financial savings [29] and if this can be tailored to meet the quality standard required for the water use, then the recycled water can be considered as a vital resource because it offers series of benefits such as useful for irrigation, cooling water for power plants and refineries, toilet flushing at household level, controlling the dust in the city, mixing and preparing cement during construction activities, beautification of artificial lakes and parks, and industrial processes. This provides us with ample reason to consider treatment of water as a sustainable solution for achieving water security.

Sources of uncertainty in water security assessment can be demonstrated by risk-based water security indicators, such as the uncertainty in the climate model outputs. The IPCC report indicated that climate change is increasing the frequency of extreme weather conditions like flood, drought, etc. More than 87% of the climate change impact will be on water related infrastructure [20]. These indicators are important in analysis of risks based on what if scenarios, which provides the essential evidence to enable options between alternative courses of action. The risk analysis explores the range of likely future situations ranging from business as usual to extremely unlikely conditions [30]. Decision making process involves choices between different courses of actions counting their benefits in terms of risk reduction or cost–benefit ratio.

Scale is critical in assessing water security [31]. National level assessments make it difficult to take action at operationalization level. Agriculture dominates water use in the Asia-Pacific region, but focuses on water productivity, which is often used as a means of improving production. Therefore, this indicator deals with the available water, and how to use, for example, an increase in water storage, can become a more urgent issue in areas with economic water shortages. There is also a question mark over use of such indicators that explains the overall picture of the country, especially if watersheds cross national borders. Watershed based water security can be defined as "sustainable access on a watershed basis to adequate quantities of water of acceptable quality to ensure human and ecosystem health" [26].

It is possible that the region can be concluded water secure or insecure but interestingly, without the explicit environmental consideration the shared conclusion from water indicator's analysis that water security is achieved by the developed part of the world such as many countries in Europe, North America, Australia, and Japan, is a kind of flaw [19]. Creating and using indicators for water security has to be directed towards some management control or assessment action. Originally, a single indicator was mainly used to assess the quality of rivers for human use. Earlier, for example, a simple parameter indicator (e.g., Biochemical oxygen demand (BOD)) was sufficient as a river pollution control measure. As the industrial revolution began, various types of chemicals started spilling into rivers requiring multiple measurements. Thus, indicators were required to measure multiple effects and combine them into single-value indicators. A recent shift to water security has called for the addition of economic and ecosystem concerns. In addition, the largest consumer of water in most countries is irrigation agriculture. With the rapid growth of the global population, the fear of a decline in the supply of water available in food production, industry, and homes has become a major problem. This has led to a search for indicators of water resource security.

Water security clearly requires an integrated approach while recognizing and accepting that tradeoffs may be needed between the different water-using sectors. Indicators tend to be heuristic, and political and water specialists need to focus on those indicators which tell the most compelling story to other sectors if they are to encourage water-smart investments. Roger Calow [26] suggested five pillars of water security where indicators are water availability and access, risk and variability, equity and livelihoods, ecosystems and biodiversity, and institutions and actors [26]. Ian Makin [26] focused on national and regional measures of water security, rather than individual basins highlighting the importance of water management issues in 49 countries across the Asia and Pacific region and the threat from many sources—population growth, urbanization, increasing water pollution, over-abstraction of groundwater, water-related disasters, and climate change.

Asian Water Development Outlook (AWDO) [1] provides a robust, pragmatic, and readily understood framework for assessing water security. This report presents a framework for the assessment of water security in five key dimensions: household water security, economic water security, urban water security, environmental water security, and resilience to water-related disasters. Status of water security in five dimensions represents inherent tension among water uses that emerge under increasing stress from competing water use sectors. These, when aggregated, provide an indicator of national and regional water security. The indicators (referred as dimensions) are seen as the means of measuring the outcomes of integrated water resources management. These provide a baseline for analyzing trends and the effects of policies and reforms that can be monitored and reported to stakeholders and offer a new way for leaders to look at the strengths and weaknesses of water resources management and service delivery. These indicators also indicate the direction and priority for increasing investment, improving governance, and expanding capacity in the water sector. Dunn and Bakker [32] developed a water security framework as a tool for improving governance for watersheds in Canada. This is in contrast to others who seek to develop national or regional indicators describing the challenge of developing and applying indicators at the watershed level that were originally designed for national or regional application and question their relevance and sensitivity for use at a community level and for including socioeconomic considerations. However, considering the importance of water quality and transboundary water resource management, this study explained water security based on seven dimensions that are shown in Figure 4.

Figure 4. Water security dimensions (based on the Appendix A).

6. Sustainable Solutions

Given global changes and its potential impacts water security is a growing concern. It touches upon all aspects of life and requires a holistic approach, which actively integrates social, cultural, and economic perspectives, scientific and technical solutions, and attention to societal dynamics. The complexities in water security point out that there will be different solutions for managing water scarcity at different spaces and times. To address water security, interdisciplinary collaborations are required across sectors, communities, and political borders to manage the competition or conflicts over water resources. Here, sustainable water management is a key which focuses on the conjunctive and efficient use of different water resources and on water allocation strategies that build the economic and social returns and enhance the water productivity. In addition, there is a need for a special focus on equity in access to water as well and social impacts of water allocation policies.

Based on the complex and intricate nature of water security, its dimension and different paradigms as mentioned above, this section advocates various sustainable solutions to achieve water security under three themes. A summary for reviewed papers supporting ideas for these three different themes of solutions to achieve water security is shown in Table 1.

Table 1. List of reviewed papers used to find water security solutions under following three different themes.

Water Security Solution	References
Governance (Adaptive and Polycentric) including Integrated Water Resource Management (IWRM)	Kumar et al., 2020; Pahl Wostl et al., 2018; Saraswat et al., 2017; Head et al., 2014; Cook and Bakker, 2012; Wiek and Larson, 2012; Uhlendahl et al., 2011; Huitema et al., 2009; Brown et al., 2008; Masago et al., 2019; Al-Jawad et al., 2019; Weather et al., 2009; Radiff 1999
Combination of Hard and Soft Approaches	Masago et al., 2019; Mo et al., 2019; Mishra et al., 2018; Tian et al., 2017; Voulvoulis et al., 2015; Grafton et al., 2011; Biswas and Tortajada, 2019; Brown et al., 2008; Mishra et al., 2020
Nature-Based Solutions	Van-Rees et al., 2019; Lindsay et al., 2019; Vorosmarty et al., 2018; Jamero et al., 2017; Chakraborty and Chakraborty 2017; Gunderson et al., 1995; Mishra et al. 2020

6.1. Polycentric and Adaptive Governance

Water governance becomes a sustainable solution when it safeguards ecosystems yet is able to consider social and economic wellbeing of people [33]. Governance refers not only to national level water legislation, regulations and institutions, but also the processes to promote stakeholder or community participation in designing water and sanitation systems and empowering communities with knowledge by increasing their ability to make decisions about the management systems. Restructuring of water governance methods refer to the social mobilization and actions designed in promotion of ownership, capacity building, coinvestment, willingness to pay for services, and incentives for participation at the community level. Van Rijswick et al. [34] recognized several factors important for sustainable water governance. These are knowledge on water systems, monitoring and enforcement, values, principles, and policy discourses, responsibility, authority and means, stakeholder involvement, conflict resolution capability, tradeoffs between social objectives such as allocation, regulation and agreements, and financial arrangements. These can be taken as a roadmap for sustainable water governance. Besides, the water governance solutions also promote gender equality in decision making [35]. It determines the relevant roles for the government entities to ensure the delivery of water and sanitation and keep checks on public and private players to meet the required needs. Building sustainable governance as solutions towards water security is a process, which is very effective in correcting market distortions, incentives, and adjustments to affordable pricing.

The water sector is considered underperforming and requires massive investment in infrastructure building and developing capacity. The complexity of the decision-making system, inability to receive timely information, conflicts of water rights (vertical and horizontal integration), poor coordination and interactions, and lack of holistic planning are the reasons and important challenges of water governance [36].

To deal with this complexity and scale of governance challenge for water security, here we would especially like to focus on polycentric governance. Polycentric means that water management plans and policies should be framed and agreed by all relevant stakeholders. In other words, both top-down and bottom-up approaches should be given weight. For adaptive governance, more emphasis will be on finding the best pathways to make robust water management plans amid rapid global changes. The benefit of such plans should reach the end users in terms of providing clean water, protection from hydrological hazards, and maintaining the health of the ecosystem.

In order to make a region more sustainable in terms of water resources, its locally available water resources must not be compromised by its socioeconomic activities. There is an urgent need for co-management, which includes the cycle of co-design, co-implementation, and co-delivery throughout the whole water cycle.

(a) Regional Circular and Ecological Sphere (RCES)—RCES is a concept that complements and supports regional resources by building broader networks, which is composed of natural connections (connections among forests; city and countryside; groundwater, rivers and the sea) and, economic connections (composed of human resources, funds, and others), thus complementing each other and generating synergy [37]. In terms of water security, this concept will lead to self-reliant and decentralized society. This will promote water reuse, reclamation, and restoration using decentralized water management systems. Additionally, it will enhance our reliability on renewable source of energy with less dependence on water resources.

(b) Participatory Watershed Land-use Management (PWLM) approach—PWLM is another very innovative and successful approach for more robust water resource management [38]. It helps to make land-use and climate change adaptation policies more effective at a local scale. This is an integrative method using both participating tactics and computer simulation modeling for the water resource management at a regional scale. The whole process is divided into four main steps: (a) scenario analysis, (b) impact assessment, (c) developing adaptation and mitigation measures and its integration in local government policies, and (d) improvement of land-use plan.

(c) Citizen science—Citizen science can be generally defined as "the engagement of non-professionals in scientific investigations" [39,40]. This is also a unique way for the general people to contribute to monitor water pollution and its management progress. This can be done by participating in "citizen science" initiatives, e.g., by reporting detections of water pollution in their locality. Normally, the citizens involved here are volunteers and unpaid, and contribute out of their own personal interest in the topic of the investigation. Although this is an old technique, it got into the limelight recently because of advances in information and communications technology (ICT) including mobile smartphones with internet, GPS, and camera capabilities [40,41]. Use of ICT has made it much easier for citizen volunteers to interact with professional scientists and pass this useful information about water quality to the decision makers.

(d) Integrated Water Resource Management (IWRM)—IWRM is a vital approach to improve water security and sustainability through better water use efficiency and conservation. Although the term or the concept IWRM, is available and frequently being used by different stakeholders since the last few decades but its ground application remains questionable [42–45]. While much progress has been made on water quality, water quantity components, this study suggests the critical elements for the success of IWRM in the ground are to prioritize the inclusion of human or socioeconomic dimension, nexus dimension (water–food–energy–health) etc., which are still unexplored in most of the regions [46,47]. In general, conventional models helps us to quantify and project plausible future of the water environment, but they cannot guarantee their goal attainment because of adaptive responses by humans and management decisions which might have unintended consequences [48–50]. Therefore, an integrated perspective in analyzing water related risk through socio-hydrological pathways is deemed essential for better understanding the action research and policy implication for sustainable water management [46].

6.2. Combination of Hard and Soft Measures

(a) Water reuse, restoration and reclamation—Considering the interwoven issue of water scarcity in the lieu of rapid global change and frequent extreme weather conditions, it is important to think holistically for both water quality and quantity. This issue becomes more acute in case of countries with arid or semiarid environments with very little freshwater available. Henceforth, the concept of 3Rs (reduction, restoration, and reclamation) is very important to achieve a water secure world. Water reuse in terms of restoration and reclamation become a matter of prime concern, which has multidimensional (financial, socioeconomic, and environmental) benefits. To achieve this, all water users and stakeholders must be aware about the use of reclaimed water, its social acceptance, and pros and cons of using this reclaimed water [51]. However, in order to make this reuse more sustainable, a sound and scientific knowledge about the source of wastewater or treated saltwater, chemical composition, and its sectoral usage is most important [52]. For reducing the water usage, several innovative ways are already in practice such as smart monitoring to minimize nonrevenue loss of water, intelligent irrigation using agriculture engineering in the field, rainwater harvesting, financing urban water usage, water tariff, etc. [53].

(b) Water conservation technologies (WCT)—WCT are of larger importance for developing countries like India with traditional irrigation schemes having efficiencies of 30–40% [54]. It is envisioned that increased use of WCT will play an important role in improving the productivity of rain-based agriculture and irrigation efficiency. Water conservation and water saving technologies are delivering benefits such as improved water productivity, improved farm profits and others, if it is well planned and managed. In addition, the WCT reduces the nonbeneficial consumptive water use such as the well-managed drip irrigation or orchard crops and enhances the distributional profits under regulatory frameworks.

(c) Modeling and Forecasting—Different hydrologic simulation models are powerful tools, which enable to assess various implications of the rapidly changing global processes on water resources variables. Although, various hydrologic models with specific features are available for simulating the water resources variables like runoff, groundwater recharge, surface water–groundwater interaction, water quality, contaminant fate, and transport etc. Based on the goals to be achieved, different models (continuous or event-based type; empirical, conceptual, and physical types) can be selected. However, these models can be effectively applied only when it has a significant amount of required input datasets and skilled people to handle it. For data scarce countries, it is important to have easy access to free available data and tools for sustainable water management. On the other hand, capacity building is also critically important for making these countries self-sufficient for doing better in analyzing future status of water resources and hence can design appropriate adaptation and mitigation strategies for water resource management.

6.3. Nature-Based Solutions

Nature-based solutions are those that use or simulate natural processes to address contemporary challenges, including those associated with water management. Its objectives are, for instance, to increase water availability (soil water retention and groundwater recharge are nature-based solutions), to improve water quality (natural and artificial wetlands, and riparian forest buffers), or to reduce water-related disasters and climate change risks (restoration, flood plains, and roof gardens). In other words, nature-based solutions are ecological processes driven by vegetation and soil in forests, pastures, humid areas, as well as in agricultural and urban landscapes, which play an important role in water movement, storage, and transformation. Nature-based solutions offer some of the most effective and sustainable ways to improve water security, and they frequently offer additional benefits for communities where they are implemented, including improved agriculture, job creation, and climate resilience. It can also play an important role in improving the supply and quality of water and reducing the impact of natural disasters.

Freshwater supply depends on healthy functioning of freshwater (riverine) ecosystems. However, studies are scarce that consider whole ecosystems such as a river [55]. Furthermore, most of the riverine systems are heavily altered anthropogenically, this is a big challenge that makes sustainable water resource management, miss vital 'pieces' or functions that relate to the diversity of the ecosystem that are needed to bring solutions that are sustainable. Freshwater ecosystems are often highly engineered and thus can undergo this act of missing the vital pieces of ecosystem diversity that is quite common, which eventually make the freshwater systems lose their resilience.

For example, freshwater systems such as rivers have undergone disconnection in the longitudinal, lateral, and vertical dimensions of the river systems [56], disrupting (and distorting) the freshwater ecosystems such as a river that continues to deliver freshwater related ecosystem benefits [56,57].

The degradation in resilience does not exclude the society that lives within the freshwater ecosystems (e.g., a river basin) causing decision-making gridlocks. For example, distorted governance on freshwater issues have occurred due to human interventions such as dams and barrages, changing of the river courses, covering the flow (and in many cases surrounding landscapes) with impervious surfaces have compromised diverse ecosystem functions and resilience in freshwater environments through decades of mismanagement. These gridlocks especially take place when high economic returns are available at the expense of functional diversity [58]. Therefore, keeping the freshwater ecosystems as diverse as possible can enhance this resilience, which can be used as a thread or anchor to bind the whole ecosystem together. Using the resilience concept as a binder can make the freshwater system undergo changes without changing its vital state and function. A key sustainable solution in this regard is binding through the available knowledge component in the freshwater landscape through stakeholder engagement that creates values in diverse ecosystem functions for capturing the dynamism and nonlinearity of the freshwater ecosystems (including its connection to other ecosystems) [59–62]. It is here that nature-based solutions (or landscape-based solutions) can work as eco-friendly technologies to enhance development of freshwater services, characterized by a number of low-cost, more flexible adaptation options in contrast to conventional 'grey' technologies, which the developing world, especially in urban areas, are now trying to implement [63].

Additionally, sustainable water management can be achieved by facilitation of a change in mindset towards landscape-based water management such as participatory watershed land use management (PWLM). The participatory approach can identify and engage champions within its offices and communities to promote long-term sustainable planning for urban water security [64–66]. It can be challenging at times due to low interest among citizens [67]. Pace of urbanization is important while introducing sustainable construction practices. Watershed management plans should be based on retrofitting models in collaboration with stakeholders and communities are essential. This should be based on physical analysis of existing water scarcity coping practices. Interactive design records are needed to adopt smart water management.

Another set of sustainable solutions include interlinked water, energy, and food components, with a balance between natural resource use and society's demand on such resources [67,68]. This forms a critical nexus essential for addressing challenges of sustainable development. Water, food, and energy are dependent on each other and share many comparable characteristics, including people's livelihoods, so a collective (i.e., a nexus) approach should be able to address this interdependency effectively.

The relationship between ecosystem and water security is mutually beneficial and a core component of sustainable solutions to water security as this ensures that sufficient and good quality of freshwater is available to support the functioning of ecosystem (and healthy functioning of the ecosystem continues to deliver good quality and quantity of water). Unfortunately, understanding of the relationship between ecosystem and water security is still a new territory but stressed as an important component to achieve overall development in the society such as Sustainable Development Goals [69,70]. As ecosystems provide water required both in terms of quality and quantity for different communities, maintaining food security as well as to mitigate hydrological hazards necessary to achieve water security at a watershed level, it is important to ensure that ecosystems are conserved.

Traditional and local adaptation strategies for an ecosystem-based approach are needed as they are often based on 'time tested' methods and adaptation strategies for water management in the whole landscape, and not only its physical and chemical parameters as modern scientific methods try to achieve. Further investigation on local adaptation strategies is also needed for achieving water security and fighting effectively against the climate change impacts is necessary for policy makers [71]. For example, every village, town, or other type of human settlement unit can have a list of traditional and local adaptation strategies so that they can deliver at the time of acute water crisis. It is reported that the water and social agreement for a water sensitive city would be significantly different to that conventional urban water approaches and integrates the normative values of environmental protection, demand–supply security, flood control, human wellbeing, and economic sustainability [22].

Nature-based solutions are now fast becoming the center stage of solving diverse and complex problems related to water security. In 2018 for the first time, Brazil hosted the world launching of the United Nations World Water Development Report, which publicly stated the importance of nature-based solutions (NBS) for water. According to Stefan Uhlenbrook, the Coordinator and Director of the UN World Water Assessment Programme (WWAP), reservoirs, irrigation canals, and water treatment plants are not the only water management instruments at our disposal. "We can't wait for nature to solve all problems by itself, but we can get inspired and use it in favor of the planet."

7. Conclusions

Flooding, drought, inadequate access to drinking water, and sanitation are some of the well-known water problems across the globe. There is a need for sustainable solutions, which are appropriate for varying contexts, local consideration to societal values, which leverage the multiple uses and make use of synergies with management of other resources. As these problems are increasing across different regions and different scales, there is an urgency to take measures for the protection of the water and to improve legislation and public awareness in this field to find the optimum way to manage, protect, and serve our limited water resources, and enforce water pollution control and the protection of water by suggesting remediation alternatives to reduce or control the influence of the contamination.

This paper explored new paradigms and evolving definitions of water security, and presented a wide range of sustainable solutions to solve various water-related problems. Water security is fundamental to achieving any kind of sustainable economic and human development. The challenges of water security are unprecedented, and a new approach to provide the underpinning science for water security management is urgently needed by the global community. An integrative framing of water security should take place at the policy and governance level, where established priorities and decisions are taken with agreement of stakeholders. An integrative approach is likely to bring good water governance and set a new approach to sustainable water management. The evolving paradigm of achieving water security will be able to provide sufficient water for socioeconomic activities for domestic, industrial, and commercial purposes, and clean drinking water to meet basic needs at an affordable price with proper sanitation, while treating and collecting wastewater to curb water pollution. Various dimensions of achieving water security extend to protecting ecosystems and increasing the role of nature to sustain its own functioning and developing ability to cope with changing conditions. It also deals with timely response to the risk of water related disasters such as droughts, floods, spreading of diseases, and pollution.

Attaining water security in a changing context is not a simple or 'one size fits all' solution. Expanding the portfolio of solutions from fully engineered systems to management systems based on capacity building, community awareness, managing the wetlands, and conservation of aquifers need to incorporate the natural processes to achieve desired results to attain a water secure future for human wellbeing. It addresses the impacts of human activities on water quantity and quality, aquatic ecosystems and climate, and the context of rapid economic growth and climate change. It can be concluded that water security has multiple and highly interconnected dimensions, and that each of these involves complex interactions between human society and the natural environment. In conclusion, achieving a water secure world requires balance between social, environmental, and economic components. It requires adequate integration of soft and hard measures such as in storing and transporting water and in protecting the resource itself.

Author Contributions: Conceptualization: B.K.M., P.K., C.S., S.C., A.G.; methodology: B.K.M., C.S., P.K., S.C.; formal analysis: B.K.M., P.K., C.S., S.C.; investigation: B.K.M., P.K., C.S., S.C.; writing—original draft preparation: B.K.M., P.K., C.S., S.C.; writing—review and editing: B.K.M., P.K., C.S., S.C., A.G. All authors have read and agreed to the published version of the manuscript.

Funding: This research received no external funding.

Institutional Review Board Statement: Not applicable.

Informed Consent Statement: Not applicable.

Conflicts of Interest: The authors declare no conflict of interest.

Appendix A

Table A1. List of reviewed papers and their relation with seven different dimensions of water security.

Reference Number	Title of the Reviewed Papers	Category (Key Topics According to the Seven Dimensions of Water Security)
[1]	Strengthening water security in Asia and the Pacific	Transboundary issues
[2]	Water: Asia's new battleground	Transboundary issues,
[3]	Coping with water scarcity. Challenge of the twenty-first century	Household, Economic, Urban, Environmental
[4]	Impact of urbanization on water shortage in face of climatic aberrations	Urban, Environmental, Transboundary issues, Safe or desirable quality, Economic
[5]	Global water crisis and future food security in an era of climate change	Economic, Environmental
[6]	Water Crisis Report	Transboundary issues, Economic, Environmental, Safe or desirable quality, Resilience to water related disasters
[7]	Water quality management: a globally neglected issue	Safe and desirable quality
[8]	Future Outlook of Urban Water Environment in Asian Cities	Economic, Urban, Environmental
[9]	Land use, water management and future flood risk	Resilience to water related disasters, Safe or desirable quality
[10]	Determinants of residential water consumption: Evidence and analysis from a 10-country household survey	Economic, Household
[11]	Responding to the challenges of water security: the Eighth Phase of the International Hydrological Programme	Resilience to water related disasters, Environmental, Safe or desirable quality, Household, Economic

Table A1. *Cont.*

Reference Number	Title of the Reviewed Papers	Category (Key Topics According to the Seven Dimensions of Water Security)
[12]	Climate Change 2014: Synthesis Report. Contribution of Working Groups I, II and III to the Fifth Assessment Report of the Intergovernmental Panel on Climate Change	Environmental
[13]	UN-Water Analytical brief: Water security and the global water agenda	Environmental, Economic, Safe or desirable quality, Resilience to water related disasters
[14]	A clash of paradigms in the water sector? Tensions and synergies between integrated water resources management and the human rights-based approach to development	Environmental, Economic, Transboundary water related issues, Safe or desirable quality
[15]	Water security: what does it mean, what may it imply?	Environmental Economic, Urban, Resilience to water related disasters
[16]	Web of Science use in published research and review paper Sciences 1997–2017: a selective, dynamic, cross-domain, content-based analysis	For method used in this paper
[17]	Water security: Debating an emerging paradigm	Integrative of the seven dimensions
[18]	China's water security: Current status, emerging challenges and future prospects. Environmental Science and Policy	Environmental, Economic, Urban, Safe or desirable quality
[19]	"Water scarcity: Fact or Fiction?"	Economic, Household, Urban, Environmental
[20]	Climate, climate change and human health in Asian Cities.	Environmental, Urban, Safe or desirable quality
[21]	Global hydrological cycles and world water resources	Environmental
[22]	Transitioning to Water Sensitive Cities: Historical, Current and Future Transition States	Urban
[23]	A comprehensive optimum integrated water resource management approach for multidisciplinary water resources management problems	Integrative of seven dimensions
[24]	Governing water: contentious transnational politics and global institution building	Environmental, Transboundary water related issues
[25]	Adaptive water governance: assessing the institutional prescriptions of adaptive (co-) management from a governance perspective and defining a research agenda	Household, Economic, Urban, Environmental, Resilience to water related disasters, Safe or desirable quality, Transboundary issues
[26]	Assessing Water Security with Appropriate Indicators	Integrative of seven dimensions
[27]	Integrated urban water management scenario modeling for sustainable water governance in Kathmandu Valley, Nepal	Urban, Economic, Environmental,
[28]	Hydrogeochemical Evolution and Appraisal of Groundwater Quality in Panna District, Central India	Safe or desirable quality, Economic, Environmental
[29]	Framework to assess sources controlling soil salinity resulting from irrigation using recycled water: an application of Bayesian Belief Network	Safe or desirable quality, Urban, Households
[30]	Risk-based principles for defining and managing water security risk-based principles for defining and managing water security	Environmental, Economic, Safe or desirable quality, Resilience to waters related hazards
[31]	Urban water security indicators: Development and pilot.	Urban
[32]	Fresh Water-Related Indicators in Canada: An Inventory and Analysis	Environmental
[33]	Water, people, and sustainability—a systems framework for analyzing and assessing water governance regimes	Integrative of seven dimensions
[34]	Ten building blocks for sustainable water governance: an integrated method to assess the governance of water	Environmental, Economic, Safe or desirable quality, Resilience to water related disasters

Table A1. *Cont.*

Reference Number	Title of the Reviewed Papers	Category (Key Topics According to the Seven Dimensions of Water Security)
[35]	Climate justice in lieu of climate change: a sustainable approach to respond to the climate change injustice and an awakening of the environmental movement	Environmental
[36]	Population growth and water supply: The future of Ghanaian cities. Megacities and Rapid Urbanization: Breakthroughs in Research and Practice	Urban
[37]	Creation of a Regional Circular and Ecological Sphere (RCES) to address local challenges	Environmental
[38]	Participatory approach for enhancing robust water resource management: case study of Santa Rosa sub-watershed near Laguna Lake	Urban, Environmental, Safe or desirable quality
[39]	The history of public participation in ecological research	Environmental
[40]	Citizen science websites/apps for invasive species sightings: An analysis of 26 ongoing initiatives.	Environmental
[41]	Citizens as sensors: The world of volunteered geography	Environmental
[42]	IWRM: what should we teach? A report on curriculum development at the University of the Western Cape, South Africa	Household, Economic, Urban, Environmental, Resilience to water related disasters, Safe or desirable quality, Transboundary issues
[43]	Integrated water resource management (IWRM): an approach to face the challenges of the next century and to avert future crises	Household, Economic, Urban, Environmental, Resilience to water related disasters, Safe or desirable quality, Transboundary issues
[44]	Good water governance and IWRM in Zambia: challenges and chances	Household, Economic, Urban, Environmental, Resilience to water related disasters, Safe or desirable quality, Transboundary issues
[45]	The water-energy-food nexus: trade-offs, thresholds and transdisciplinary approaches to sustainable development	Economic, Urban, Environmental, Resilience to water related disasters, Safe or desirable quality
[46]	Debates—Perspectives on socio-hydrology: Capturing feedbacks between physical and social processes	Resilience to water related disasters
[47]	Stochastic multi-objective modelling for optimization of water-food-energy nexus of irrigated agriculture	Economic, Environmental
[48]	Socio-hydrology: use-inspired water sustainability science for the Anthropocene	Environmental, Resilience to water related disasters
[49]	Earth systems: Model human adaptation to climate change	Economic, Urban, Environmental
[50]	Small-island communities in the Philippines prefer local measures to relocation in response to sea-level rise	Household, Resilience to water related disaster
[51]	Solutions to water resource scarcity: water reclamation and reuse	Economic, Urban, Environmental
[52]	Occurrence and risk assessment of emerging contaminants in a water reclamation and ecological reuse project	Environmental, Safe or desirable water quality
[53]	The potential of water reuse as a management option for water security under the ecosystem services approach	Economic, Environmental, Safe or desirable water quality
[54]	Post-adaptation behaviour of farmers towards soil and water conservation technologies of watershed management in India	Economic, Environmental
[55]	How is ecosystem health defined and measured? A critical review of freshwater and estuarine studies	Economic, Environmental
[56]	Rivers as Socioecological Landscapes	Environmental dimension
[57]	Connectivity in rivers	Environmental dimension

Table A1. Cont.

Reference Number	Title of the Reviewed Papers	Category (Key Topics According to the Seven Dimensions of Water Security)
[58]	Barriers and Bridges to Renewal of Ecosystems and Institutions	Environmental and Transboundary dimension
[59]	Ecological stakeholder analogs as intermediaries between freshwater biodiversity conservation and sustainable water management	Environmental and Transboundary dimension
[60]	Integrating the social, hydrological and ecological dimensions of freshwater health: the Freshwater Health Index	Environmental, Safe or desirable quality, and Resilience to water related disasters dimension
[61]	Exploring spatial variations in the relationships between residents' recreation demand and associated factors: a case study in Texas	Household, Economic, Urban, Environmental, Resilience to water related disasters, Safe or desirable quality dimension
[62]	Resilience and thresholds in river ecosystems	Environmental, Resilience to water related disasters dimension
[63]	Managing urban water crises: adaptive policy responses to drought and flood in Southeast Queensland, Australia	Household, Economic, Urban, Environmental, Resilience to water related disasters, Safe or desirable quality dimension
[64]	Participatory watershed land-use management approach: Interfacing science, policy and participation for effective decision making	Environmental, Safe or desirable quality, and Resilience to water related disasters dimension
[65]	The Role of Community Champions in Long-Term Sustainable Urban Water Planning	Household, Economic, Urban, Environmental, Resilience to water related disasters, Safe or desirable quality dimension
[66]	Public Participation in Natural Hazard Mitigation Policy Formation: Challenges for Comprehensive Planning	Household, Economic, Environmental, Resilience to water related disasters, Safe or desirable quality dimension
[67]	Sustainable development and the water–energy–food nexus: A perspective on livelihoods	Household, Economic, Urban, Environmental, Resilience to water related disasters, Safe or desirable quality, Transboundary dimension
[68]	Considering the energy, water and food nexus: towards an integrated modelling approach	Household, Economic, Urban, Environmental, Resilience to water related disasters, Safe or desirable quality, Transboundary dimension
[69]	Ecosystem-based water security and the sustainable development goals.	Environmental, Resilience to water related disasters, Safe or desirable quality dimension
[70]	UN - United Nations. Sustainable Development Goal 6 Synthesis Report 2018 on Water and Sanitation	Household, Economic, Urban, Environmental, Resilience to water related disasters, Safe or desirable quality, Transboundary dimension
[71]	Global analysis of urban surface water supply vulnerability	Household, Economic, Urban, Environmental, Resilience to water related disasters, Safe or desirable quality, Transboundary dimension

References

1. Asian Water Development Outlook (AWDO). Strengthening Water Security in Asia and the Pacific. 2016. Available online: https://www.adb.org/publications/asian-water-development-outlook-2016 (accessed on 27 August 2018).
2. Chellaney, B. *Water: Asia's New Battleground*; Georgetown University Press: Washington, DC, USA, 2011.
3. UN-Water. *Coping with Water Scarcity. Challenge of the Twenty-First Century*; UN-Water; FAO: Geneva, Switzerland, 2007; p. 23.
4. Majumder, M. *Impact of Urbanization on Water Shortage in Face of Climatic Aberrations*; Springer Briefs in Water Science and Technology; Springer: Singapore, 2015; p. 98. ISSN 2194-7244.
5. Hanjra, M.A.; Qureshi, M.E. Global water crisis and future food security in an era of climate change. *Food Policy* **2010**, *35*, 365–377. [CrossRef]
6. Guppy, L.; Anderson, K. *Water Crisis Report*; United Nations University Institute for Water, Environment and Health: Hamilton, ON, Canada, 2017.

7.   Biswas, A.K.; Tortajada, C. Water quality management: A globally neglected issue. *Int. J. Water Resour. Dev.* **2019**, *35*, 913–916. [CrossRef]

8.   Masago, Y.; Mishra, B.K.; Jalilov, S.; Kefi, M.; Kumar, P.; Dilley, M.; Fukushi, K. *Future Outlook of Urban Water Environment in Asian Cities*; United Nations University: Tokyo, Japan, 2019.

9.   Wheater, H.; Evans, E. Land use, water management and future flood risk. *Land Use Policy* **2009**, *26*, S251–S264. [CrossRef]

10. Grafton, R.Q.; Ward, M.B.; To, H.; Kompas, T. Determinants of residential water consumption: Evidence and analysis from a 10-country household survey. *Water Resour. Res.* **2011**, *47*, W08537. [CrossRef]

11. Jimenez-Cisneros, B. Responding to the challenges of water security: The Eighth Phase of the International Hydrological Programme, 2014–2021. *Proc. IAHS* **2015**, *366*, 10–19. [CrossRef]

12. IPCC. *Climate Change 2014: Synthesis Report. Contribution of Working Groups I, II and III to the Fifth Assessment Report of the Intergovernmental Panel on Climate Change*; Pachauri, R.K., Meyer, L.A., Eds.; IPCC: Geneva, Switzerland, 2014; p. 151.

13. United Nations University (UNU). UN-Water Analytical Brief: Water Security and the Global Water Agenda. Hamilton, ON, Canada. 2013, p. 47. Available online: http://inweh.unu.edu/wp-content/uploads/2013/11/water-security-and-the-global-water-agenda (accessed on 16 August 2018).

14. Tremblay, H. A clash of paradigms in the water sector? Tensions and synergies between integrated water resources management and the human rights-based approach to development. (1 August 2010). *Nat. Resour. J.* **2011**, *51*, 307–356.

15. Schultz, B.; Uhlenbrook, S. Water security: What does it mean, what may it imply? In Proceedings of the International Symposium "Water for a Changing World Developing Local Knowledge and Capacity", Delft, The Netherlands, 13–15 June 2007.

16. Li, K.; Rollins, J.; Yan, E. Web of Science use in published research and review papers 1997–2017: A selective, dynamic, cross-domain, content-based analysis. *Scientometrics* **2018**, *115*, 1–20. [CrossRef]

17. Cook, C.; Bakker, K. Water security: Debating an emerging paradigm. *Glob. Environ. Chang.* **2012**, *22*, 94–102. [CrossRef]

18. Jiang, Y. China's water security: Current status, emerging challenges and future prospects. *Environ. Sci. Policy* **2015**, *54*, 106–125. [CrossRef]

19. Rijsberman, F.R. Water scarcity: Fact or Fiction? *Agric. Water Manag.* **2006**, *80*, 5–22. [CrossRef]

20. Kovats, S.; Akhtar, R. Climate, climate change and human health in Asian Cities. *Environ. Urban.* **2008**, *20*, 165–175. [CrossRef]

21. Oki, T.; Kanae, S. Global hydrological cycles and world water resources. *Science* **2006**, *313*, 1068–1072. [CrossRef]

22. Brown, R.; Keath, N.; Wong, T. Transitioning to Water Sensitive Cities: Historical, Current and Future Transition State. In Proceedings of the 11th International Conference on Urban Drainage, Edinburgh, UK, 31 August–5 September 2008.

23. Al-Jawad, J.Y.; Alsaffar, H.M.; Bertram, D.; Kalin, R.M. A comprehensive optimum integrated water resource management approach for multidisciplinary water resources management problems. *J. Environ. Manag.* **2019**, *239*, 211–224. [CrossRef]

24. Gareau, B.J.; Crow, B. *Ken Conca. Governing Water: Contentious Transnational Politics and Global Institution Building*; International Environmental Agreements; MIT Press: Cambridge, MA, USA, 2006; p. 466.

25. Huitema, D.; Mostert, E.; Egas, W.; Moellenkamp, S.; Pahl-Wostl, C.; Yalcin, R. Adaptive water governance: Assessing the institutional prescriptions of adaptive (co-) management from a governance perspective and defining a research agenda. *Ecol. Soc.* **2009**, *14*, 26. [CrossRef]

26. GWP. *Assessing Water Security with Appropriate Indicators*; Global Water Partnership: Stockholm, Sweden, 2014.

27. Saraswat, C.; Mishra, B.K.; Kumar, P. Integrated urban water management scenario modeling for sustainable water governance in Kathmandu Valley, Nepal. *Sustain. Sci.* **2017**, *12*, 1037–1053. [CrossRef]

28. Kumar, P.; Kumar, A.; Singh, C.K.; Saraswat, C.; Avtar, R.; Ramanathan, A.L.; Herath, S. Hydrogeochemical Evolution and Appraisal of Groundwater Quality in Panna District, Central India. *Expo. Health* **2016**, *8*, 19–30. [CrossRef]

29. Rahman, M.M.; Hagare, D.; Maheshwari, B. Framework to assess sources controlling soil salinity resulting from irrigation using recycled water: An application of Bayesian Belief Network. *J. Clean. Prod.* **2014**, *105*, 406–419. [CrossRef]

30. Hall, J.W.; Borgomeo, E. Risk-based principles for defining and managing water security risk-based principles for defining and managing water security. *Philos. Trans. R. Soc. A* **2013**, *371*, 20120407. [CrossRef] [PubMed]

31. Jensen, O.; Wu, H. Urban water security indicators: Development and pilot. *Environ. Sci. Policy* **2018**, *83*, 33–45. [CrossRef]

32. Dunn, G.; Bakker, K. Fresh Water-Related Indicators in Canada: An Inventory and Analysis. *Can. Water Resour. J.* **2011**, *36*, 135–148. [CrossRef]

33. Wiek, A.; Larson, K.L. Water, people, and sustainability—A systems framework for analyzing and assessing water gover-nance regimes. *Water Resour Manag.* **2012**, *26*, 3153–3171. [CrossRef]

34. Van-Rijswick, M.; Edelenbos, J.; Hellegers, P.; Kok, M.; Kuks, S. Ten building blocks for sustainable water governance: An integrated method to assess the governance of water. *Water Int.* **2014**, *39*, 725–742. [CrossRef]

35. Saraswat, C.; Kumar, P. Climate justice in lieu of climate change: A sustainable approach to respond to the climate change injustice and an awakening of the environmental movement. *Energy Ecol. Environ.* **2016**, *1*, 67–74. [CrossRef]

36. Cobbinah, P.B.; Okyere, D.K.; Gaisie, E. Population growth and water supply: The future of Ghanaian cities. In *Megacities and Rapid Urbanization: Breakthroughs in Research and Practice*; IGI Global: Pasadena, CA, USA, 2020; pp. 96–117.

37. Ministry of Environment. Creation of a Regional Circular and Ecological Sphere (RCES) to Address Local Challenges. 2018. Available online: https://www.env.go.jp/policy/kihon_keikaku/plan/plan_5 (accessed on 20 August 2020).

38. Kumar, P.; Johnson, B.A.; Dasgupta, R.; Avtar, R.; Chakraborty, S.; Masayuki, K.; Macandog, D. Participatory approach for enhancing robust water resource management: Case study of Santa Rosa sub-watershed near Laguna Lake, Philippines. *Water* **2020**, *12*, 1172. [CrossRef]

39. Rushing, M.; Primack, R.; Bonney, R. The history of public participation in ecological research. *Front. Ecol. Environ.* **2012**, *10*, 285–290. [CrossRef]

40. Johnson, B.; Mader, A.; Dasgupta, R.; Kumar, P. Citizen science websites/apps for invasive species sightings: An analysis of 26 ongoing initiatives. *Glob. Ecol. Conserv.* **2019**, *21*, e00812. [CrossRef]

41. Goodchild, M.F. Citizens as sensors: The world of volunteered geography. *GeoJournal* **2007**, *69*, 211–221. [CrossRef]

42. Jonker, L. IWRM: What should we teach? A report on curriculum development at the University of the Western Cape, South Africa. *Phys. Chem. Earth Parts A/B/C* **2005**, *30*, 881–885. [CrossRef]

43. Radif, A.A. Integrated water resource management (IWRM): An approach to face the challenges of the next century and to avert future crises. *Desalination* **1999**, *124*, 145–153. [CrossRef]

44. Uhlendahl, T.; Salian, P.; Casarotto, C.; Doetsch, J. Good water governance and IWRM in Zambia: Challenges and chances. *Water Policy* **2011**, *13*, 845–862. [CrossRef]

45. Kurian, M. The water-energy-food nexus: Trade-offs, thresholds and transdisciplinary approaches to sustainable development. *Environ. Sci. Policy* **2017**, *68*, 97–106. [CrossRef]

46. Baldassarre, D.; Viglione, G.A.; Carr, G.; Kuil, L.; Yan, K.; Brandimarte, L.; Bloschl, G. Debates—Perspectives on socio-hydrology: Capturing feedbacks between physical and social processes. *Water Resour. Res.* **2015**, *51*, 4770–4781. [CrossRef]

47. Mo, L.; Qiang, F.; Singh, V.P.; Dong, L.; Tianxiao, L. Stochastic multi-objective modelling for optimization of water-food-energy nexus of irrigated agriculture. *Adv. Water Resour.* **2019**, *127*, 209–224.

48. Sivapalan, M.; Konar, M.; Srinivasan, V.; Chhatre, A.; Wutich, A.; Scott, C.A.; Wescoat, J.L.; Iturbe, I.R. Socio-hydrology: Use-inspired water sustainability science for the Anthropocene. *Earths Future* **2014**, *2*, 225–230. [CrossRef]

49. Palmer, P.I.; Smith, M.J. Earth systems: Model human adaptation to climate change. *Nature* **2014**, *7515*, 365–366. [CrossRef]

50. Jamero, M.L.; Onuki, M.; Esteban, M.; Billones-Sensano, X.K.; Tan, N.; Nellas, A.; Takagi, H.; Thao, N.D.; Valenzuela, V.P. Small-island communities in the Philippines prefer local measures to relocation in response to sea-level rise. *Nat. Clim. Chang.* **2017**, *7*, 581–586. [CrossRef]

51. Tian, Y.; Hu, H.; Zhang, J. Solutions to water resource scarcity: Water reclamation and reuse. *Environ. Sci. Pollut. Res.* **2017**, *24*, 5095–5097. [CrossRef] [PubMed]

52. Lin, X.; Xu, J.; Keller, A.A.; He, L.; Gu, Y.; Zheng, W.; Sun, D.; Lu, Z.; Huang, J.; Huang, X.; et al. Occurrence and risk assessment of emerging contaminants in a water reclamation and ecological reuse project. *Sci. Total Environ.* **2020**, *744*, 140977. [CrossRef]

53. Voulvoulis, N. The potential of water reuse as a management option for water security under the ecosystem services approach. *Desalination Water Treat.* **2015**, *53*, 3263–3271. [CrossRef]

54. Bagdi, G.L.; Mishra, P.K.; Kurothe, R.S.; Arya, S.L.; Patil, S.L.; Singh, A.K.; Bihari, B.; Prakash, O.; Kumar, A.; Sundarambal, P. Post-adaptation behaviour of farmers towards soil and water conservation technologies of watershed management in India. *Inter. Soil & Water Cons. Res.* **2015**, *3*, 161–169.

55. O'Brien, A.; Townsend, K.; Hale, R.; Sharley, D.; Pettigrove, V. How is ecosystem health defined and measured? A critical review of freshwater and estuarine studies. *Ecol. Ind.* **2016**, *8*, 722–729. [CrossRef]

56. Chakraborty, A.; Chakraborty, S. Rivers as Socioecological Landscapes. In *Rivers and Society: Landscapes Governance and Livelihoods*; Earthscan Studies in Water Resource Management; Cooper, M., Chakraborty, A., Chakraborty, S., Eds.; Earthscan; Routledge: London, UK, 2017; Chapter 2; pp. 9–26.

57. Wohl, E. Connectivity in rivers. *Prog. Phys. Geogr. Earth Environ.* **2017**, *41*. [CrossRef]

58. Gunderson, L.H.; Holling, C.S.; Light, S. *Barriers and Bridges to Renewal of Ecosystems and Institutions*; Columbia University Press: New York, NY, USA, 1995.

59. Van-Rees, C.B.; Cañizares, J.R.; Garcia, G.M.; Reed, J.M. Ecological stakeholder analogs as intermediaries between freshwater biodiversity conservation and sustainable water management. *Environ. Pol. Gov.* **2019**, *29*, 303–312. [CrossRef]

60. Vollmer, D.; Shaad, K.; Souter, N.J.; Farrell, T.; Dudgeon, D.; Sullivan, C.A.; Fauconnier, I.; MacDonald, G.M.; McCartney, M.P.; Power, A.G.; et al. Integrating the social, hydrological and ecological dimensions of freshwater health: The Freshwater Health Index. *Sci. Total Environ.* **2018**, *627*, 304–313. [CrossRef]

61. Lee, K.H.; Schuett, M.A. Exploring spatial variations in the relationships between residents' recreation demand and associated factors: A case study in Texas. *Appl. Geogr.* **2014**, *53*, 213–222. [CrossRef]

62. Parsons, M.; Thoms, M.; Capon, T.; Capon, S.; Reid, M. *Resilience and Thresholds in River Ecosystems*; National Water Commission; Waterlines: Canberra, Australia, 2009.

63. Head, B.W. Managing urban water crises: Adaptive policy responses to drought and flood in Southeast Queensland, Australia. *Ecol. Soc.* **2014**, *19*, 33. [CrossRef]

64. Endo, I.; Didham, R.J. *Participatory Watershed Land-Use Management Approach: Interfacing Science, Policy and Participation for Effective Decision Making*; IGES: Hayama, Japan, 2016; Available online: https://www.iges.or.jp/en/pub/participatory-watershed-land-use-management-0/en (accessed on 2 September 2020).

65. Lindsay, J.; Rogers, B.C.; Church, E.; Gunn, A.; Hammer, K.; Dean, A.J.; Fielding, K. The Role of Community Champions in Long-Term Sustainable Urban Water Planning. *Water* **2019**, *11*, 476. [CrossRef]

66. Godschalk, D.R.; Brody, S.; Burby, R. Public Participation in Natural Hazard Mitigation Policy Formation: Challenges for Comprehensive Planning. *J. Environ. Plan. Manag.* **2010**, *46*, 733–754. [CrossRef]
67. Biggs, E.M.; Bruce, E.; Boruff, B.; Duncan, J.M.A.; Horsley, J.; Pauli, N.; McNeil, K.; Neef, A.; Ogtrop, F.V.; Curnow, J.; et al. Sustainable development and the water–energy–food nexus: A perspective on livelihoods. *Environ. Sci. Policy* **2015**, *54*, 389–397. [CrossRef]
68. Bazilian, M.; Rogner, H.; Hoells, M.; Hermann, S.; Arent, D.; Gielen, D.; Steduto, P.; Mueller, A.; Komor, P.; Tol, R.S.J.; et al. Considering the energy, water and food nexus: Towards an integrated modelling approach. *Energy Policy* **2011**, *39*, 7896–7906. [CrossRef]
69. Vörösmarty, C.J.; Osuna, V.R.; Cak, A.; Bhaduri, A.; Bunn, S.E.; Corsi, F.; Gastelumendi, J.; Green, P.; Harrison, I.; Lawford, R.; et al. Ecosystem-based water security and the sustainable development goals. *Ecohydrol. Hydrobiol.* **2018**. [CrossRef]
70. UN—United Nations. *Sustainable Development Goal 6 Synthesis Report 2018 on Water and Sanitation*; United Nations Publishers: New York, NY, USA, 2018.
71. Padowski, J.C.; Gorelick, S.M. Global analysis of urban surface water supply vulnerability. *Environ. Res. Lett.* **2014**, *9*, 104004. [CrossRef]

MDPI
St. Alban-Anlage 66
4052 Basel
Switzerland
Tel. +41 61 683 77 34
Fax +41 61 302 89 18
www.mdpi.com

Water Editorial Office
E-mail: water@mdpi.com
www.mdpi.com/journal/water

9 783036 518695